T0402085

Mechanosensing Biology

Masaki Noda
Editor

Mechanosensing Biology

 Springer

Editor
Masaki Noda M.D., Ph.D.
Professor, Director
Medical Research Institute
Tokyo Medical and Dental University
1-5-45 Yushima, Bunkyo-ku
Tokyo 113-8510, Japan
noda.mph@mri.tmd.ac.jp

ISBN 978-4-431-89756-9   e-ISBN 978-4-431-89757-6
DOI 10.1007/978-4-431-89757-6
Springer Tokyo Dordrecht Heidelberg London New York

Library of Congress Control Number: 2010941751

Cover illustration: Mechanisms involved in the response of osteoblasts to mechanical forces. From Chapter 8, courtesy of P. J. Marie. Figures were produced using Servier Medical Art (www.servier.com).

Printed on acid-free paper

Springer is part of Springer Science+Business Media (www.springer.com)

Preface

Mechanical stress is known to regulate body function, evidence of which can be seen in many tissues such as bone, muscle, heart, and vessels. Bedridden patients lose bone when they are immobilized for a long time. Astronauts also experience muscle and bone loss during space flight. The heart functions to pump blood, causing mechanical stress to itself and to vascular tissue. The effects of mechanical stress can be observed not only in adults but also in developmental periods of life. Even the earliest establishment of primordial tissues required microenvironmental stress that would later play a role in the maintenance of cell structure and the shape of organs.

The function of certain membrane channels is regulated by mechanical stress. In conjunction with local mechanical stimuli, systemic regulatory events such as endocrine and neurological controls work interactively. To respond to its environment, the body requires the signals of mechanical stress in the skeletal tissue cells. It has been suggested that multiple signaling pathways operate in diverse types of cells by responding in different ways to mechanical stress. In muscle cells, membrane proteins have been shown to maintain their localization and functions under loading conditions, whereas loss of mechanical stress can lead to rapid loss of membrane proteins such as dystrophin. Homeostasis is impaired upon loss of mechanical stress, leading to pathological conditions such as osteopenia, muscle atrophy, and vascular tissue dysfunction. It is important, therefore, to understand the mechanisms of such signaling induced by mechanical stress in the maintenance of homeostasis. These mechanical signaling events are believed to maintain the functioning of the body and must be considered in contemplating new approaches to treating dysfunction and disease.

In this monograph, mechanical stress is discussed by experts in the field with respect to the molecular, cellular, and tissue aspects in close connection with medicine. Taking these aspects together, the book provides the most up-to-date information on cutting-edge advancements in the field of mechanobiology. In elderly populations, such mechanical pathophysiology, as well as the mechanical activities of locomotor and cardiovascular systems, is important because skeletal and heart functions decline and cause various diseases in other organs. For this reason, concern about mechanical stress-related health problems of elderly patients has been rapidly increasing. This book provides a timely contribution to research into locomotor and circulatory diseases that are major problems in contemporary society.

Masaki Noda

Contents

Contributors

Corinne Albiges-Rizo (Chapter 14)
Institut Albert Bonniot, INSERM U823, CNRS ERL3148 Université Joseph
Fourier, Equipe DySAD, Site Santé, BP 170, 38042 Grenoble Cedex 9, France

Joji Ando (Chapter 2)
Laboratory of Biomedical Engineering, School of Medicine, Dokkyo Medical
University, 880 Kita-kobayashi, Mibu, Tochigi 321-0293, Japan

Makoto Asashima (Chapter 3)
Research Center for Stem Cell Engineering, National Institute of Advanced
Industrial Science and Technology (AIST), Tsukuba Central 4, 1-1-1 Higashi,
Tsukuba, Ibaraki 305-8562, Japan
and
Department of Life Sciences (Biology), Graduate School of Arts and Sciences,
The University of Tokyo, 3-8-1 Komaba, Meguro-ku, Tokyo 153-8902, Japan

Anne Blangy (Chapter 14)
Montpellier University, Centre de Recherche de Biochimie Macromoléculaire,
CNRS, 1919 route de Mende, 34297 Montpellier Cedex 5, France

Lynda Faye Bonewald (Chapter 10)
Department of Oral Biology, University of Missouri at Kansas City School
of Dentistry, 650 East 25th Street, Kansas City, MO 64108-2784, USA

Yoichi Ezura (Chapter 6)
Department of Molecular Pharmacology, Division of Advanced Molecular
Medicine, Medical Research Institute, Tokyo Medical and Dental University,
1-5-45 Yushima, Bunkyo-ku, Tokyo 113-8510, Japan

Kimihide Hayakawa (Chapter 1)
ICORP/ORST, Cell Mechanosensing Project, Japan Science and Technology
Agency, Nagoya, Aichi 466-8550, Japan

Tadayoshi Hayata (Chapter 6)
Department of Molecular Pharmacology, Division of Advanced Molecular
Medicine, Medical Research Institute, Tokyo Medical and Dental University,
1-5-45 Yushima, Bunkyo-ku, Tokyo 113-8510, Japan

Kyoji Ikeda (Chapter 9)
Department of Bone and Joint Disease, National Center for Geriatrics and
Gerontology (NCGG), 35 Gengo, Morioka, Obu, Aichi 474-8511, Japan

Yuzuru Ito (Chapter 3)
Research Center for Stem Cell Engineering, National Institute of Advanced
Industrial Science and Technology (AIST), Tsukuba Central 4, 1-1-1 Higashi,
Tsukuba, Ibaraki 305-8562, Japan

Shinsuke Kido (Chapter 11)
Department of Medicine and Bioregulatory Sciences, The University of
Tokushima Graduate School of Medical Sciences, 3-18-15 Kuramoto-cho,
Tokushima 770-8503, Japan

Rika Kuriwaka-Kido (Chapter 11)
Department of Medicine and Bioregulatory Sciences, The University of
Tokushima Graduate School of Medical Sciences, 3-18-15 Kuramoto-cho,
Tokushima 770-8503, Japan

Pierre J. Marie (Chapter 8)
Laboratory of Osteoblast Biology and Pathology, Inserm U606, Hopital
Lariboisiere, 2 rue Ambroise Pare, 75475 Paris Cedex 10, France
and
University Paris Diderot, Paris, France

Toshio Matsumoto (Chapter 11)
Department of Medicine and Bioregulatory Sciences, The University of
Tokushima Graduate School of Medical Sciences, 3-18-15 Kuramoto-cho,
Tokushima 770-8503, Japan

Tatsuo Michiue (Chapter 3)
Department of Life Sciences (Biology), Graduate School of Arts and Sciences,
The University of Tokyo, 3-8-1 Komaba, Meguro-ku, Tokyo 153-8902, Japan

Yuko Miyagoe-Suzuki (Chapter 4)
Department of Molecular Therapy, National Institute of Neuroscience,
National Center of Neurology and Psychiatry, 4-1-1 Ogawa-Higashi
Kodaira, Tokyo 187-8502, Japan

Atsuko Mizuno (Chapter 7)
Department of Pharmacology, Jichi Medical University, 3311-1 Yakushiji
Shimotsuke, Tochigi 329-0498, Japan

Yoshiro Nakajima (Chapter 3)
Research Center for Stem Cell Engineering, National Institute of Advanced
Industrial Science and Technology (AIST), Tsukuba Central 4, 1-1-1 Higashi
Tsukuba, Ibaraki 305-8562, Japan

Tetsuya Nakamoto (Chapter 6)
Global Center of Excellence Program "International Research Center for Tooth
and Bone Diseases", Tokyo Medical and Dental University, 1-5-45 Yushima,
Bunkyo-ku, Tokyo 113-8510, Japan
and
Department of Molecular Pharmacology, Medical Research Institute,
Tokyo Medical and Dental University, 1-5-45 Yushima, Bunkyo-ku,
Tokyo 113-8510, Japan

Toshitaka Nakamura (Chapter 12)
Department of Orthopaedic Surgery, School of Medicine, University of
Occupational and Environmental Health, 1-1 Iseigaoka, Yahatanishi-ku,
Kitakyushu 807-8555, Japan

Takeshi Nikawa (Chapter 5)
Department of Nutritional Physiology, Institute of Health Biosciences,
University of Tokushima Graduate School, 3-18-15 Kuramoto-cho,
Tokushima 770-8503, Japan

Masaki Noda (Chapter 6)
Global Center of Excellence Program "International Research Center for Tooth
and Bone Diseases", Tokyo Medical and Dental University, 1-5-45 Yushima,
Bunkyo-ku, Tokyo 113-8510, Japan
and
Department of Molecular Pharmacology, Division of Advanced Molecular
Medicine, Medical Research Institute, Tokyo Medical and Dental University,
1-5-45 Yushima, Bunkyo-ku, Tokyo 113-8510, Japan
and
Hard Tissue Genome Research Center, Tokyo Medical and Dental University,
1-5-45 Yushima, Bunkyo-ku, Tokyo 113-8510, Japan
and
Department of Orthopedic Surgery, Tokyo Medical and Dental University,
1-5-45 Yushima, Bunkyo-ku, Tokyo 113-8510, Japan

Takuya Notomi (Chapter 6)
Global Center of Excellence Program "International Research Center for Tooth
and Bone Diseases", Tokyo Medical and Dental University, 1-5-45 Yushima,
Bunkyo-ku, Tokyo 113-8510, Japan
and
Department of Molecular Pharmacology, Medical Research Institute,
Tokyo Medical and Dental University, 1-5-45 Yushima,
Bunkyo-ku, Tokyo 113-8510, Japan

Kiyoshi Ohnuma (Chapter 3)
Top Runner Incubation Center for Academia-Industry Fusion, Nagaoka University
of Technology, 1603-1 Kamitomiokamachi, Nagaoka, Niigata 940-2188, Japan

Yuushi Okumura (Chapter 5)
Department of Nutritional Physiology, Institute of Health Biosciences,
University of Tokushima Graduate School, 3-18-15 Kuramoto-cho,
Tokushima 770-8503, Japan

Géraldine Pawlak (Chapter 14)
Montpellier University, Centre de Recherche de Biochimie Macromoléculaire,
CNRS, 1919 route de Mende, 34297 Montpellier Cedex 5, France
and
Institut Albert Bonniot, INSERM U823, CNRS ERL3148 Université Joseph
Fourier, Equipe DySAD, Site Santé, BP 170, 38042 Grenoble Cedex 9, France

Emmanuelle Planus (Chapter 14)
Institut Albert Bonniot, INSERM U823, CNRS ERL3148 Université Joseph
Fourier, Equipe DySAD, Site Santé, BP 170, 38042 Grenoble Cedex 9, France

Akinori Sakai (Chapter 12)
Department of Orthopaedic Surgery, School of Medicine, University of
Occupational and Environmental Health, 1-1 Iseigaoka, Yahatanishi-ku,
Kitakyushu 807-8555, Japan

Tim Skerry (Chapter 13)
Mellanby Bone Centre, School of Medicine and Biomedical Sciences,
University of Sheffield, Beech Hill Road, Sheffield S10 2RX, UK

Masahiro Sokabe (Chapter 1)
Department of Physiology, Nagoya University Graduate School of Medicine,
65 Tsurumai, Showa-ku, Nagoya, Aichi 466-8550, Japan
and
ICORP/ORST, Cell Mechanosensing Project, Japan Science and Technology
Agency, Nagoya, Aichi 466-8550, Japan

Makoto Suzuki (Chapter 7)
Edogawabashi-clinic, DSD building 4F, 348 Yamabuki-cho, Shinjuku-ku, Tokyo
162-0801, Japan

Shin'ichi Takeda (Chapter 4)
Department of Molecular Therapy, National Institute of Neuroscience, National
Center of Neurology and Psychiatry, 4-1-1 Ogawa-Higashi, Kodaira, Tokyo
187-8502, Japan

Hitoshi Tatsumi (Chapter 1)
Department of Physiology, Nagoya University Graduate School of Medicine,
65 Tsurumai, Showa-ku, Nagoya, Aichi 466-8550, Japan

Virginie Vives (Chapter 14)
Montpellier University, Centre de Recherche de Biochimie Macromoléculaire,
CNRS, 1919 route de Mende, 34297 Montpellier Cedex 5, France

Ken Watanabe (Chapter 9)
Department of Bone and Joint Disease, National Center for Geriatrics and
Gerontology (NCGG), 35 Gengo, Morioka, Obu, Aichi 474-8511, Japan

Kimiko Yamamoto (Chapter 2)
Laboratory of System Physiology, Department of Biomedical Engineering,
Graduate School of Medicine, The University of Tokyo, 7-3-1 Hongo, Bunkyo-ku,
Tokyo 113-0033, Japan

Part I
Cells and Signals

Chapter 1
Nanotechnology in Mechanobiology: Mechanical Manipulation of Cells and Organelle While Monitoring Intracellular Signaling

Hitoshi Tatsumi, Kimihide Hayakawa, and Masahiro Sokabe

1.1 Introduction

Cell migration requires a regulated interplay of actin-filament dynamics and turnover of cell-matrix adhesions (Pollard and Borisy 2003; Ridley et al. 2003). These processes are mechanically coupled by a large complex of cytoplasmic proteins linking adhesive molecules to cytoskeletons (Zamir et al. 1999; Szaszak et al. 2002). The response of vascular endothelial cells to cyclic stretching is an excellent example of mechano-induced morphological change. Endothelial cells in vivo show a spindle-like shape aligning their long axis parallel to the vessel running. When the cells are cultured on an elastic sheet, they show a polygonal shape with random orientation. However, when they are subjected to uni-axial cyclic stretch, they start to change their shape from a cobble-like shape to a spindle-like shape, aligning their long axis perpendicular to the stretch axis (Shirinsky et al. 1989; Naruse et al. 1998b) as seen in the vessel wall in vivo. Focal contacts (FCs) and cytoskeletons play crucial roles to maintain the shape of the endothelial cells (Drake et al. 1992; Kawakami et al. 2001). Redistributions of FCs and cytoskeletons are necessary for the shape remodeling of cells. Mechanosensing machinery followed by intracellular signaling and redistribution of FCs must be elucidated to understand the molecular process underlying the shape remodeling of the cell in response to mechanical stimuli. To explore the molecular mechanisms underlying the force-dependent cell shape remodeling, it is crucial to apply precisely controlled mechanical stimuli (ranging from pN to nN) to FCs and to observe the response with a high spatial–temporal resolution (ranging from 100 nm to 1 μm, and 1 ms to 1 min) in living cells. In this chapter, we

H. Tatsumi (✉) and M. Sokabe
Department of Physiology, Nagoya University Graduate School of Medicine,
65 Tsurumai, Showa-ku, Nagoya, Aichi 466-8550, Japan
e-mail: tatsumi@med.nagoya-u.ac.jp

K. Hayakawa and M. Sokabe
ICORP/ORST, Cell Mechanosensing Project, Japan Science and Technology Agency,
Nagoya, Aichi 466-8550, Japan

M. Noda (ed.), *Mechanosensing Biology*,
DOI 10.1007/978-4-431-89757-6_1, © Springer 2011

describe a variety of methods for controlled mechanical stimulations as well as the laboratory systems for monitoring the mechanosensing, and the mechano-induced signaling, e.g., activation of mechanosensitive (MS) channels, intracellular calcium ion concentration ($[Ca^{2+}]_i$) increases, adhesive contact formation and mechano-induced reorganization of cytoskeletons in live cells including neuronal growth cones and human umbilical vein endothelial cells (HUVECs).

1.2 A Variety of Methods for Applying Forces to Cells and Analyses of Mechanosensing and Signaling

1.2.1 Stretching Cells Cultured on Elastic Sheets

Two types of methods for stretching cells on an elastic sheets have been reported; one is a flexible (elastic)-bottomed cultured plate (Fig. 1.1a) (Iba and Sumpio 1991) and the other is a square elastic sheet for uni-axial stretching (Naruse et al. 1998b) (Fig. 1.1c). Cells are cultured on an elastic sheet coated with extracellular matrix like fibronectin, laminin or collagen, and subjected to stretch. In the first type, negative pressure (e.g., 150 mmHg pressure) is applied to the elastic plate to expand the cell substratum, and cells are stretched. The experimental setup is simple, and the amplitude and duration of stretching are controlled by changing the amplitude and duration of applied negative pressure. It should be noted that the direction and amplitude of stretching are dependent on the place of a cell in the plate; e.g., cells in the center of a plate are stretched in all directions, while the cells near 12 o'clock are stretched in the 6–12 o'clock direction (see Fig. 1.1b). The amplitude of stretching is also dependent on the place of a cell in the plate; the largest stretching (e.g., 24%) is applied to the cells in the periphery and it declines as cells are located towards the center of the plate (e.g., 1%) (Iba and Sumpio 1991). A merit of this technique is that we can immediately get a variety of responses depending on the place of the cells if we correctly distinguish the differences. At the same time, however, this merit turns out to be a demerit when we make a biochemical mass analysis, since the cells in a dish give an average of mixed responses.

A square elastic sheet is used for uni-axial stretching. The machinery as shown in Fig. 1.1c stretches the chamber. Cells are stretched in the same direction, and most of the cells are stretched in the same amplitude; however, slightly less stretching is applied to the cells at the margin of the chamber. With this type of machinery, FCs aligning their long axis parallel to the stretch axis disappear and those perpendicular to the stretch axis elongate during the cyclic stretching (Naruse et al. 1998a). Cells can be observed before and after cyclic stretching with an objective lens, as shown in the Fig. 1.1d, which also enables monitoring the response, e.g., intracellular Ca ion concentration ($[Ca^{2+}]_i$) changes. These methods enable slow and creeping analyses of mechanical stimuli-dependent cell-remodeling, and the biochemical process behind it (see more details in Shi et al. 2007; Wang et al. 2001).

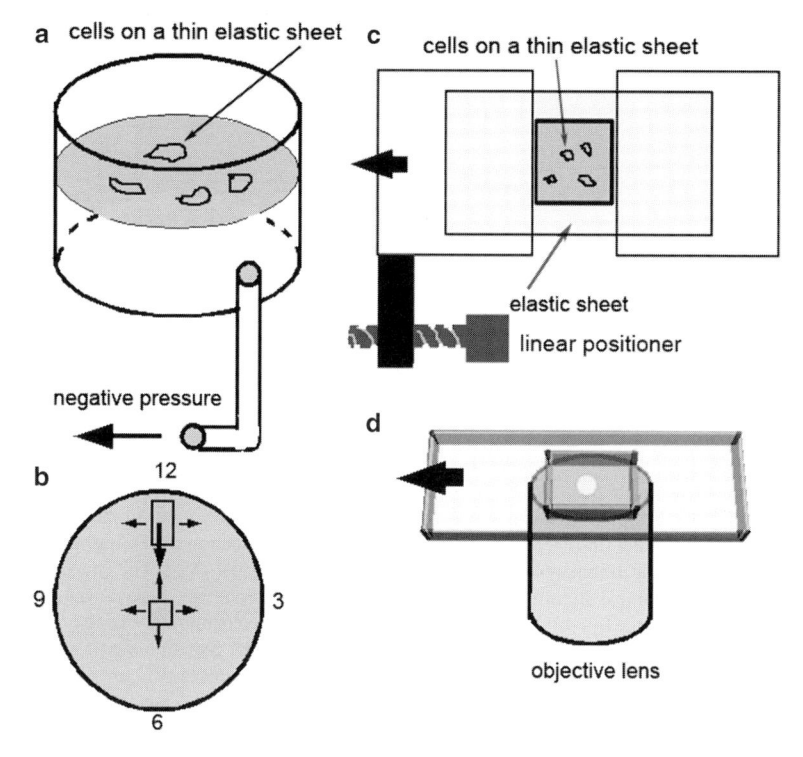

Fig. 1.1 Methods for stretching cells cultured on an elastic sheet. A flexible (elastic)-bottomed culture sheet (**a**) and an elastic sheet for uni-axial stretching (**c**) are shown schematically. Cells are cultured on elastic sheets and subjected to stretch. (**a**) Negative pressure is applied to the elastic plate to expand the cell substratum. Cells in the center of the plate are stretched in all directions, while the cells, e.g., in the 12 o'clock are stretched in the 6–12 o'clock direction (**b**). The *arrows* show the direction and amplitude of the stretching. (**c**) A square elastic sheet is used for uni-axial stretching. Cells are cultured on a square chamber slide with a thin elastic substratum. A thick elastic sheet is attached on the rigid materials; the *left side* of the elastic sheet is stretched with a linear positioner driven by a stepping motor. Elastic plates and cells are placed above an objective lens (**d**). These stretching systems are commercially available from Flexcell Corp. (http://www.flexcellint.com/) and Scholertec (http://www2.odn.ne.jp/stec/)

1.2.2 Localized Force Application by Dragging a Pipette While Recording Intracellular Signaling

Force can be applied to a cell by dragging a pipette as shown in Fig. 1.2a. The highly precise mechanical stimulus can be applied to different parts of dorsal root ganglion (DRG) neurons, using a piezo-driven glass microcapillary whose movement is computer-controlled (Imai et al. 2000). The DRG neurons are primary afferent neurons, which carry sensory signals from the skin of the limbs and the trunk, muscle and visceral organs to the spinal cord. Various mechanical, thermal, chemical, and/or noxious events are known to generate action potentials at the peripheral nerve

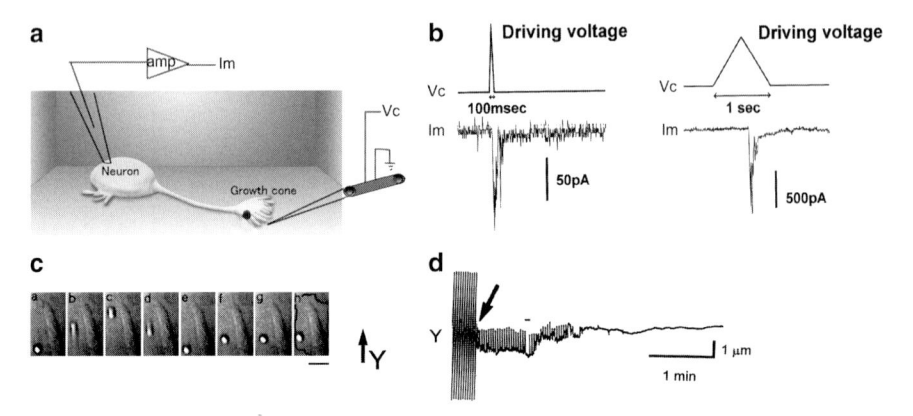

Fig. 1.2 Schematic drawing of a piezo-driven pipette and optical trapping of a bead for applying force and analyzing the mechano-induced response in growth cones. (**a**) A neuron with a growth cone and a piezo-driven pipette. (**b**) Driving voltage (triangular pulses) of various durations (100 ms and 1 s) was applied to piezo-actuators to make horizontal movements of the tip of a glass microcapillary. After the tip reached its maximal displacement at the top of the triangular pulse, the tip returned to the initial position. The control signal is shown (Vc). The inward current induced by the mechanical stimuli is shown in trace (Im). The current was evoked near the maximum displacement of the tip. (**c**) A polarized beam from a laser was focused on a bead on a growth cone to trap and drag the bead in Y direction. The bead stays on the growth cone (f, g, h panels) after electrical stimulation. *Horizontal bar* denotes 2 μm. (**d**) Analysis of optical tweezers-induced movement of beads before and after the adhesion on the growth cone membrane. The bead attached on the growth cone and was slightly displaced along the Y-axis by optical trapping (*arrows*). The maximum distance of bead movement by the optical trapping decreased and finally became almost zero after 10 min of observation, indicating that optical trapping could no longer move the beads. Some of the data are based on our previous reports (Tatsumi and Katayama 1999; Imai et al. 2000); they are modified and presented in panels (**b**)–(**d**)

endings of DRG neurons. It is plausible that the peripheral nerve endings of DRG neurons possess a variety of membrane sensory receptors activated by mechanical, thermal, chemical, or noxious stimuli. There is a wide consensus that the membrane of growth cones formed at the nerve endings undergoes large changes in tension during extension and retraction of neurites. It is assumed that these regions are endowed with MS machinery such as stretch-activated channels. However, the mechanism of mechanoelectric transduction at the peripheral nerve endings of DRG neurons is poorly understood. We examined the whole cell current induced by mechanical stimulation, and explored the distribution of sensing machinery.

The schematic drawing in Fig. 1.2 shows the design of the pipette attached to the piezo-actuator and the voltage signal for controlling the piezo-actuator. A driving voltage (triangular pulses) of various amplitudes and durations (100 ms to 1 s) was applied to the piezo actuators to make horizontal movement of the tip of the glass microcapillary in the order of 0–3.9 μm. The tip reached its maximal displacement at the top of the triangular pulse, and subsequently the tip returned to the initial position. The mechanical stimulation induced an inward Cl⁻ current (Im) in the case of DRG growth cones, as shown in Fig. 1.2. Detailed analyses of the current can be found in our report (Imai et al. 2000). The tips used for mechanical stimulation are heat-polished and rounded with tip diameters of 1–2 μm. The image of the experiment and the control signal are shown in Fig. 1.2. Filopodia as well as lamellipodia of growth cones were most sensitive to mechanical stimuli. However, when the neurite or soma of DRG cells was stimulated in the same way, electrical responses were hardly recorded, suggesting that the MS Cl⁻ channels were not distributed uniformly in dorsal root neurons. More detailed analyses of the current are reported elsewhere (Imai et al. 2000).

Force application by dragging pipette is also used to examine the role of tension in the formation of actin filament bundles (Hirata et al. 2004). A bent glass micropipette coated with glutaraldehyde was attached horizontally on an intact lamella of a semi-intact cell and displaced in parallel to the cell's longitudinal axis. The traction force causes formation of actin bundles parallel to the force direction. The micropipette was treated with 3-aminopropyltriethoxysilane, dried, coated with 20% glutaraldehyde for 5 min at room temperature, and washed before use, which makes a tight contact between the pipette and the cell surface. The horizontally placed outer wall of the bent tip of the micropipette was attached on the surface lamella of a semi-intact cell, and then the lamella could be dragged as the pipette moved horizontally.

1.2.3 Mechanical Force Application to the Cell Surface by Optically Dragging a Bead for Analyzing the Contact Formation

Optical trapping and dragging a bead can be used for applying forces to the cell (Dai and Sheetz 1995, 1999; Tatsumi and Katayama 1999; Giannone et al. 2003). These studies applied forces to the beads to measure the properties of the cell membrane or adhesive structures.

Growth cone migration depends on interactions between cell-surface adhesion receptors and components of the extracellular matrix. It has also been suggested that cell migration results from the generation of traction forces generated in the cytoskeleton at the sites of cell adhesion (Dembo and Wang 1999). The neuronal growth cone is important for axonal guidance and elongation. Contact between a growth cone and the target cell initiates synapse formation, and the importance of cell–cell adhesive molecules for path finding and synapse formation has also been suggested. However, cell–cell adhesive contacts at the growth cone of living neurons have been observed in only a few studies. Activity-dependent transmitter release and adhesive cell–cell interactions have been studied separately; however, the relationship between these events has been examined in mammalian central neurons by us (Tatsumi and Katayama 1999). The properties of the adhesive contact at the growth cones of the diagonal band of Broca (DBB), a well-known cholinergic center in the mammalian brain, were studied by optically manipulating a latex bead attached to the growth cone surface. To examine the cellular response just after the formation of adhesive contact, small beads which were optically trapped were repeatedly attached to and detached from the growth cone membrane (Fig. 1.2c). While this procedure of attachment and detachment was repeated, electrical stimulation was applied to the DBB cell body. The bead attached to and stayed on the growth cone within 2 min of the stimulation due to the contact formation.

This maintained contact was observed in 40% of beads examined. The optical force moved the bead to the limit of dragging because of the contact formation, and the bead was released from the optical trap (see Fig. 1.2d), suggesting that a sub-membrane barrier (e.g., cytoskeletons) disturbs the dragging of the bead by the trapping force as reported previously. Then, the bead returned to the original position (or attached point). In all cases, the movable distance (denoted in Fig. 1.2d by an arrow) of the bead by the laser force gradually decreased, and finally the optical force did not move beads after 10 min of observation, suggesting that surface structures bound to the bead began to be associated with rigid cytoskeletal structures. Recent progress using knockout cell lines shows that talin1 is involved in the force-dependent reinforcement of initial integrin–cytoskeleton bonds in human melanoma lines (Giannone et al. 2003).

Optical trapping and dragging of a bead (shown in Fig. 1.2c) were made as follows (see also Fig. 1.5a): a polarized beam from a 50-mW red (643 nm) Argon–Krypton laser (model 643-100RS; Omnichrome, CA) or 250 mW near-infrared (1,064 nm) YAG laser (model DPY321; Coherent, CA) was focused through a water-immersion objective lens (×63, NA 1.2; Zeiss, Germany; or ×100 NA 1.45; Nikon, Japan) on a small Latex bead (0.52 μm diameter, carboxylate microspheres; Polysciences, PA) on the growth cone (Fig. 1.2a). A remote mirror steering system (Nanomover; Melles Griot, Japan) was placed to steer the laser trapping point. The maximum particle retention force of the laser optical trapping was 0.7 pN in our setup (Block 1990). Latex beads were treated with bovine serum albumin (5 mg/ml) and were applied to growth cones. These beads have also been used as cell surface markers in experiments using molluscan neuronal growth cones. A single bead was trapped and slowly displaced by moving the trapping point on the lamellipodium of the growth cone.

1.2.4   Force Application Via a Bead Attached on the Surface of a Cell for the Analyses with High Spatial–Temporal Resolution

High-speed, single-cell analyses at the molecular level are crucial to elucidate the mechanism underlying the mechanosensing and mechano-induced remodeling of cells. However, detailed analyses could not be made during cyclic stretching of the cell substrate, since cells escape from the view of the microscope during the stretching (Fig. 1.1d). We have developed a new assay system to apply localized mechanical stress to an endothelial cell by using a fibronectin-conjugated bead (FN-bead), which enables imaging of individual FCs during the mechanical stimulation and recording mechano-induced responses. The values mentioned in this section are used to analyze and explore the molecular mechanism underlying the

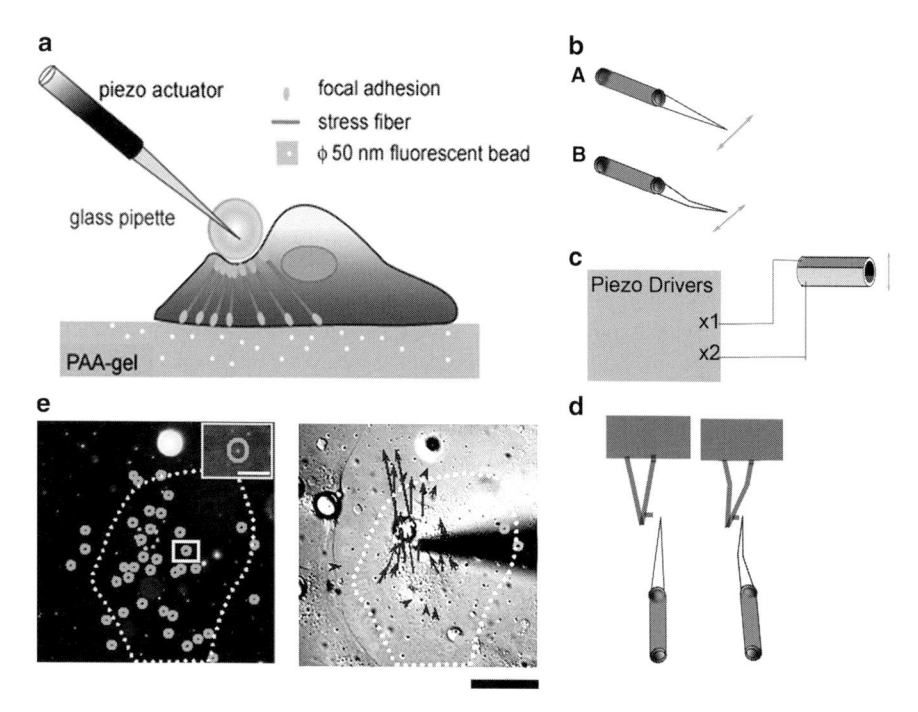

Fig. 1.3 Mechanical stimulation by displacing a FN-bead. (**a**) Schematic drawing of a bead on a cell. A FN-coated bead was displaced by a piezo-driven glass pipette, which gives localized mechanical force to the cell surface and the FCs. A glass micropipette is attached at the tip of a piezo-actuator as shown in (**b**); straight type (**A**) and bent type (**B**). The bent type was suitable for a displacement in parallel to the cell substratum. (**c**) The amplitude of the force applied to the bead is controlled with piezo-drivers (Actuator driver, ENP-5041BS; ECHO Electronics). The output X1 and X2 are applied to the half of cylindrical piezo actuators (Fuji Ceramics, Tokyo Japan), respectively, in the anti-phase; e.g., one +50 V and the other −50 V from a basal bias voltage of 100 V. (**d**) Stiffness of the tip of the glass capillary is estimated by bending the tip of an AFM probe (SN-AF01; Olympus, Japan). (**e**) An example of gel deformation caused by the mechanical stimulation. An FN-bead (shown by a *circle* in the DIC image in the *right*) on a HUVEC (denoted by the *white broken line* that denotes the cell perimeter) was displaced by a piezoelectric-driven glass pipette (a *shadow* in the DIC image). *Small gray circles* indicate the position of 50 nm fluorescent beads (*left panel*). Insets show fluorescent beads at higher magnification (a fluorescent spot at the center of the circle; *scale bar*, 20 μm). Each *arrow* in the *bottom panel* indicates the direction and relative amplitude of fluorescent particle displacement (*longest arrow* corresponds to 0.76 μm displacement) when the FN-bead was moved 4 μm in the direction shown by *arrows*. Data in panel (**e**) are based on our previous report (Hayakawa et al. 2008)

mechano-induced channel activation and intracellular calcium ion concentration increases (see Sects. 1.2.6 and 1.2.7).

FN-beads are plated on HUVECs (Fig. 1.3a). The FN-beads adhere to the cell surface within a few minutes after their plating as in the case of growth cones (Fig. 1.2). Integrin, paxillin, and vinculin were accumulated beneath the beads within 5 min, showing that FCs were formed beneath the beads (Hayakawa et al. 2008). These FCs are generally connected to the basal FCs via actin stress fibers within 30 min. Glass beads (Φ 10 μm; Duke Scientific, USA) are conjugated with fibronectin as described previously (Jacobson et al. 1978).

By displacing a bead with a glass capillary localized mechanical force was applied to the basal FCs via actin stress fibers that connect the FCs beneath the bead and the FCs at the bottom of the cell, as schematically drawn in Fig. 1.3a. A FN-coated bead can be displaced precisely by a piezo-driven glass pipette (approximately 1 μm) for 100 ms. The force required to displace the FN bead for 1 μm was estimated at approximately 30 nN using a calibrated micropipette in our experimental setup (10 μm bead on endothelial cells). The pipette was made from glass capillary (1 mm; Narishige, Japan) with a programmable puller (Model P-97; Sutter Instruments, USA) that could reproduce pipettes with almost the same shaped tip under controlled temperature environments. The stiffness of the tip of the glass capillary was estimated by bending the tip using a calibrated tip of an AFM probe (SN-AF01; Olympus, Japan); the displacement of an FN-bead and bending of the tip of the capillary were monitored with a CCD camera to estimate the force. The force applied to the bead was estimated from the relation between the displacement of the FN-bead and the bending of the tip of the glass capillary (Fig. 1.3d). The force required for 1 μm displacement of the FN bead was estimated at 35.7 ± 5 nN ($n = 5$) in our conditions.

To elucidate the area where the force was transmitted by FN-bead displacement, we employed a polyacrylamide gel substrate culture system based on (Munevar et al. 2001). HUVECs are cultured on a fibronectin-conjugated polyacrylamide gel containing fluorescent particles (Φ 50 nm) for time-lapse imaging of the deformation of the gel (Fig. 1.3a). The time-dependent changes in the mechanical force generated by displacement of the FN bead can be examined by time-lapse recording of the fluorescent particles embedded in the substrate; translocation of beads reflects the deformation of the substrate. The amplitude and direction of deformation in the gel upon displacement of a FN bead by 4 μm were estimated (Fig. 1.3e). The image analysis of movements of the fluorescent particles showed that the mechanical force was transmitted to the area approximately 20 μm from the bead, where the substrate was moved in the direction of bead displacement.

Apical surface of the ells pretreated with cytochalasin D (cytoD) (100 nM, 30 min) could be deformed with a similar degree to that of the control FN-bead displacement, but the force required for the same displacement of the bead on the cytoD-treated cells was estimated at only 2.3 ± 0.5 nN ($n = 6$). This small amount is probably due to the disassembly of mechanically resistive stress fibers by cytoD, implying that the mechanical force was transmitted to the cell bottom via actin stress fibers.

1.2.5 Mechanical Stretching of Actin Stress Fibers by Displacing a FN-Bead to Activate MS Channels in HUVECs

To date, two mechanisms have been proposed for MS channel activation. One is that tension development in the lipid bilayer directly activates MS channels.

This is mostly based on biophysical analyses of the bacterial large conductance MS channel, MscL (Sukharev et al. 1999), and TRPC1 (Maroto et al. 2005) reconstituted in cytoskeleton-free liposomes. The second mechanism proposes that the extracellular matrix (Du et al. 1996) and cytoskeleton (Fukushige et al. 1999; Sokabe et al. 1991) are involved in channel activation; more specifically, tension in the cytoskeleton activates MS channels. A couple of studies (Byfield et al. 2004; Evans and Waugh 1977) have shown that a membrane interacting with cytoskeletons is stiffer than the membrane alone, implying that cytoskeletons would make the membrane a more efficient force-transmitting device. In fact, a certain MS channel shows a loss of mechanosensitivity in a cytoskeleton-free membrane (Zhang et al. 2000). It has also been reported that the cytoskeleton by itself will work as a force-focusing structure (Hu et al. 2003) owing to its linear structure with a relatively high elastic modulus (Kojima et al. 1994). Unfortunately, however, this idea is based on indirect evidence from electrophysiological and genetic analyses in eukaryotic cells (Corey et al. 2004). It is imperative to design and conduct experiments in which one can directly manipulate cytoskeletons while recording MS channel activities, preferably in intact cells.

We used intact cultured HUVECs to demonstrate that MS channels are activated by stress in the cytoskeleton. When HUVECs are subjected to uni-axial stretch, MS channels are activated, followed by an increase in the intracellular Ca^{2+} concentration ($[Ca^{2+}]_i$) because of Ca^{2+} influx through the channels (Naruse and Sokabe 1993). Stretching the extracellular substrate generates stress in the actin cytoskeleton (Pourati et al. 1998), which has been postulated to activate MS channels. However, detailed electrophysiological and imaging analyses of the channel activation is technically very difficult with this type of mechanical stimulation because of the movement of the cells upon stretching. We employed a new assay system that enables us to apply localized mechanical stresses onto focal adhesions to activate MS channels by dragging fibronectin-conjugated beads adhering to the dorsal cell surface or by injecting phalloidin-conjugated beads into cells (see Sect. 1.2.7), which then bind to actin stress fibers (and/or actin filaments). We demonstrated that direct mechanical stimulation (stretching) of the actin cytoskeleton using optical tweezers can activate MS channels in cultured HUVECs.

Immediately after a displacement of an FN bead, $[Ca^{2+}]_i$ increased rapidly and spread over the cell in a few seconds (Fig. 1.4c), then it returned to the resting $[Ca^{2+}]_i$ level in 3 min. The amplitude of the $[Ca^{2+}]_i$ increased monotonically with the degree of displacement (0.15–0.9 μm). The minimal response was induced by as short as 0.15 μm displacement (ca. 5.3 nN). The transient increase in $[Ca^{2+}]_i$ was largely inhibited (88.1 ± 4.3%, $n=5$) by 20 μM Gd^{3+} or in Ca^{2+}-free medium, but was not affected by thapsigargin (1 μM), suggesting that the $[Ca^{2+}]_i$ increase was mediated by the influx of Ca^{2+} through MS channels.

We examined whether cytoskeletal structures were involved in the activation of the MS channels by FN-bead displacement. The FN-bead-dependent $[Ca^{2+}]_i$ increase was nearly perfectly abolished (95.1 ± 5.5%, $n=3$) by the F-actin-disrupting agent cytoD (100 nM, 30 min). By contrast, the $[Ca^{2+}]_i$ increase was not affected by the microtubule-disrupting agent colcemid (1 μM, 30 min), suggesting that actin stress fibers are essential to induce the $[Ca^{2+}]_i$ increase by FN-bead displacement.

Electrophysiology should provide a better means than Ca^{2+} imaging to make quantitative analyses of the MS channel properties. For this purpose, we built a fully electronically controlled setup, shown in Fig. 1.4a, and made whole-cell patch-clamp recordings to monitor mechanically-induced channel currents. FN-bead

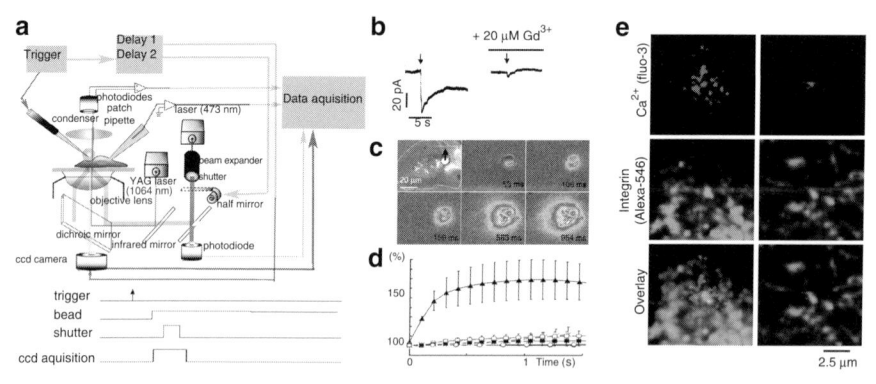

Fig. 1.4 High-speed imaging of $[Ca^{2+}]_i$ and the electrophysiological response induced by mechanical stimulation. (**a**) An electro-optical system for data acquisition. The trigger signal commands a piezo-driven glass pipette to move; the same signal triggers a focal plane shutter of a camera (α sweet; Minolta, Japan) after passing through a delay circuit. We put in a delay (e.g., 5 ms), since the piezo-driven glass pipette takes time to move. The signal also starts the acquisition of $[Ca^{2+}]_i$ images with a CCD camera under the control of software (MetaMorph; Universal Imaging, USA). The focal plane shutter enables brief (0.1 ms to 1 s) evanescent illumination of the specimen via a high NA (more than 1.4) lens; laser light is projected to the periphery of the objective lens to produce total internal reflection at the surface of the coverglass. We use evanescent (TIRF) illumination because it illuminates only the thin intracellular area near the bottom of the cell, which enables the recording of the rapid $[Ca^{2+}]_i$ increase beneath the membrane. To monitor the precise time of the illumination, half the laser power is projected to a photodiode. Signals of the bead movement, of laser illumination and of $[Ca^{2+}]_i$ image are fed into a data acquisition system. The time sequence of the trigger and the data acquisition is presented at *bottom* of the panel; trigger (an *arrow*), bead (bead movement detected with two photodiodes), shutter (laser illumination detected with diode) and CCD image acquisition. (**b**) Displacement of the bead induced an inward current. Gd^{3+} reduced the amplitude of the inward current. *Arrows* denote the displacement of the FN-bead. (**c**) Displacement of the FN-bead (1 μm for 100 ms) evoked an increase in $[Ca^{2+}]_i$. The *red circle* in the DIC image denotes the FN-bead. The *black arrow* indicates the direction of displacement. (**d**) $[Ca^{2+}]_i$ transients induced by bead displacement in different conditions. FN-bead displacement (1 μm for 100 ms) did not induce $[Ca^{2+}]_i$ increase in nominally Ca^{2+}-free solution (*asterisks*), and in solution contained 20 μM Gd^{3+} (*open squares*) or 100 nM cytochalasin D (*closed squares*). No $[Ca^{2+}]_i$ increase was also observed by the displacement of a bead conjugated with anti-LDL antibody (*open circles*) or with non-activating anti-integrin antibody (K-20) (*closed circles*) in normal solution. $[Ca^{2+}]_i$ increases were induced only when the FN-bead was displaced in the normal solution (*closed triangles*). Symbols except closed *triangles* are overlapped due to the $[Ca^{2+}]_i$ transients are low. (**e**) $[Ca^{2+}]_i$ microdomains 2–4 ms after the onset of the stimulation are shown as *red spots*. These spots are located in the vicinity of integrin clusters (*green*). The *upper left panel* in (**e**) shows approximately 10 $[Ca^{2+}]_i$ microdomains. The *middle left panel* shows integrin and the *lower left panel* shows an overlay of the $[Ca^{2+}]_i$ microdomain and the integrin. The *right-hand panels* in (**e**) show another example of the $[Ca^{2+}]_i$ microdomains 0–2 ms after the onset of the stimulation. Data in panels (**b**)–(**e**) are based on our previous report (Hayakawa et al. 2008)

displacement induced an inward current (23.4 ± 4 pA, $n = 15$) in less than 10 ms, and the current peaked within 100 ms (Fig. 1.4b) then gradually declined. This transient inward current was also significantly reduced by 20 μM Gd^{3+} ($91.5 \pm 3.1\%$ inhibition, $n = 4$; Fig. 1.4b), suggesting that they are from the same origin – MS channels. The short delay (<10 ms) implies that MS channels were activated directly not via biochemical processes (e.g., second messengers or G-proteins).

1.2.6 Direct Mechanical Stretching of Actin Stress Fibers by Dragging Beads Optically to Activate MS Channels

We developed an innovative technique to test the hypothesis directly that mechanical stress in the actin cytoskeleton activates MS channels. Beads that were coated with the actin cytoskeleton binding agent phalloidin were microinjected to cells through a glass micropipette. The injected beads and actin stress fibers were imaged by confocal microscopy (Fig. 1.5b), showing that most of the beads (or aggregates of beads) and stress fibers were overlapped in the same optical slice. In addition, live imaging of the preparation showed that the beads stayed in the same position, suggesting that the beads were closely associated with stress fibers and/or actin networks that connect stress fibers.

A traction force to actin cytoskeleton was applied by dragging a bead aggregate (approximately 400 nm in diameter) with laser tweezers. MS channel currents were recorded with the whole-cell patch-clamp technique while monitoring the position of the trapping point of laser tweezers (Fig. 1.5a). When the trapping point passed the aggregate, a transient inward current was evoked (Fig. 1.5c) and returned to a basal level in less than 1 s. For a second stimulation, the direction of the movement of the trapping point was reversed and passed the same aggregate, which again evoked an inward current with similar amplitude (Fig. 1.5c, right arrow). The average amplitude of the inward current was 7.4 ± 1.1 pA ($n = 22$), which was 32% of that induced by FN-bead displacement. The amplitude of the force of the laser trapping was estimated at 5.5 pN. When the laser-trapping focus passed an aggregate of phalloidin-conjugated beads, the $[Ca^{2+}]_i$ increased, suggesting that the Ca^{2+}-permeable MS channels are activated by direct mechanical manipulation of actin cytoskeleton. When a nonlabeled bead was trapped and moved by the optical tweezers, force would be transmitted to the substrate in a similar way without activating MS channels; these beads were displaced slightly and were probably stopped by their interaction with subcellular structures, such as microtubules, intermediate filaments or internal membranes. The stress in these structures was not transmitted to the MS channels or perhaps was not large enough to activate them.

The above results, showing that MS channels are activated by the stress in actin stress fibers terminated at basal FCs, leads to the hypothesis that MS channels in HUVECs locate near FCs. To test this idea, we examined the hypothetical colocalization of MS channels and FCs by imaging the $[Ca^{2+}]_i$ microdomain that comprises

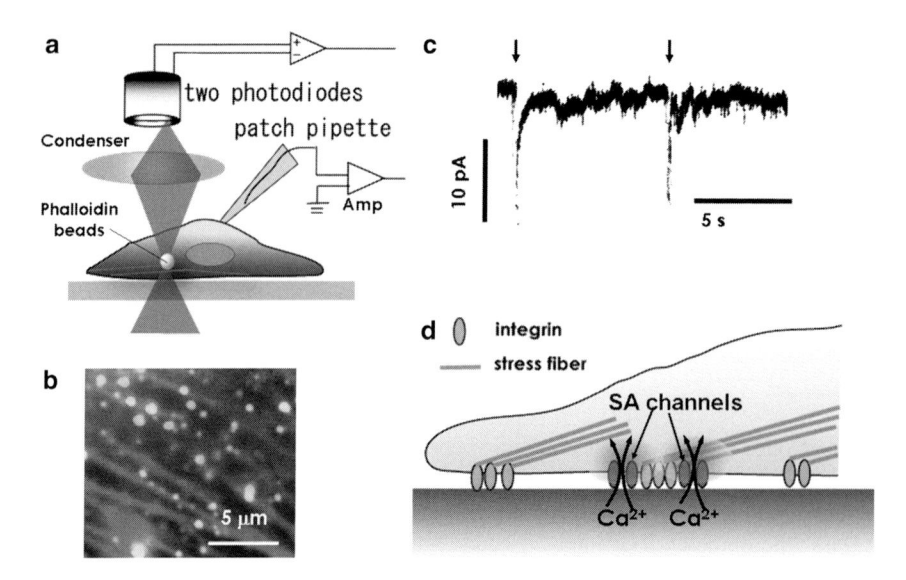

Fig. 1.5 Activation of MS channels by applying mechanical force to beads attached to actin stress fibers. (**a**) Phalloidin-conjugated 40 nm fluorescent beads were microinjected into HUVECs. These beads bound to the actin stress fibers and were trapped by laser optical tweezers. The movement of the trapping point was monitored with two photodiodes. Whole-cell patch-clamp recordings were made from the same cell. (**b**) Phalloidin-conjugated *green* fluorescent beads were located along the actin stress fibers (*red*). The cell was fixed 30 min after microinjection and stained with Rhodamine–phalloidin in this case. (**c**) A transient inward current was induced when the optical tweezers transiently passed an aggregate of phalloidin-coated beads (shown by the *two arrows*). (**d**) A schematic drawing of distribution of integrin clusters (*green*) and MS channels (*red*). The mechanical stress in actin stress fibers activates MS channels in the vicinity of the integrin clusters and induces Ca^{2+} influx to cytosol (*arrows*). This ionic flow is detected by the patch-clamp amplifier illustrated in panel (**a**). The data in panels (**b**) and (**c**) are based on our previous report (Hayakawa et al. 2008)

high-$[Ca^{2+}]_i$ regions at the cytoplasmic vestibule of Ca^{2+}-permeable channels (Zenisek et al. 2003). We imaged $[Ca^{2+}]_i$ microdomains using high-speed evanescent field microscopy, while monitoring the distribution of FCs as described in the following section.

1.2.7 $[Ca^{2+}]_i$ Microdomains in the Vicinity of the FCs Apparently Correspond to the MS Ca^{2+} Permeable Channels

In order to visualize the locus of the $[Ca^{2+}]_i$ microdomains precisely, it is necessary to acquire a fast snapshot (~milliseconds) of the $[Ca^{2+}]_i$ increase immediately after the onset of mechanical stimulation. We imaged the distribution of β1 integrin in living HUVECs, as previously described (Kawakami et al. 2001). In this experiment, $[Ca^{2+}]_i$

transients under the bead were imaged with a single brief (2 ms duration) evanescent laser illumination at, or 2 ms after, the onset of mechanical stimulation (Fig. 1.4e). A mean of 4 ± 0.94 ($\pm$SEM; $n=8$) $[Ca^{2+}]_i$ microdomains appeared in the vicinity of FCs underneath the FN bead with 2 ms of illumination; the diameter of the microdomain was 0.96 ± 0.07 μm ($n=33$), which agrees with the value in a previous report (Zou et al. 2002) and also with theoretical estimations. A recent study demonstrates that a Ca^{2+} microdomain reflects the activation of a single MS channel (Zou et al. 2002). The diameter of the Ca^{2+} microdomain (2.2 μm, 15 ms after channel opening) is similar to that imaged in this study, supporting the idea that it is possible to image the activation of a single MS channel or a small number of MS channels in the cluster. A theoretical estimation of the size of a Ca^{2+} microdomain also supports the above idea. Using the diffusion constant of Ca^{2+} in cytoplasm (0.6×10^{-6} cm^2/s; Hodgkin and Keynes 1957), the diameter of the Ca^{2+} microdomain 4 ms after the opening of a Ca^{2+} channel is calculated to be 1.1 μm, a value close to 0.96 μm in this study.

The number of $[Ca^{2+}]_i$ microdomains also principally agrees with the number of MS channels estimated from our whole-cell current measurements; the peak amplitude of the whole-cell MS current (23 pA) evoked by displacing a FN bead corresponds to 14–24 MS channels, given that the conductance of a single MS channel in HUVECs ranges from 24 to 40 pS (Lansman et al. 1987; Popp et al. 1992; Yao et al. 2001). If half of them were located on the basal surface of the cell, we would expect to see 7–12 channels (or microdomains) under these recording conditions. This is in the same order of the number in the above observation (approximately 4). All these results support the idea that the $[Ca^{2+}]_i$ microdomain in this study represents a single activated MS channel or a single cluster of a few MS channels. Interestingly, the center of individual $[Ca^{2+}]_i$ microdomains did not overlap with FCs in all $[Ca^{2+}]_i$ microdomains examined, but rather lay ca. 800 nm from the center of FCs (Fig. 1.4e); the area of integrin clusters did not overlap that of $[Ca^{2+}]_i$ microdomains. These results suggest that MS channels in the vicinity of FCs were activated by the force exerted along certain structures, possibly submembranous cytoskeletons, which may link MS channels with the FCs.

The importance of the cytoskeleton for activating MS channels has been repeatedly suggested, based on electrophysiological analyses of single MS channel activities (Sachs 1991; Sokabe et al. 1991; Sukharev et al. 1999; Yao et al. 2001). However, evidence from the studies on cytoskeleton involvement in cell mechanotransduction was indirect. The present study has succeeded in providing direct evidence on this hypothesis by a combination of electrophysiological, advanced imaging and laser trapping techniques.

In HUVECs, the force generated by the laser trapping (5.5 pN) was most likely transmitted through an actin filament to MS channels, where it activated them. The resulting inward current was ~7.4 pA at –40 mV, which corresponded to an activation of 5–8 MS channels. This suggests that a sub-pN force could activate a single MS channel in intact cells. It should be noted that, although there are a large number of channels within a cell, only a very limited number of channels (5–8 in the above case) could be activated by the optical manipulation of actin filaments.

We speculate that these relatively small forces (sub-pN) work on the MS channels of intact HUVECs, compared with activation forces in MscL of *Escherichia coli* (Sukharev et al. 1999; Gullingsrud and Schulten 2003), where approximately 40 pN was used in MD simulations, suggesting that cytoskeletal structures work as an efficient force-transmitting and -focusing structure to confer higher mechano-sensitivity on the MS channel in eukaryotic cells. This complex mechanosensing machinery would evolve from simple prokaryotes to complex eukaryotes. The molecular basis of the sensitivity increase in mechanosensing by cytoskeleton channel complex is not yet clarified.

A very small fraction of the force applied to the cell surface was still sufficient to activate MS channels. The force applied to the cell surface by the FN bead (35 nN) induced an inward current of 23 pA, whereas the optical manipulation (5.5 pN) of the actin cytoskeleton induced an inward current of 7.4 pA. These results suggest that most of the force on the FAs (or integrins) would be transmitted to the extracellular substrate via cytoskeleton, organelles and FAs. A very small fraction of the force acti-vates MS channels. The specific molecular components that transmit the force from an actin stress fiber to the MS channels presumably form a structural linkage, but they are not yet identified, remaining an interesting subject for future studies.

1.3   Roles of Mechanosensing in Cell Migration

The studies mentioned above not only provide direct evidence that development of tension in actin stress fibers can activate MS channels in intact cells but also reveal a new function of the actin cytoskeleton as a force-transmitting and -focusing device between integrins and MS channels. Such a concept has already been suggested in previous studies based on indirect observations (Doyle and Lee 2005; Hu et al. 2003; Munevar et al. 2004). Taken together, these studies along with ours open the door to a complex but intriguing world of cellular mechanotransduction involving remote activation and regulation of MS channels, since stress in the actin cytoskel-eton changes during a variety of mechano-related cell behaviors, such as morpho-genesis and locomotion. A good example can be seen in the locomotion of keratocytes (Doyle and Lee 2005), in which increased traction stresses induce a Gd^{3+}-sensitive $[Ca^{2+}]_i$ transient change, resulting in adhesion disassembly and rapid retraction at the rear of the cell. It is suggested that stress in the actin cytoskeleton will be focused on the MS channel to induce its activation followed by Gd^{3+}-sensitive Ca^{2+} entry.

The relationship between MS channels and cell adhesion has also been dis-cussed in a recent report (Doyle and Lee 2005). A range of cell behaviors, including cell migration (Doyle and Lee 2005; Munevar et al. 2004; Tanaka et al. 2005) and morphogenesis (Naruse et al. 1998b) are inhibited by the potential MS channel blocker Gd^{3+}. Stretching forces applied through flexible substrata also induced increases in both $[Ca^{2+}]_i$ and traction forces in NIH3T3 fibroblasts (Munevar et al. 2004). They suggest that stretch-activated Ca^{2+} entry in the frontal region regulates the organization of focal adhesions and the output of mechanical forces. The mechanotransduction mechanism proposed in this study, exerted by a molecular complex of MS channel, cytoskeleton, and cell adhesion, evidently plays an important role in these cellular functions.

In nerve growth cones, the role of MS channels in migration has not been totally clarified. TRPC1, a problematic MS channel candidate (Maroto et al. 2005; Gottlieb et al. 2008), may be involved in turning (Wang and Poo 2005).

1.4 Future Perspectives

Cells sense stress in the membrane and/or cytoskeletons. At present, stress is estimated from the strain of materials with known elastic modulus. However, the quantitative estimation of stress in the subcellular structures is generally difficult. In the case of membrane voltage, it is directly measured with microelectrodes or patch-clamp recording techniques; membrane voltage is also estimated with voltage-sensitive dyes. A technique for estimating the stress in the subcellular structures is truly required. FRET (fluorescence resonance energy transfer) is introduced to estimate conformational changes in proteins in response to amphipaths (chemical compounds to induce stress in the membrane or proteins) (Corry et al. 2005; Machiyama et al. 2009). This technique will also be applicable to the detection of stress in the subcellular structures. Development of methods for detecting stress in molecules will promote a new field of science, mechanobiology, which will open a way to explore the mechano-stress-mediated interactions between biomolecules, intracellular signaling, cell remodeling, and cell migration.

Acknowledgments This work was supported in part by Grants-in-aid for General Scientific Research (#13480216 to M.S. and #14580769 to H.T.), Scientific Research on Priority Areas (#15086270 to M.S.) and Creative Research (#16GS0308 to M.S.) from the Ministry of Education Science Sports and Culture and a grant from Japan Space Forum (to M.S. and H.T.).

References

Block SM (1990) Optical tweezers: a new tool for biophysics. In: Noninvasive techniques in cell biology. Wiley-Liss Inc, New York, pp 375–402
Byfield FJ, Aranda-Espinoza H, Romanenko VG, Rothblat GH, Levitan I (2004) Cholesterol depletion increases membrane stiffness of aortic endothelial cells. Biophys J 87 : 3336–3343
Corey DP, Garcia-Anoveros J, Holt JR, Kwan KY, Lin SY, Vollrath MA, Amalfitano A, Cheung ELM, Derfler BH, Duggan A, Geleoc GSG, Gray PA, Hoffman MP, Rehm HL, Tamasauskas D, Zhang DS (2004) TRPA1 is a candidate for the mechanosensitive transduction channel of vertebrate hair cells. Nature 432 : 723–730
Corry B, Rigby P, Liu ZW, Martinac B (2005) Conformational changes involved in MscL channel gating measured using FRET spectroscopy. Biophys J 89 : L49–L51
Dai J, Sheetz MP (1995) Mechanical properties of neuronal growth cone membranes studied by tether formation with laser optical tweezers. Biophys J 68 : 988–996

Dai J, Sheetz MP (1999) Membrane tether formation from blebbing cells. Biophys J 77:3363–3370

Dembo M, Wang YL (1999) Stresses at the cell-to-substrate interface during locomotion of fibroblasts. Biophys J 76:2307–2316

Doyle AD, Lee J (2005) Cyclic changes in keratocyte speed and traction stress arise from Ca^{2+}-dependent regulation of cell adhesiveness. J Cell Sci 118:369–379

Drake CJ, Davis LA, Hungerford JE, Little CD (1992) Perturbation of beta 1 integrin-mediated adhesions results in altered somite cell shape and behavior. Dev Biol 149:327–338

Du HP, Gu GQ, William CM, Chalfie M (1996) Extracellular proteins needed for C-elegans mechanosensation. Neuron 16:183–194

Evans EA, Waugh R (1977) Osmotic correction to elastic area compressibility measurements on red-cell membrane. Biophys J 20:307–313

Fukushige T, Siddiqui ZK, Chou M, Culotti JG, Gogonea CB, Siddiqui SS, Hamelin M (1999) MEC-12, an alpha-tubulin required for touch sensitivity in C-elegans. J Cell Sci 112:395–403

Giannone G, Jiang G, Sutton DH, Critchley DR, Sheetz MP (2003) Talin1 is critical for force-dependent reinforcement of initial integrin-cytoskeleton bonds but not tyrosine kinase activation. J Cell Biol 163:409–419

Gottlieb P, Folgering J, Maroto R, Raso A, Wood TG, Kurosky A, Bowman C, Bichet D, Patel A, Sachs F, Martinac B, Hamill OP, Honore E (2008) Revisiting TRPC1 and TRPC6 mechano-sensitivity. Pflugers Archiv Eur J Physiol 455:1097–1103

Gullingsrud J, Schulten K (2003) Gating of MscL studied by steered molecular dynamics. Biophys J 85:2087–2099

Hayakawa K, Tatsumi H, Sokabe M (2008) Actin stress fibers transmit and focus force to activate mechanosensitive channels. J Cell Sci 121:496–503

Hirata H, Tatsumi H, Sokabe M (2004) Tension-dependent formation of stress fibers in fibroblasts: a study using semi-intact cells. JSME Int J Ser C 47:962–969

Hodgkin AL, Keynes RD (1957) Movements of labelled calcium in squid giant axons. J Physiol (Lond) 138:253–281

Hu SH, Chen JX, Fabry B, Numaguchi Y, Gouldstone A, Ingber DE, Fredberg JJ, Butler JP, Wang N (2003) Intracellular stress tomography reveals stress focusing and structural anisotropy in cytoskeleton of living cells. Am J Physiol Cell Physiol 285:C1082–C1090

Iba T, Sumpio BE (1991) Morphological response of human endothelial cells subjected to cyclic strain in vitro. Microvasc Res 42:245–254

Imai K, Tatsumi H, Katayama Y (2000) Mechanosensitive chloride channels on the growth cones of cultured rat dorsal root ganglion neurons. Neuroscience 97:347–355

Jacobson BS, Cronin J, Branton D (1978) Coupling polylysine to glass beads for plasma membrane isolation. Biochim Biophys Acta 506:81–96

Kawakami K, Tatsumi H, Sokabe M (2001) Dynamics of integrin clustering at focal contacts of endothelial cells studied by multimode imaging microscopy. J Cell Sci 114:3125–3135

Kojima H, Ishijima A, Yanagida T (1994) Direct measurement of stiffness of single actin filaments with and without tropomyosin by in vitro nanomanipulation. Proc Natl Acad Sci USA 91:12962–12966

Lansman JB, Hallam TJ, Rink TJ (1987) Single stretch-activated ion channels in vascular endothelial cells as mechanotransducers? Nature 325:811–813

Machiyama H, Tatsumi H, Sokabe M (2009) Structural changes in the cytoplasmic domain of the mechanosensitive channel MscS during opening. Biophys J 97:1048–1057

Maroto R, Raso A, Wood TG, Kurosky A, Martinac B, Hamill OP (2005) TRPC1 forms the stretch-activated cation channel in vertebrate cells. Nat Cell Biol 7:179–185

Munevar S, Wang Y, Dembo M (2001) Traction force microscopy of migrating normal and H-ras transformed 3T3 fibroblasts. Biophys J 80:1744–1757

Munevar S, Wang YL, Dembo M (2004) Regulation of mechanical interactions between fibroblasts and the substratum by stretch-activated Ca^{2+} entry. J Cell Sci 117:85–92

Naruse K, Sokabe M (1993) Involvement of stretch-activated ion channels in Ca^{2+} mobilization to mechanical stretch in endothelial cells. Am J Physiol 264:C1037–C1044

Naruse K, Sai X, Yokoyama N, Sokabe M (1998a) Uni-axial cyclic stretch induces c-src activation and translocation in human endothelial cells via SA channel activation. FEBS Lett 441:111–115

Naruse K, Yamada T, Sokabe M (1998b) Involvement of SA channels in orienting response of cultured endothelial cells to cyclic stretch. Am J Physiol 274:H1532–H1538

Pollard TD, Borisy GG (2003) Cellular motility driven by assembly and disassembly of actin filaments. Cell 112:453–465

Popp R, Hoyer J, Meyer J, Galla HJ, Gogelein H (1992) Stretch-activated non-selective cation channels in the antiluminal membrane of porcine cerebral capillaries. J Physiol 454:435–449

Pourati J, Maniotis A, Spiegel D, Schaffer JL, Butler JP, Fredberg JJ, Ingber DE, Stamenovic D, Wang N (1998) Is cytoskeletal tension a major determinant of cell deformability in adherent endothelial cells? Am J Physiol 274:C1283–C1289

Ridley AJ, Schwartz MA, Burridge K, Firtel RA, Ginsberg MH, Borisy G, Parsons JT, Horwitz AR (2003) Cell migration: integrating signals from front to back. Science 302:1704–1709

Sachs F (1991) Mechanical transduction by membrane ion channels: a mini review. Mol Cell Biochem 104:57–60

Shi F, Chiu YJ, Cho YS, Bullard TA, Sokabe M, Fujiwara K (2007) Down-regulation of ERK but not MEK phosphorylation in cultured endothelial cells by repeated changes in cyclic stretch. Cardiovasc Res 73:813–822

Shirinsky VP, Antonov AS, Birukov KG, Sobolevsky AV, Romanov YA, Kabaeva NV, Antonova GN, Smirnov VN (1989) Mechano-chemical control of human endothelium orientation and size. J Cell Biol 109:331–339

Sokabe M, Sachs F, Jing Z (1991) Quantitative video microscopy of patch clamped membranes stress, strain, capacitance, and stretch channel activation. Biophys J 59:722–728

Sukharev SI, Sigurdson WJ, Kung C, Sachs F (1999) Energetic and spatial parameters for gating of the bacterial large conductance mechanosensitive channel, MscL. J Gen Physiol 113:525–540

Szaszak M, Gaborik Z, Turu G, McPherson PS, Clark AJ, Catt KJ, Hunyady L (2002) Role of the proline-rich domain of dynamin-2 and its interactions with Src homology 3 domains during endocytosis of the AT1 angiotensin receptor. J Biol Chem 277:21650–21656

Tanaka K, Naruse K, Sokabe M (2005) Effects of mechanical stresses on the migrating behavior of endothelial cells. In: Wada H (ed) Biomechanics at micro and nanoscale levels, vol I. World Scientific Publishing, Singapore, pp 75–87

Tatsumi H, Katayama Y (1999) Growth cones exhibit enhanced cell–cell adhesion after neurotransmitter release. Neuroscience 99:855–865

Wang GX, Poo MM (2005) Requirement of TRPC channels in netrin-1-induced chemotropic turning of nerve growth cones. Nature 434:898–904

Wang JG, Miyazu M, Matsushita E, Sokabe M, Naruse K (2001) Uniaxial cyclic stretch induces focal adhesion kinase (FAK) tyrosine phosphorylation followed by mitogen-activated protein kinase (MAPK) activation. Biochem Biophys Res Commun 288:356–361

Yao X, Kwan H, Huang Y (2001) Stretch-sensitive switching among different channel sublevels of an endothelial cation channel. Biochim Biophys Acta 1511:381–390

Zamir E, Katz BZ, Aota S, Yamada KM, Geiger B, Kam Z (1999) Molecular diversity of cell-matrix adhesions. J Cell Sci 112(Pt 11):1655–1669

Zenisek D, Davila V, Wan L, Almers W (2003) Imaging calcium entry sites and ribbon structures in two presynaptic cells. J Neurosci 23:2538–2548

Zhang Y, Gao F, Popov VL, Wen JW, Hamill OP (2000) Mechanically gated channel activity in cytoskeleton-deficient plasma membrane blebs and vesicles from *Xenopus* oocytes. J Physiol (Lond) 523:117–130

Zou H, Lifshitz LM, Tuft RA, Fogarty KE, Singer JJ (2002) Visualization of Ca^{2+} entry through single stretch-activated cation channels. Proc Natl Acad Sci USA 99:6404–6409

Chapter 2
Molecular Mechanisms Underlying Mechanosensing in Vascular Biology

Kimiko Yamamoto and Joji Ando

2.1 Introduction

The endothelial cells (ECs) lining blood vessels have a variety of functions and play a central role in the homeostasis of the circulatory system. Biochemical mediators, including hormones, cytokines and neurotransmitters, have long been thought to control EC functions. Recently, however, it has become clear that biomechanical forces generated by blood flow and blood pressure regulates EC functions. ECs are constantly exposed to shear stress, a frictional force generated by flowing blood, and to cyclic strain, which is caused by pulsatile changes in blood pressure. A number of recent studies have demonstrated that ECs have the ability to sense shear stress and cyclic strain as signals and transmit them into the cell interior, where they cause the ECs to change their morphology, functions and gene expression (Kamiya and Ando 1994; Kakisis et al. 2004; Chien 2007; Wang and Thampatty 2008). The EC responses to biomechanical forces are critical to maintaining normal vascular functions, and impairment of EC responses leads to the development of vascular diseases, including hypertension, thrombosis, aneurysms, and atherosclerosis. Immature cells, such as endothelial progenitor cells (EPCs) and embryonic stem (ES) cells, as well as mature ECs, have been shown to respond to biomechanical forces by undergoing changes in their proliferation, differentiation and organogenetic activity (Yamamoto et al. 2003b, 2005; Shimizu et al. 2008; Obi et al. 2009). This article will focus on fluid shear stress and review the molecular mechanisms underlying shear stress mechanotransduction in ECs.

K. Yamamoto
Laboratory of System Physiology, Department of Biomedical Engineering,
Graduate School of Medicine, The University of Tokyo, 7-3-1 Hongo, Bunkyo-ku,
Tokyo 113-0033, Japan

J. Ando (✉)
Laboratory of Biomedical Engineering, School of Medicine, Dokkyo Medical University,
880 Kita-kobayashi, Mibu, Tochigi 321-0293, Japan
e-mail: jo-ji@umin.ac.jp

M. Noda (ed.), *Mechanosensing Biology*,
DOI 10.1007/978-4-431-89757-6_2, © Springer 2011

2.2 EC Responses to Shear Stress

Blood flow generates a frictional force, shear stress, in ECs, and the following formula can be used to calculate its intensity (τ): $\tau = \mu\,du/dr$, where μ is blood viscosity, u is blood flow velocity, r is the radius of the blood vessel, and du/dr is the flow velocity gradient. Under physiological conditions, arterial ECs are exposed to a shear stress of around 20 dynes/cm^2, and venous ECs to shear stress ranging from 1.5 to 6 dynes/cm^2 (Kamiya et al. 1984).

It has been well established that ECs are sensitive to shear stress. When cultured ECs are exposed to shear stress in fluid-dynamically designed flow-loading devices, the ECs change their morphology and functions. ECs are polygonal under static culture conditions, but become elongated with their long axis oriented in the direction of flow in response to shear stress (Krueger et al. 1971; Dewey et al. 1981). EC functions also change in response to flow. For example, ECs increase production of various vasodilating substances, including nitric oxide (NO) (Buga et al. 1991; Ohno et al. 1993; Korenaga et al. 1994), prostacyclin (Frangos et al. 1985), C-type natriuretic peptide (Okahara et al. 1995), and adrenomedullin (Chun et al. 1997) in response to shear stress, and decrease the production of vasoconstricting factors, including endothelin (Sharefkin et al. 1991) and angiotensin-converting enzyme (Rieder et al. 1997). Shear stress also results in an increase in the antithrombotic activity and fibrinolytic activity of ECs by stimulating production of thrombomodulin (Takada et al. 1994) and plasminogen activators (Diamond et al. 1989). Shear stress also affects EC synthesis of growth factors (Hsieh et al. 1991; Mitsumata et al. 1993; Morita et al. 1993; Ohno et al. 1995; Malek et al. 1993), cytokines (Sterpetti et al. 1993), and reactive oxygen species (ROS) (Laurindo et al. 1994; De Keulenaer et al. 1998; Silacci et al. 2001; Hwang et al. 2003), which are involved in EC apoptosis and adhesive interactions with leukocytes.

When shear stress modulates EC functions, it usually affects the expression of related genes. Our DNA microarray analysis showed that approximately 3% of all EC genes examined showed some kind of response to shear stress (Ohura et al. 2003). Assuming that ECs express around 20,000 genes, this finding suggests that more than 600 genes are shear stress-responsive. Many studies, including our own, have demonstrated that shear stress regulates endothelial gene expression transcriptionally and/or posttranscriptionally (Ando et al. 1999), and various transcription factors, including AP-1, NFκB, Sp1, GATA6, Egr-1, and KLF2, and their binding sites in gene promoters (shear stress response elements), have been shown to be responsible for shear stress-mediated gene responses (Shyy et al. 1995; Khachigian et al. 1997; Korenaga et al. 1997, 2001; Lin et al. 1997; Sokabe et al. 2004; Huddleson et al. 2006). On the other hand, shear stress also regulates gene expression through mRNA stabilization, e.g., in the genes encoding endothelial nitric oxide synthase (eNOS) (Weber et al. 2005), cyclooxygenase 2 (Inoue et al. 2002), granulocyte–macrophage colony stimulating factor (Kosaki et al. 1998), and urokinase-type plasminogen activator (Sokabe et al. 2004).

2.3 Shear Stress Mechanotransduction

The fact that ECs respond to shear stress by changing their morphology, function, and gene expression indicates that ECs recognize shear stress as a signal and transmit it into the cell interior. A vast number of studies have been conducted to clarify the mechanisms underlying shear stress mechanotransduction, but much remains unclear (Davies 1995; Resnick et al. 2003). The following section will address shear stress signaling pathways and candidates for shear stress sensors.

2.3.1 Shear Stress Signaling Pathways

Based on the results of numerous studies, multiple pathways appear to be involved in the shear stress signal transduction (Fig. 2.1). When ion channels are activated, a variety of ions, including Ca^{2+}, K^+, Cl^-, and Na^+, enter or exit cells, and mechanical signals are transduced into changes in membrane potential and intracellular ion concentrations. Activation of G protein coupling receptors increases the activity of adenylate cyclase (AC) and phospholipase C (PLC), which induces second messengers, including cAMP, Ca^{2+}, IP_3, and diacylglyceride (DG). cAMP then activates cAMP-dependent protein kinase (PKA; protein kinase A), which catalyzes phosphorylation of a variety of proteins. Cytosolic Ca^{2+} binds to calmodulin and activates calmodulin-dependent kinase. IP_3 binds to its receptors expressed on the endoplasmic reticulum, where Ca^{2+} is stored, and the binding triggers Ca^{2+} release into the cytoplasm, and DG activates protein kinase C (PKC). Tyrosine kinase-type receptors transmit signals through phosphatidyl inositol-3 (PI3) kinase and phosphorylation of small G proteins, such as Ras, and leads to activation of MAP kinases. On the other hand, shear stress signals can enter via adhesion molecules, such as integrins located at focal contacts, and phosphorylate focal adhesion kinase (FAK). There is also a signaling pathway mediated by ROS, which are produced in ECs exposed to shear stress. Activation of these signal transduction pathways leads to activation of transcription factors and alterations of EC functions.

2.3.2 Shear Stress Sensors

Although shear stress activates multiple signal transduction pathways, which of the pathways are primary and which are secondary remains unclear, because the initial sensing mechanism or sensors that recognize shear stress have not been identified. Thus far, various membrane molecules and cellular microdomains, including ion channels, growth factor receptors, G proteins, caveolae, adhesion proteins, the cytoskeleton, the glycocalyx, and primary cilia, have been shown to play important roles in the shear stress sensing mechanism (Fig. 2.2).

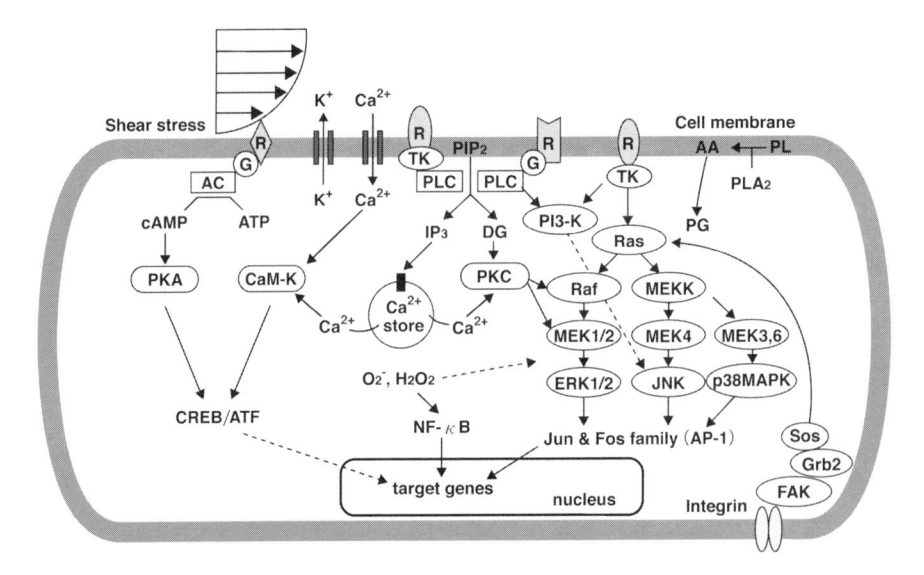

Fig. 2.1 Signal transduction factors that are capable of being activated by shear stress. Multiple pathways are involved in the shear stress signal transduction that leads to alterations in EC morphology and function and to activation of various transcription factors. It remains unclear which pathways are primary and which are secondary, and the initial sensor that recognizes shear stress has not been identified. It seems possible that shear stress activates several pathways simultaneously. R, receptor; G, G protein; AC, adenylate cyclase; cAMP, cyclic adenosine 3′,5′-monophosphate; ATP, adenosine 5′-triphosphate; TK, tyrosine kinase; PL, phospholipids; PLA2, phospholipase A2; AA, arachidonic acid; PG, prostaglandin; PLC, phospholipase C; IP3, inositol1,4,5-triphosphate; DG, 1,2-diacylglycerol; PKA, protein kinase A; CaM-K, calmodulin kinase; PKC, protein kinase C; PI3-K, phosphatidylinositol 3-phosphate; Ras and Raf, small G proteins; MEK, MAP kinase-ERK kinase; MEKK, MAP kinase-ERK kinase kinase; ERK, extracellular signal-regulated kinase; JNK, c-jun N-terminal kinase; p38 MAPK, maitogen activated kinase; Sos and Grb2, adaptor proteins; FAK, focal adhesion kinase; O_2^- and H_2O_2, reactive oxygen species; CREB/ATP, cAMP-responsive element-binding protein/activating transcription factor; NFkB, nuclear factor kappa B; Jun & Fos family (AP-1), transcription factors

2.3.2.1   Ion Channels

Various types of ion channels have been listed as candidates for shear stress sensors. Potassium ion channels open in response to shear stress, and their opening results in hyperpolarization of the plasma membrane, whereas activation of chloride ion channels by shear stress induces membrane depolarization (Nakache and Gaub 1988; Olesen et al. 1988; Barakat et al. 1999; Lieu et al. 2004; Gautam et al. 2006). Some types of Ca^{2+}-permeable cation channels have been shown to be shear stress-responsive. For example, P2X purinoceptors and transient receptor potential (trp) channels, both of which are Ca^{2+}-permeable channels expressed by ECs, open in response to shear stress and mediate the influx of extracellular Ca^{2+} across the plasma membrane (Yamamoto et al. 2000b; O'Neil and Heller 2005; Kohler et al. 2006; Oancea et al. 2006). The Ca^{2+} influx triggers subsequent Ca^{2+}-dependent

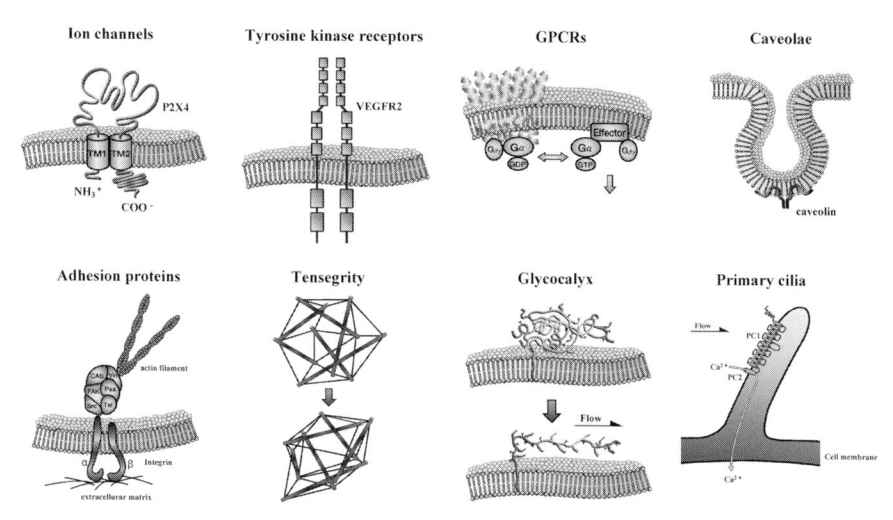

Fig. 2.2 Candidates for shear stress sensors. *Ion channels*, various types of ion channels including K+ channels, Cl− channels, and Ca²⁺ channels have been shown to be shear stress-responsive. P2X₄, a subtype of ATP-operated cation channel P2X purinoceptor, is one such ion channel, and in the form of a trimer functions as Ca²⁺ permeable channel. *Tyrosine kinase receptor*, rapid phosphorylation of vascular endothelial growth factor receptor 2 (VEGFR2) occurs in response to shear stress in a ligand-independent manner. *GPCRs*, G-protein-coupled receptors and G-protein itself have been shown to be activated by shear stress. *Caveolae*, caveolae, membrane microdomains containing a variety of receptors, ion channels, and signaling molecules, and their component protein, caveolin-1, have been demonstrated to be involved in shear stress sensing and response mechanisms. *Adhesion proteins*, various adhesion proteins located at sites of cell–cell and cell–matrix attachment are subjected to tension under shear stress and respond to it, resulting in the activation of downstream signal transduction pathways. Integrins are one class of such adhesion proteins. *Tensegrity*, tensegrity is a form of architecture that self-stabilizes its structure through the use of isolated compression-resistant elements (struts) and tensile elements (cables). When this model is applied to cells, microtubules and actin filaments are regarded as the struts and cables, respectively. It seems possible that these cytoskeleton components directly sense mechanical forces that deform cells. *Glycocalyx*, the random-coiled glycocalyx unfolds into a filament structure under flow conditions, and this conformational change affects the local concentrations and transport of ions, amino acids, and growth factors, or triggers signal transduction via the intracellular core protein of the glycocalyx. *Primary cilia*, the bending of primary cilia induced by flow activates polycystin-2 (PC2), which leads to an influx of extracellular Ca²⁺ via polycystin-1 (PC1), a member of the transient receptor potential (trp) ion channel family

signaling pathways that lead to EC responses to shear stress. The next section will review what has been discovered about P2X purinoceptor-mediated Ca²⁺ signaling of shear stress.

P2X₄ Channel-Mediated Ca²⁺ Signaling of Shear Stress

When cultured ECs were subjected to shear stress, the intracellular Ca²⁺ concentration increased in a dose-dependent manner (Fig. 2.3a) (Ando et al. 1988, 1993;

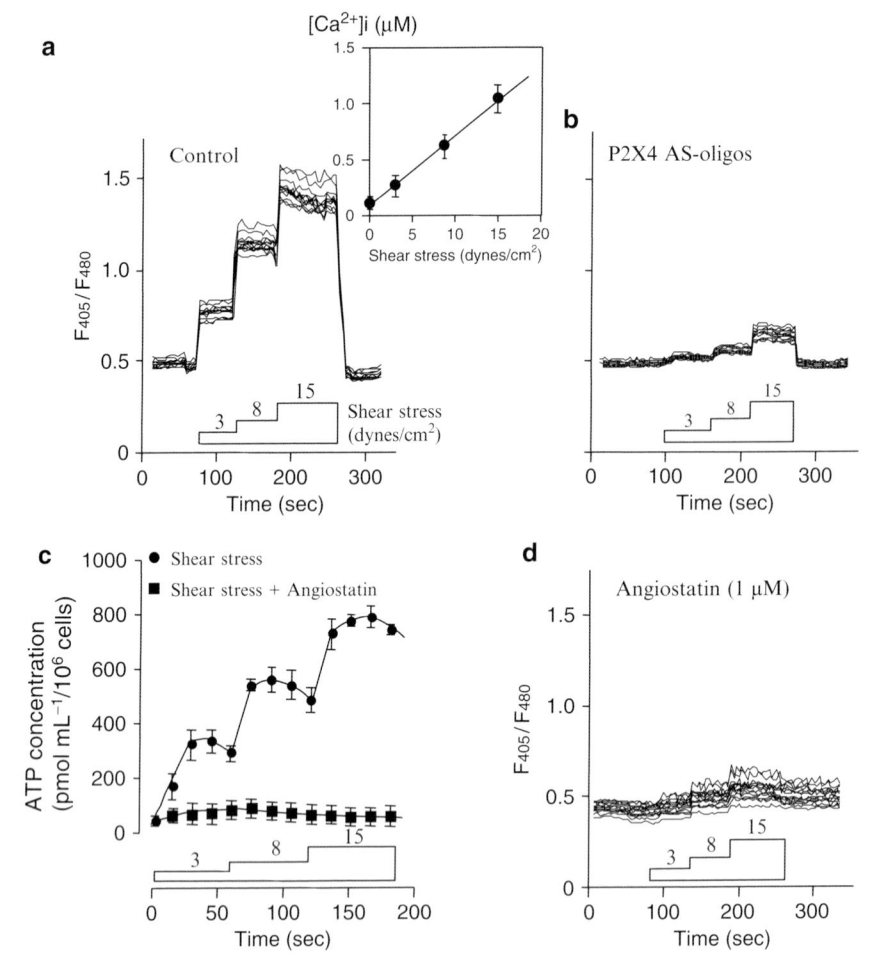

Fig. 2.3 Shear stress-dependent Ca^{2+} influx via $P2X_4$ channels. (**a**) Shear stress-induced Ca^{2+} response. Intracellular Ca^{2+} concentrations ($[Ca^{2+}]_i$) increased in a stepwise manner when cultured human pulmonary artery ECs (HPAECs) were exposed to stepwise increases in shear stress, and a linear relationship was found between the Ca^{2+} concentration and shear stress, indicating that ECs are capable of accurately converting information on shear stress into changes in Ca^{2+} concentration. The Ca^{2+} response was attributable to an influx of extracellular Ca^{2+}, because it did not occur in the absence of extracellular Ca^{2+}. The ratio of the emitted light of the fluorescent Ca^{2+} indicator Indo-1/ AM at 405 nm (F405) and 480 nm (F480) reflects $[Ca^{2+}]_i$. (**b**) Involvement of $P2X_4$ in the Ca^{2+}influx. Antisense-oligonucleotides (AS-oligos) targeted against $P2X_4$ that knockout $P2X_4$ expression in HPAECs markedly suppressed the shear stress-dependent Ca^{2+} responses. (**c**) ATP release in response to shear stress. HPAECs released ATP in a shear stress-dependent manner, and the ATP releasing response was completely blocked with angiostatin, a membrane-impermeable ATP synthase inhibitor, suggesting the involvement of cell surface ATP synthase in the shear stress-induced ATP release. (**d**) Involvement of ATP release in shear stress-dependent Ca^{2+} influx. Angiostatin, almost completely blocked the Ca^{2+}response to shear stress, suggesting that ECs have shear stress mechanotransduction mechanisms in which shear stress stimulates ECs to release ATP via cell surface ATP synthase, which leads to $P2X_4$ activation followed by a Ca^{2+} influx

Yamamoto et al. 2000a). The Ca^{2+} response was due to an influx of extracellular Ca^{2+} via $P2X_4$, a subtype of ATP-operated cation channel P2X purinoceptor. Treatment of ECs with an antisense oligonucleotide targeted to their $P2X_4$ channels blocked the shear stress-induced Ca^{2+} influx (Fig. 2.3b). Activation of $P2X_4$ required ATP, which was supplied in the form of endogenous ATP released by the ECs (Yamamoto et al. 2003a). The ECs released ATP dose-dependently in response to shear stress (Fig. 2.3c), and suppression of the ATP release with the ATP synthase inhibitor angiostatin abolished the shear stress-induced Ca^{2+} responses (Fig. 2.3d). These findings suggest that ECs are capable of accurately converting information regarding shear stress intensity into changes in intracellular Ca^{2+} concentrations through ATP release and $P2X_4$ activation. Although the mechanism responsible for the ATP release in response to shear stress remains unclear, several possibilities have been suggested: shear stress may increase ATP release through vesicular exocytosis or ATP binding cassette transporters, or it may activate cell surface ATP synthase to catalyze the synthesis of ATP (Bodin and Burnstock 2001; Reigada and Mitchell 2005; Yamamoto et al. 2007).

Roles of Shear Stress Ca^{2+} Signaling in Control of Circulatory System

Our study in $P2X_4$ gene knockout mice ($P2X_4$ KO mice) revealed physiological roles of $P2X_4$-mediated shear stress signal transduction in the circulatory system (Yamamoto et al. 2006). The $P2X_4$ KO mice did not exhibit normal EC responses to shear stress, such as a Ca^{2+} influx and subsequent production of NO. The vasodilation induced by acute increases in blood flow in situ was much weaker in the $P2X_4$ KO mice, and the $P2X_4$ KO mice had higher blood pressure than wild-type mice. No adaptive vascular remodeling, i.e., decrease in vessel size in response to a chronic decrease in blood flow, was observed in the $P2X_4$ KO mice (Kuhlencordt et al. 2004). The impaired vascular remodeling resembled that observed in eNOS KO mice. These findings suggest that Ca^{2+} signaling of shear stress via $P2X_4$ plays a crucial role in the control of vascular tone, and in blood flow-dependent vasodilation and vascular remodeling, through endothelial NO production.

2.3.2.2 Tyrosine Kinase Receptors

Activation of tyrosine kinase receptors, including vascular endothelial growth factor receptor 2 (VEGFR2) and angiopoietin receptor Tie-2, occurs in ECs exposed to shear stress, and the activation is assumed to be ligand-independent, because it occurs in the absence of VEGF or angiopoietin (Chen et al. 1999; Shay-Salit et al. 2002; Jin et al. 2003; Lee and Koh 2003). We showed that shear stress induces a ligand-independent phosphorylation of VEGFR2 in ES-cell-derived VEGFR2-positive cells. Although the mechanisms by which mechanical forces activate tyrosine kinase receptors are not well understood, mechanical forces may trigger dimerization of VEGFR2 monomers by affecting their spatial distribution in the cell membrane, or they may activate the receptors by changing their conformation and promoting the binding of tyrosine kinases, such as Src, that are capable of phosphorylating the receptors. Phosphorylation of these tyrosine kinase receptors leads to activation of various protein kinases, including ERK, JNK, PI3-kinase, and Akt, which result in eNOS activation and inhibition of apoptosis.

2.3.2.3 G Proteins

G-protein-coupled receptors (GPCRs) have been postulated to play a role in shear stress signal transduction (Gudi et al. 1996; Jo et al. 1997). GPCR conformational dynamics in a single EC were detected by real-time molecular imaging using fluorescence resonance energy transfer (FRET), and shear stress was found to cause a conformational transition of bradykinin B2 GPCRs that led to activation of the receptors (Chachisvilis et al. 2006). Shear stress was also demonstrated to activate purified G proteins reconstituted in liposomes in the absence of receptor proteins, suggesting that G proteins themselves act as a primary mechanotransducer (Gudi et al. 1998).

2.3.2.4 Caveolae

Caveolae are membrane microdomains measuring around 50–100 nm in length that are visible as flask-shaped invaginations below the surface of cells (Fig. 2.2), and they contain many signaling molecules, including receptors, ion channels, and protein kinases (Anderson 1993; Shaul and Anderson 1998). It has been well documented that caveolae play an important role in shear stress signal transduction (Rizzo et al. 1998, 2003; Boyd et al. 2003). We have observed that flow-induced Ca^{2+} responses in ECs start at caveolae and propagate through the entire cell in the form of a Ca^{2+} wave (Isshiki et al. 1998). The Ca^{2+} increase occurring in the vicinity of caveolae causes them to rapidly liberate eNOS in the cytoplasm, where the activated eNOS catalyzes the production of NO. Treatment of ECs with an antibody against caveolin-1, a constitutive protein of caveolae, has been shown to prevent shear stress-mediated ERK activation (Park et al. 2000). It was also demonstrated that shear stress activates sphingomyelinase located in caveolae, causing them to produce ceramide, and leading to ERK activation and Akt-mediated eNOS activation (Czarny and Schnitzer 2004). The roles of caveolae and caveolin-1 in shear stress-mediated regulation of vascular functions have been assessed in caveolin-1 KO mice, which exhibit complete absence of caveolae in vessel walls (Yu et al. 2006). The caveolin-1 KO mice were characterized by impaired blood flow-dependent vasodilation and vascular remodeling in comparison with wild-type mice. These impairments were rescued by reconstituting caveolin-1 into the endothelium of the KO mice, suggesting that caveolae and caveolin-1 are involved in shear stress-mediated control of vascular functions.

2.3.2.5 Adhesion Proteins

It has been proposed that shear stress is transmitted from the apical surface of ECs through the cytoskeleton to points of attachment at cell–cell and cell–matrix adhesions, and if that is true, adhesion proteins may serve as mechanotransducers.

Integrins are transmembrane glycoproteins composed of α and β subunits (Fig. 2.2). Their extracellular domain binds directly to extracellular matrix proteins, and their cytoplasmic domains interact with many proteins aggregated at focal contacts, including both signaling molecules, such as FAK, Src family protein kinases, Fyn, and p130[CAS], and cytoskeletal proteins, such as α-actinin, vinculin, talin, tensin, and paxillin (Ruoslahti 1991; Burridge and Chrzanowska-Wodnicka 1996). Evidence has been found that shear stress activates integrins. When integrins are activated by shear stress, FAK, paxillin, c-Src, Fyn, and p130[CAS] are rapidly activated, thereby leading to the activation of Ras-ERK pathways (Ishida et al. 1996; Li et al. 1997). The results of experiments in which a magnetic twisting device was used to apply shear stress directly to cell surface integrins suggested that integrins are capable of functioning as mechanosensors and transmitting shear stress signals to the cytoskeleton (Wang et al. 1993; Chen et al. 2001). Indeed, when integrins were twisted by magnetic microbeads coated with antibodies against integrins, cytoskeletal filaments became reoriented and a force-dependent cell stiffening response occurred. It has also been demonstrated that activation of integrins by shear stress induces microfilament reorganization through activation of RohA small GTPase and phosphorylation of an actin-regulatory protein coffilin, and that shear stress causes integrins to translocate to caveolae, leading to activation of caveolin-1, Src, and myosin light chain kinase, which results in the formation of stress fibers (Shyy and Chien 2002). In addition, there is evidence implicating integrin signaling in the activation of VEGFR2 induced by shear stress but not by VEGF (Wang et al. 2002).

Platelet endothelial cell adhesion molecule-1 (PECAM-1), a member of the immunoglobulin superfamily, is localized to the cell–cell borders of ECs, where it mediates the leukocytes extravasation during the inflammatory response. A novel mechanosignaling pathway via PECAM-1 has been proposed (Harada et al. 1995; Osawa et al. 2002; Fujiwara 2006). PECAM-1 is tyrosine phosphorylated within 30 s of the start of exposure to shear stress, and as a result Ras signaling pathway is activated, leading to ERK activation. Similar signaling events occurred when magnetic beads coated with antibodies against PECAM-1 were used to directly apply tugging force to PECAM-1 molecules on the EC surface. These results seem to indicate that PECAM-1 is a mechanosensitive molecule.

Cadherins are the major proteins of the adherens junctions that mediate cell–cell adhesions. The cytoplasmic domains of cadherins are linked to the actin cytoskeleton via catenin proteins, including β-catenin and plakoglobin (Dejana et al. 1999). It has recently been proposed that vascular endothelial cadherin (VE-cadherin), which is specific for ECs, forms a mechanosensory complex with PECAM-1 and VEGFR2 that plays a critical role in shear stress signal transduction (Tzima et al. 2005). In this system, it is assumed that PECAM-1 directly transduces mechanical forces, that VEGFR2 activates PI3 kinase, which in turn, mediates integrin activation, and that VE-cadherin functions as an adaptor to form signaling complexes. ECs in which VE-cadherin or PECAM-1 had been knocked out failed to show normal responses to shear stress, such as activation of PI3 kinase, Akt, and integrin, or an alignment of actin filaments in the direction of flow.

2.3.2.6 Cytoskeleton

Living cells stabilize their structure and shape by means of an interconnected network of cytoskeleton components that include microfilaments, microtubules, and inter-mediate filaments. A tensegrity cell model has been proposed to explain how mechanical forces are transduced into a biochemical response (Fig. 2.2) (Ingber 1991, 1997, 2006; Wang et al. 2001). The cell model is constructed with a series of isolated compression-resistant sticks that resist the pull of surrounding tensile strings and thereby create an internal prestress that stabilizes the entire network. When mechanical forces are applied to the tensegrity model, the structural elements rearrange without undergoing any topographical disruption or loss of tensional continuity, which may directly activate signaling molecules that associate with the cytoskeleton. The cytoskeleton has been shown to play an important role in shear stress-induced NO production and ICAM-1 gene expression by ECs (Knudsen and Frangos 1997; Imberti et al. 2000).

2.3.2.7 Glycocalyx

The surface of ECs is covered with a layer of membrane-bound macromolecules that constitute the glycocalyx. The thickness of the EC surface glycocalyx ranges from 0.05 to 1 μm as vessels decrease in size from arterial and venous macrovessels to microvessels (Secomb et al. 2001). The glycocalyx has been considered a pos-sible shear stress sensor, because it is located between flowing blood and the cell membrane (Weinbaum et al. 2003; Tarbell and Pahakis 2006). Involvement of the glycocalyx in EC responses to shear stress has been demonstrated by the finding that degradation of hyaluronic acid glycosaminoglycans within the glycocalyx with hyaluronidase significantly decreases flow-induced NO production in isolated canine femoral arteries (Mochizuki et al. 2003), and by the finding that enzymatic removal of heparan sulfate with heparinase completely inhibits NO production in response to shear stress in bovine aortic ECs (Florian et al. 2003). Two possible mechanisms have been proposed to explain how the EC glycocalyx mediates shear stress mechanotransduction. One possible mechanism is that heparan sulfate pro-teoglycan is present as a random coil under no-flow conditions, but with increasing flow becomes unfolded into a filament structure (Fig. 2.2) (Siegel et al. 1996). This conformational change is accompanied by an increase in binding sites for Na^+ ions, and the Na^+ binding may trigger the signal transduction. In addition to glycocalyx-mediated regulation of the local concentration gradient and transport of ions, amino acids, and growth factors, it is also possible that shear stress is transmitted to the cell interior through the actin cytoskeleton or intracellular signaling molecules that directly associate with the core protein of the glycocalyx.

2.3.2.8 Primary Cilia

The presence of primary cilia having a rod-like, nonmotile structure and protruding from the apical cell membranes has been reported in embryonic ECs, HUVECs, and human aortic ECs (Bystrevskaya et al. 1988; Iomini et al. 2004; Van der Heiden et al. 2008). Recent studies provide evidence that the primary cilia mediate the mechanism by which ECs sense and respond to shear stress (Singla and Reiter 2006; Hierck et al. 2008). Since primary cilia are physically connected to cytoskeletal microtubules, their bending by flow is assumed to transmit shear stress signals into the cells through the cytoskeleton. The bending of the primary cilia may also activate Ca^{2+}-permeable ion channels and trigger Ca^{2+} signaling. It has recently been shown that polycystin-1, an 11-transmembrane protein with a long extracellular domain, and polycysitn-2, a member of a superfamily of trp channels, are localized on the cilia of ECs and together involved in shear stress sensing (Fig. 2.2) (Nauli et al. 2008; Aboualaiwi et al. 2009). ECs in which polycystin-1 and polycystin-2 have been knocked out are unable to transduce shear stress into changes in intracellular Ca^{2+} concentration or to produce NO in response to shear stress.

2.4 Conclusion

A large amount of work has been done on shear stress-dependent endothelial mechanobiology during the past 30 years, and it has revealed how ECs recognize shear stress and respond to it. Nevertheless, shear stress sensing mechanisms are still not fully understood. A striking feature of shear stress mechanotransduction is that the shear stress activates a variety of membrane molecules and microdomains almost simultaneously, leading to signal transduction through multiple pathways. It should be noted in regard to shear stress sensing by ECs that they are simultaneously exposed to both shear stress and cyclic strain in vivo. As a physical force, cyclic strain is roughly 10,000 times greater than shear stress, and causes a large deformation of ECs that ranges from 5 to 10% at a frequency of around 1 Hz (Dobrin 1978). Since this means that ECs would have to recognize shear stress while being greatly deformed by cyclic strain, it seems unlikely that ECs sense shear stress through integrins, the cytoskeleton, or unfolding of molecules directly caused by mechanical force. Unknown principles may be at work in the shear stress sensing mechanisms. Recent studies have provided data suggesting that changes in the properties of the plasma membrane are involved in shear stress mechanotransduction. Plasma membrane fluidity has been shown to increase in response to shear stress (Haidekker et al. 2000; Butler et al. 2001, 2002; Gojova and Barakat 2005;

Dangaria and Butler 2007). The response occurred within 10 s of application of shear stress and is assumed to lead to activation of GPCRs. It has also been demonstrated that depleting cholesterol from the plasma membrane abolishes EC responses to shear stress, including ERK activation and eNOS activation (Park et al. 1998; Lungu et al. 2004). Shear stress may first modify the properties of the plasma membrane and then affect ion channels, receptors, adhesion proteins, and phosphorylation status of signaling molecules. Elucidation of shear stress sensing mechanisms should lead to a better understanding of how blood flow regulates the circulatory system, how physical exercise exerts a beneficial effect on the human body (Green et al. 2004; Laughlin et al. 2008), and how hemodynamic factors are involved in the pathophysiology of vascular diseases, including hypertension, thrombosis, aneurysms, and atherosclerosis. The invention of novel means of manipulating endothelial shear stress sensing may lead to the development of new therapies for the above vascular diseases. Furthermore, the increase in knowledge about shear stress mechanotransduction should be helpful in better understanding the vital phenomena that occur based on interactions between genetic information and environmental factors, including mechanical forces, as a whole.

Acknowledgments This work was partly supported by grants-in-aid for Scientific Research (A18200030, B19300155, and S21220011) and Scientific Research on Priority Areas from the Japanese Ministry of Education, Culture, Sports, Science and Technology (17076002). The authors acknowledge Dr. Akira Kamiya for valuable support and guidance for our work.

References

Aboualaiwi WA, Takahashi M et al (2009) Ciliary polycystin-2 is a mechanosensitive calcium channel involved in nitric oxide signaling cascades. Circ Res 104:860–869
Anderson RG (1993) Caveolae: where incoming and outgoing messengers meet. Proc Natl Acad Sci USA 90:10909–10913
Ando J, Komatsuda T et al (1988) Cytoplasmic calcium response to fluid shear stress in cultured vascular endothelial cells. In Vitro Cell Dev Biol 24:871–877
Ando J, Ohtsuka A et al (1993) Wall shear stress rather than shear rate regulates cytoplasmic Ca++ responses to flow in vascular endothelial cells. Biochem Biophys Res Commun 190:716–723
Ando J, Korenaga R et al (1999) Flow-induced endothelial gene regulation. In: Lelkes PI (ed) Mechanical forces and the endothelium. Harwood Academic Publishers, Singapore, pp 111–126
Barakat AI, Leaver EV et al (1999) A flow-activated chloride-selective membrane current in vascular endothelial cells. Circ Res 85:820–828
Bodin P, Burnstock G (2001) Evidence that release of adenosine triphosphate from endothelial cells during increased shear stress is vesicular. J Cardiovasc Pharmacol 38:900–908
Boyd NL, Park H et al (2003) Chronic shear induces caveolae formation and alters ERK and Akt responses in endothelial cells. Am J Physiol Heart Circ Physiol 285:H1113–H1122
Buga GM, Gold ME et al (1991) Shear stress-induced release of nitric oxide from endothelial cells grown on beads. Hypertension 17:187–193
Burridge K, Chrzanowska-Wodnicka M (1996) Focal adhesions, contractility, and signaling. Annu Rev Cell Dev Biol 12:463–518
Butler PJ, Norwich G et al (2001) Shear stress induces a time- and position-dependent increase in endothelial cell membrane fluidity. Am J Physiol Cell Physiol 280:C962–C969

Butler PJ, Tsou TC et al (2002) Rate sensitivity of shear-induced changes in the lateral diffusion of endothelial cell membrane lipids: a role for membrane perturbation in shear-induced MAPK activation. FASEB J 16:216–218

Bystrevskaya VB, Lichkun VV et al (1988) An ultrastructural study of centriolar complexes in adult and embryonic human aortic endothelial cells. Tissue Cell 20:493–503

Chachisvilis M, Zhang YL et al (2006) G protein-coupled receptors sense fluid shear stress in endothelial cells. Proc Natl Acad Sci USA 103:15463–15468

Chen KD, Li YS et al (1999) Mechanotransduction in response to shear stress. Roles of receptor tyrosine kinases, integrins, and Shc. J Biol Chem 274:18393–18400

Chen J, Fabry B et al (2001) Twisting integrin receptors increases endothelin-1 gene expression in endothelial cells. Am J Physiol Cell Physiol 280:C1475–C1484

Chien S (2007) Mechanotransduction and endothelial cell homeostasis: the wisdom of the cell. Am J Physiol Heart Circ Physiol 292:H1209–H1224

Chun TH, Itoh H et al (1997) Shear stress augments expression of C-type natriuretic peptide and adrenomedullin. Hypertension 29:1296–1302

Czarny M, Schnitzer JE (2004) Neutral sphingomyelinase inhibitor scyphostatin prevents and ceramide mimics mechanotransduction in vascular endothelium. Am J Physiol Heart Circ Physiol 287:H1344–H1352

Dangaria JH, Butler PJ (2007) Macrorheology and adaptive microrheology of endothelial cells subjected to fluid shear stress. Am J Physiol Cell Physiol 293:C1568–C1575

Davies PF (1995) Flow-mediated endothelial mechanotransduction. Physiol Rev 75:519–560

De Keulenaer GW, Chappell DC et al (1998) Oscillatory and steady laminar shear stress differentially affect human endothelial redox state: role of a superoxide-producing NADH oxidase. Circ Res 82:1094–1101

Dejana E, Bazzoni G et al (1999) Vascular endothelial (VE)-cadherin: only an intercellular glue? Exp Cell Res 252:13–19

Dewey CF Jr, Bussolari SR et al (1981) The dynamic response of vascular endothelial cells to fluid shear stress. J Biomech Eng 103:177–185

Diamond SL, Eskin SG et al (1989) Fluid flow stimulates tissue plasminogen activator secretion by cultured human endothelial cells. Science 243:1483–1485

Dobrin PB (1978) Mechanical properties of arteries. In: Caro CG, Pedley TJ, Schroter RC, Seed WA (eds) The mechanics of the circulation. Oxford University Press, New York, pp 397–460

Florian JA, Kosky JR et al (2003) Heparan sulfate proteoglycan is a mechanosensor on endothelial cells. Circ Res 93:e136–e142

Frangos JA, Eskin SG et al (1985) Flow effects on prostacyclin production by cultured human endothelial cells. Science 227:1477–1479

Fujiwara K (2006) Platelet endothelial cell adhesion molecule-1 and mechanotransduction in vascular endothelial cells. J Intern Med 259:373–380

Gautam M, Shen Y et al (2006) Flow-activated chloride channels in vascular endothelium. Shear stress sensitivity, desensitization dynamics, and physiological implications. J Biol Chem 281:36492–36500

Gojova A, Barakat AI (2005) Vascular endothelial wound closure under shear stress: role of membrane fluidity and flow-sensitive ion channels. J Appl Physiol 98:2355–2362

Green DJ, Maiorana A et al (2004) Effect of exercise training on endothelium-derived nitric oxide function in humans. J Physiol 561:1–25

Gudi SR, Clark CB et al (1996) Fluid flow rapidly activates G proteins in human endothelial cells. Involvement of G proteins in mechanochemical signal transduction. Circ Res 79:834–839

Gudi S, Nolan JP et al (1998) Modulation of GTPase activity of G proteins by fluid shear stress and phospholipid composition. Proc Natl Acad Sci USA 95:2515–2519

Haidekker MA, L'Heureux N et al (2000) Fluid shear stress increases membrane fluidity in endothelial cells: a study with DCVJ fluorescence. Am J Physiol Heart Circ Physiol 278:H1401–H1406

Harada N, Masuda M et al (1995) Fluid flow and osmotic stress induce tyrosine phosphorylation of an endothelial cell 128 kDa surface glycoprotein. Biochem Biophys Res Commun 214:69–74

Hierck BP, Van der Heiden K et al (2008) Primary cilia sensitize endothelial cells for fluid shear stress. Dev Dyn 237:725–735

Hsieh HJ, Li NQ et al (1991) Shear stress increases endothelial platelet-derived growth factor mRNA levels. Am J Physiol 260:H642–H646

Huddleson JP, Ahmad N et al (2006) Up-regulation of the KLF2 transcription factor by fluid shear stress requires nucleolin. J Biol Chem 281:15121–15128

Hwang J, Ing MH et al (2003) Pulsatile versus oscillatory shear stress regulates NADPH oxidase subunit expression: implication for native LDL oxidation. Circ Res 93:1225–1232

Imberti B, Morigi M et al (2000) Shear stress-induced cytoskeleton rearrangement mediates NF-kappaB-dependent endothelial expression of ICAM-1. Microvasc Res 60:182–188

Ingber D (1991) Integrins as mechanochemical transducers. Curr Opin Cell Biol 3:841–848

Ingber DE (1997) Tensegrity: the architectural basis of cellular mechanotransduction. Annu Rev Physiol 59:575–599

Ingber DE (2006) Cellular mechanotransduction: putting all the pieces together again. FASEB J 20:811–827

Inoue H, Taba Y et al (2002) Transcriptional and posttranscriptional regulation of cyclooxygenase-2 expression by fluid shear stress in vascular endothelial cells. Arterioscler Thromb Vasc Biol 22:1415–1420

Iomini C, Tejada K et al (2004) Primary cilia of human endothelial cells disassemble under laminar shear stress. J Cell Biol 164:811–817

Ishida T, Peterson TE et al (1996) MAP kinase activation by flow in endothelial cells. Role of beta 1 integrins and tyrosine kinases. Circ Res 79:310–316

Isshiki M, Ando J et al (1998) Endothelial Ca^{2+} waves preferentially originate at specific loci in caveolin-rich cell edges. Proc Natl Acad Sci USA 95:5009–5014

Jin ZG, Ueba H et al (2003) Ligand-independent activation of vascular endothelial growth factor receptor 2 by fluid shear stress regulates activation of endothelial nitric oxide synthase. Circ Res 93:354–363

Jo H, Sipos K et al (1997) Differential effect of shear stress on extracellular signal-regulated kinase and N-terminal Jun kinase in endothelial cells. Gi2- and Gbeta/gamma-dependent signaling pathways. J Biol Chem 272:1395–1401

Kakisis JD, Liapis CD et al (2004) Effects of cyclic strain on vascular cells. Endothelium 11:17–28

Kamiya A, Ando J (1994) Fluid shear stress and vascular endothelial cell biomechanics. In: Hirasawa Y, Sledge CB et al (ed) Clinical Biomechanics and Related Research. Springer-Verlag Tokyo, pp 255–271

Kamiya A, Bukhari R et al (1984) Adaptive regulation of wall shear stress optimizing vascular tree function. Bull Math Biol 46:127–137

Khachigian LM, Anderson KR et al (1997) Egr-1 is activated in endothelial cells exposed to fluid shear stress and interacts with a novel shear-stress-response element in the PDGF A-chain promoter. Arterioscler Thromb Vasc Biol 17:2280–2286

Knudsen HL, Frangos JA (1997) Role of cytoskeleton in shear stress-induced endothelial nitric oxide production. Am J Physiol 273:H347–H355

Kohler R, Heyken WT et al (2006) Evidence for a functional role of endothelial transient receptor potential V4 in shear stress-induced vasodilatation. Arterioscler Thromb Vasc Biol 26:1495–1502

Korenaga R, Ando J et al (1994) Laminar flow stimulates ATP- and shear stress-dependent nitric oxide production in cultured bovine endothelial cells. Biochem Biophys Res Commun 198:213–219

Korenaga R, Ando J et al (1997) Negative transcriptional regulation of the VCAM-1 gene by fluid shear stress in murine endothelial cells. Am J Physiol 273:C1506–C1515

Korenaga R, Yamamoto K et al (2001) Sp1-mediated downregulation of P2X4 receptor gene transcription in endothelial cells exposed to shear stress. Am J Physiol Heart Circ Physiol 280:H2214–H2221

Kosaki K, Ando J et al (1998) Fluid shear stress increases the production of granulocyte-macrophage colony-stimulating factor by endothelial cells via mRNA stabilization. Circ Res 82:794–802

Krueger JW, Young DF et al (1971) An in vitro study of flow response by cells. J Biomech 4:31–36

Kuhlencordt PJ, Rosel E et al (2004) Role of endothelial nitric oxide synthase in endothelial activation: insights from eNOS knockout endothelial cells. Am J Physiol 286:C1195–C1202

Laughlin MH, Newcomer SC et al (2008) Importance of hemodynamic forces as signals for exercise-induced changes in endothelial cell phenotype. J Appl Physiol 104:588–600

Laurindo FR, de Pedro M A et al (1994) Vascular free radical release. Ex vivo and in vivo evidence for a flow-dependent endothelial mechanism. Circ Res 74:700–709

Lee HJ, Koh GY (2003) Shear stress activates Tie2 receptor tyrosine kinase in human endothelial cells. Biochem Biophys Res Commun 304:399–404

Li S, Kim M et al (1997) Fluid shear stress activation of focal adhesion kinase. Linking to mitogen-activated protein kinases. J Biol Chem 272:30455–30462

Lieu DK, Pappone PA et al (2004) Differential membrane potential and ion current responses to different types of shear stress in vascular endothelial cells. Am J Physiol Cell Physiol 286:C1367–C1375

Lin MC, Almus-Jacobs F et al (1997) Shear stress induction of the tissue factor gene. J Clin Invest 99:737–744

Lungu AO, Jin ZG et al (2004) Cyclosporin A inhibits flow-mediated activation of endothelial nitric-oxide synthase by altering cholesterol content in caveolae. J Biol Chem 279:48794–48800

Malek AM, Gibbons GH et al (1993) Fluid shear stress differentially modulates expression of gene encoding basic fibroblast growth factor and platelet-derived growth factor B chain in vascular endothelium. J Clin Invest 92:2013–2021

Mitsumata M, Fishel RS et al (1993) Fluid shear stress stimulates platelet-derived growth factor expression in endothelial cells. Am J Physiol 265:H3–H8

Mochizuki S, Vink H et al (2003) Role of hyaluronic acid glycosaminoglycans in shear-induced endothelium-derived nitric oxide release. Am J Physiol Heart Circ Physiol 285:H722–H726

Morita T, Yoshizumi M et al (1993) Shear stress increases heparin-binding epidermal growth factor-like growth factor mRNA levels in human vascular endothelial cells. Biochem Biophys Res Commun 197:256–262

Nakache M, Gaub HE (1988) Hydrodynamic hyperpolarization of endothelial cells. Proc Natl Acad Sci USA 85:1841–1843

Nauli SM, Kawanabe Y et al (2008) Endothelial cilia are fluid shear sensors that regulate calcium signaling and nitric oxide production through polycystin-1. Circulation 117:1161–1171

O'Neil RG, Heller S (2005) The mechanosensitive nature of TRPV channels. Pflugers Arch 451:193–203

Oancea E, Wolfe JT et al (2006) Functional TRPM7 channels accumulate at the plasma membrane in response to fluid flow. Circ Res 98:245–253

Obi S, Yamamoto K et al (2009) Fluid shear stress induces arterial differentiation of endothelial progenitor cells. J Appl Physiol 106:203–211

Ohno M, Gibbons GH et al (1993) Shear stress elevates endothelial cGMP. Role of a potassium channel and G protein coupling. Circulation 88:193–197

Ohno M, Cooke JP et al (1995) Fluid shear stress induces endothelial transforming growth factor beta-1 transcription and production. Modulation by potassium channel blockade. J Clin Invest 95:1363–1369

Ohura N, Yamamoto K et al (2003) Global analysis of shear stress-responsive genes in vascular endothelial cells. J Atheroscler Thromb 10:304–313

Okahara K, Kambayashi J et al (1995) Shear stress induces expression of CNP gene in human endothelial cells. FEBS Lett 373:108–110

Olesen SP, Clapham DE et al (1988) Haemodynamic shear stress activates a K+ current in vascular endothelial cells. Nature 331:168–170

Osawa M, Masuda M et al (2002) Evidence for a role of platelet endothelial cell adhesion molecule-1 in endothelial cell mechanosignal transduction: is it a mechanoresponsive molecule? J Cell Biol 158:773–785

Park H, Go YM et al (1998) Plasma membrane cholesterol is a key molecule in shear stress-dependent activation of extracellular signal-regulated kinase. J Biol Chem 273:32304–32311

Park H, Go YM et al (2000) Caveolin-1 regulates shear stress-dependent activation of extracellular signal-regulated kinase. Am J Physiol Heart Circ Physiol 278:H1285–H1293

Reigada D, Mitchell CH (2005) Release of ATP from retinal pigment epithelial cells involves both CFTR and vesicular transport. Am J Physiol Cell Physiol 288:C132–C140

Resnick N, Yahav H et al (2003) Fluid shear stress and the vascular endothelium: for better and for worse. Prog Biophys Mol Biol 81:177–199

Rieder MJ, Carmona R et al (1997) Suppression of angiotensin-converting enzyme expression and activity by shear stress. Circ Res 80:312–319

Rizzo V, McIntosh DP et al (1998) In situ flow activates endothelial nitric oxide synthase in luminal caveolae of endothelium with rapid caveolin dissociation and calmodulin association. J Biol Chem 273:34724–34729

Rizzo V, Morton C et al (2003) Recruitment of endothelial caveolae into mechanotransduction pathways by flow conditioning in vitro. Am J Physiol Heart Circ Physiol 285:H1720–H1729

Ruoslahti E (1991) Integrins. J Clin Invest 87:1–5

Secomb TW, Hsu R et al (2001) Effect of the endothelial surface layer on transmission of fluid shear stress to endothelial cells. Biorheology 38:143–150

Sharefkin JB, Diamond SL et al (1991) Fluid flow decreases preproendothelin mRNA levels and suppresses endothelin-1 peptide release in cultured human endothelial cells. J Vasc Surg 14:1–9

Shaul PW, Anderson RG (1998) Role of plasmalemmal caveolae in signal transduction. Am J Physiol 275:L843–L851

Shay-Salit A, Shushy M et al (2002) VEGF receptor 2 and the adherens junction as a mechanical transducer in vascular endothelial cells. Proc Natl Acad Sci USA 99:9462–9467

Shimizu N, Yamamoto K et al (2008) Cyclic strain induces mouse embryonic stem cell differentiation into vascular smooth muscle cells by activating PDGF receptor beta. J Appl Physiol 104:766–772

Shyy JY, Chien S (2002) Role of integrins in endothelial mechanosensing of shear stress. Circ Res 91:769–775

Shyy JY, Lin MC et al (1995) The cis-acting phorbol ester "12-O-tetradecanoylphorbol 13-acetate"-responsive element is involved in shear stress-induced monocyte chemotactic protein 1 gene expression. Proc Natl Acad Sci USA 92:8069–8073

Siegel G, Walter A et al (1996) Anionic biopolymers as blood flow sensors. Biosens Bioelectron 11:281–294

Silacci P, Desgeorges A et al (2001) Flow pulsatility is a critical determinant of oxidative stress in endothelial cells. Hypertension 38:1162–1166

Singla V, Reiter JF (2006) The primary cilium as the cell's antenna: signaling at a sensory organelle. Science 313:629–633

Sokabe T, Yamamoto K et al (2004) Differential regulation of urokinase-type plasminogen activator expression by fluid shear stress in human coronary artery endothelial cells. Am J Physiol Heart Circ Physiol 287:H2027–H2034

Sterpetti AV, Cucina A et al (1993) Shear stress increases the release of interleukin-1 and interleukin-6 by aortic endothelial cells. Surgery 114:911–914

Takada Y, Shinkai F et al (1994) Fluid shear stress increases the expression of thrombomodulin by cultured human endothelial cells. Biochem Biophys Res Commun 205:1345–1352

Tarbell JM, Pahakis MY (2006) Mechanotransduction and the glycocalyx. J Intern Med 259:339–350

Tzima E, Irani-Tehrani M et al (2005) A mechanosensory complex that mediates the endothelial cell response to fluid shear stress. Nature 437:426–431

Van der Heiden K, Hierck BP et al (2008) Endothelial primary cilia in areas of disturbed flow are at the base of atherosclerosis. Atherosclerosis 196:542–550

Wang JH, Thampatty BP (2008) Mechanobiology of adult and stem cells. Int Rev Cell Mol Biol 271:301–346

Wang N, Butler JP et al (1993) Mechanotransduction across the cell surface and through the cytoskeleton. Science 260:1124–1127

Wang N, Naruse K et al (2001) Mechanical behavior in living cells consistent with the tensegrity model. Proc Natl Acad Sci USA 98:7765–7770

Wang Y, Miao H et al (2002) Interplay between integrins and FLK-1 in shear stress-induced signaling. Am J Physiol Cell Physiol 283:C1540–C1547

Weber M, Hagedorn CH et al (2005) Laminar shear stress and 3′ polyadenylation of eNOS mRNA. Circ Res 96:1161–1168

Weinbaum S, Zhang X et al (2003) Mechanotransduction and flow across the endothelial glyco-calyx. Proc Natl Acad Sci USA 100:7988–7995

Yamamoto K, Korenaga R et al (2000a) Fluid shear stress activates Ca(2+) influx into human endothelial cells via P2X4 purinoceptors. Circ Res 87:385–391

Yamamoto K, Korenaga R et al (2000b) P2X(4) receptors mediate ATP-induced calcium influx in human vascular endothelial cells. Am J Physiol Heart Circ Physiol 279:H285–H292

Yamamoto K, Sokabe T et al (2003a) Endogenously released ATP mediates shear stress-induced Ca²⁺ influx into pulmonary artery endothelial cells. Am J Physiol Heart Circ Physiol 285:H793–H803

Yamamoto K, Takahashi T et al (2003b) Proliferation, differentiation, and tube formation by endothelial progenitor cells in response to shear stress. J Appl Physiol 95:2081–2088

Yamamoto K, Sokabe T et al (2005) Fluid shear stress induces differentiation of Flk-1-positive embryonic stem cells into vascular endothelial cells in vitro. Am J Physiol Heart Circ Physiol 288(4):H1915–H1924

Yamamoto K, Sokabe T et al (2006) Impaired flow-dependent control of vascular tone and remodeling in P2X4-deficient mice. Nat Med 12:133–137

Yamamoto K, Shimizu N et al (2007) Involvement of cell surface ATP synthase in flow-induced ATP release by vascular endothelial cells. Am J Physiol Heart Circ Physiol 293:H1646–H1653

Yu J, Bergaya S et al (2006) Direct evidence for the role of caveolin-1 and caveolae in mechan-otransduction and remodeling of blood vessels. J Clin Invest 116:1284–1291

Chapter 3
Mechanobiology During Vertebrate Organ Development

Makoto Asashima, Tatsuo Michiue, Kiyoshi Ohnuma, Yoshiro Nakajima, and Yuzuru Ito

3.1 Regulation of Neural Crest Cell Migration

Neural crest cells are one of many migrating cell types found in vertebrate tissues. Neural crest specification occurs between the neural plate and epidermal region during vertebrate embryogenesis, and is regulated at the gene level by appropriate concentrations of cell signaling proteins such as bone morphogenetic protein (BMP), fibroblast growth factor (FGF), and Wnt (reviewed by Meulemans and Bronner-Fraser 2004). Many genes participate in this regional network to specify the neural crest. After neural tube closure, neural crest cells start migrating to the ventral half of the embryo, and the resultant epithelial-to-mesenchymal transformation dissociates the neural crest cells from the epidermis. The dissociated neural crest cells then start to move ventrally.

How this directional migration of neural crest cells is regulated remains largely unknown. We know that chemoattractants and the extracellular matrix (ECM) control the ventral migration pathway. For example, neural crest cells expressing Eph3B receptor migrate ventrally, and are blocked from migrating into the anterior somitic region containing the ephrinB1 ligand (Krull et al. 1997). In a second example, migrating cells expressing the receptor protein neuropilin recognize the

M. Asashima (✉), Y. Nakajima, and Y. Ito
Research Center for Stem Cell Engineering, National Institute of Advanced
Industrial Science and Technology (AIST), Tsukuba Central 4, 1-1-1 Higashi,
Tsukuba, Ibaraki 305-8562, Japan
e-mail: asashi@bio.c.u-tokyo.ac.jp

M. Asashima and T. Michiue
Department of Life Sciences (Biology), Graduate School of Arts and Sciences,
The University of Tokyo, 3-8-1 Komaba, Meguro-ku,
Tokyo 153-8902, Japan

K. Ohnuma
Top Runner Incubation Center for Academia-Industry Fusion,
Nagaoka University of Technology, 1603-1 Kamitomiokamachi, Nagaoka,
Niigata 940-2188, Japan

M. Noda (ed.), *Mechanosensing Biology*,
DOI 10.1007/978-4-431-89757-6_3, © Springer 2011

ligand protein semaphorin in posterior Sclerotome cells (He and Tessier-Lavigne 1997; Kolodkin et al. 1997). In both cases, receptor proteins are expressed on neural crest cells, whose migration is then regulated according to the concentrations of ligands in surrounding tissues. ECM proteins such as fibronectin and laminin are also necessary for neural crest cell migration (Goodman and Newgreen 1985; Duband et al. 1986; Bronner-Fraser 1986), mediated via interactions with integrin proteins expressed on the neural crest cells.

Contact inhibition of locomotion was recently implicated in the directional migration of neural crest cells (Carmona-Fontaine et al. 2008). Neural crest explants will never invade other neural crest explants. Indeed, when two neural crest cells contact each other, lamellipodia are collapsed and migration stops. This result suggested that this type of neural cell recognizes the presence of other cells and regulates its locomotion accordingly. Moreover, our in vitro experiments showed that neural cells determine their migrating behavior by directly contacting their surrounding environment, as described in the next section.

3.2 Integration of Mechanical Information by Neuronal Cells In Vitro

Long-distance, directional migration of neuronal cells is a critical feature of the developing and regenerating nervous system. However, physical barriers such as connective tissue prevent the cells from migrating freely towards their destination (Fig. 3.1a). Thus, the cells not only mechanically sense the surrounding geometry but also integrate the mechanical information to migrate directionally (Fig. 3.1b) (Ingber 2003), although this relationship between the geometry and cell migration has not been well studied. Recent advances in the soft-lithography technique developed by Whitesides and co-workers allow various geometries to be simulated in a micrometer-size cell culture chamber (Kane et al. 1999). Thus, the relationships between geometry and cell response can be studied. In this way, spindle orientation,

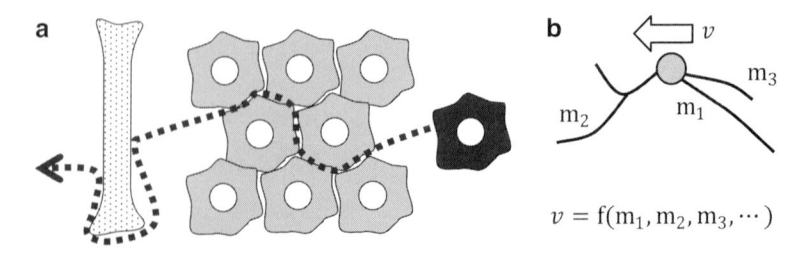

Fig. 3.1 Schematic illustrations of the mechanical information based on the surrounding geometry and cell migration. (**a**) The cells migrate while avoiding physical barriers. (**b**) The cells migrate with verbosity v by integrating mechanical information (m_1, m_2, m_3, …)

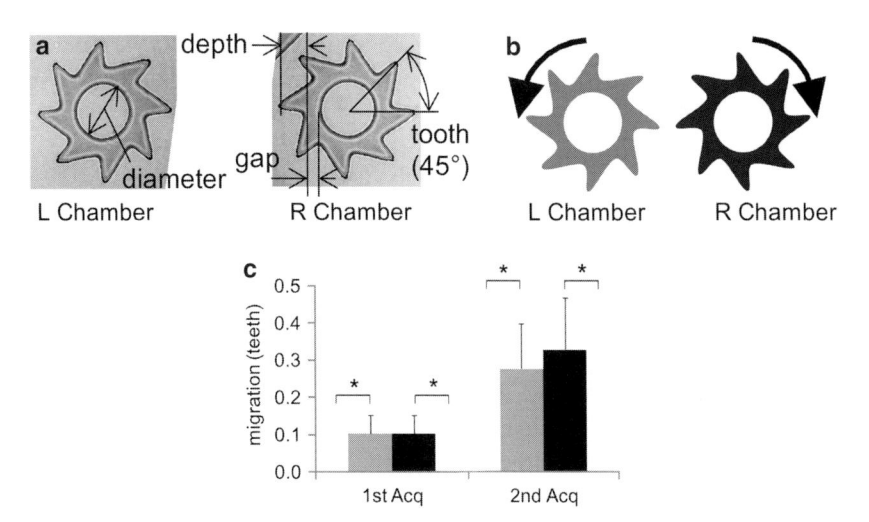

Fig. 3.2 Ratchet wheel-shaped microchambers. (**a**) PDMS stamp of the ratchet wheel-shaped microchambers. The *left* and the *right* chambers are line-symmetrical to one another and are termed the L and R chambers, respectively. (**b**) Definition of the positive direction in the L and R chamber. The cells attach to the shading area (*gray* or *black*). (**c**) The mean tangential migration distance during the first and second acquisitions for the L chamber (*gray*) and R chamber (*black*). * $P<0.05$ (*t* test)

growth, differentiation, and migration were associated with the shape of the culture vessel (Brock et al. 2003; Jiang et al. 2005; Recknor et al. 2006). We used this technique in a simple in vitro experiment. We focused on periodic structure, which is abundant in vivo, in a ratchet-shaped microchamber to study the migration of PC12 cells (Fig. 3.2a). Our working hypothesis is that neuronal cells migrate directionally on a periodic scaffold structure if the periodic unit is asymmetric. Polydimethylsiloxane (PDMS) agent was applied to collagen-coated culture dishes to make microchambers with central cores. Cells were attached only to the collagen-coated areas between the central core and the ratchet-shaped outer frame. The cells migrated virtually one-dimensionally in a tangential direction. The left (L) and right (R) microchambers shown in Fig. 3.2a were line-symmetrical to one another and arranged alternately to serve as control chambers for each other. The gap between the teeth and core was equivalent to the size of the cell body. The core diameter, tooth depth, and gap between the teeth and core were approximately 100, 40, and 30 μm, respectively. The PC12 cells were plated in the L and R chambers in neurite outgrowth medium in which the cells extend long neurites over about 10 days (Greene et al. 1982; Ohnuma et al. 2006). Because neuronal cell migration follows this neurite extension (Hatten 2002), we expected that cell migration in the chambers would differ between the cells with short neurites and those with long neurites. Time-lapse micrographs were therefore acquired twice: for 70 h from 1 day after plating (first acquisition), when the cells have short neurites; and for

70 h from 12 days after plating (second acquisition), when the cells have long neurites. We defined the cell migration direction in which the ratchet teeth were tapered as a positive direction (counterclockwise migration in the L chamber or clockwise migration in the R chamber; Fig. 3.2b).

We found that the PC12 cells in the L and R chambers migrated the same distance in a positive direction during both acquisitions (Fig. 3.2c); in other words, migration in the opposite direction in the opposite-geometry chambers. These results suggested that the direction of migration is dependent upon chamber geometry, and support our working hypothesis that the periodic nature of the scaffold determines the directional migration if the periodic unit is asymmetric.

To elucidate the relationship between the chamber geometry and the shape of neurite-extending PC12 cells, we analyzed the position of the cell body relative to the neurite tips. Time-course tracing of cell body migration showed that the neurite tips remained around the tips of the ratchet teeth. We also found a tendency for the cells to migrate in a positive direction with the extended neurites about one tooth ahead of the cell body, by placing the neurite tip at the tip of the tooth, and then passing the cell body alongside the neurite tip, which remains at the tooth tip (illustrated in Fig. 3.3).

Although it was not clear from this experiment whether the observed dynamics between neurite tips and cell body caused or resulted from the directional migration, we propose a possible simplified mechanical model for directional migration of neuronal cells. Since the positively-directed neurites were guided along the tooth edge and the negatively-directed neurites extended in a straight line, the tangential component of the maximal tension of the positive-direction neurite should be higher than that of the negative-direction neurite. Thus, the probability of continuous forward migration is higher than that of backward migration. Mahmud et al. (2009) reported similar results using fibroblasts and cancer cells including melanoma.

In conclusion, we designed an experimental model of neuronal cell migration guided by mechanical information about the surrounding geometry. Currently, there is no evidence that such a scaffold structure exists in vivo. However, both repetitive structures, including the somite and the cortical layer, and asymmetric protein distribution are abundant in in vivo phenomena. The present study provided a new

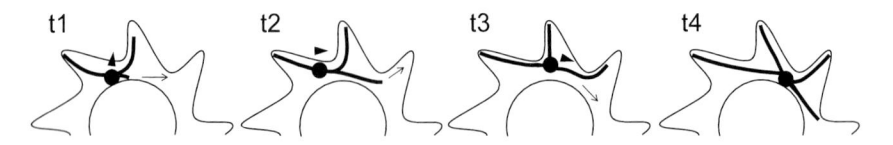

Fig. 3.3 Schematic illustrations of the characteristic tendency in the relationship between neurite tips and cell body. The cell body (*circle*) and neurites (*solid thick lines*) are represented. The positively migrating cells tended to extend a neurite to one ratchet tooth in front of the cell body and to leave a few neurites behind at the tooth tips. *Arrow* direction of neurite growth. *Arrowhead* direction of the cell body migration

concept for neuronal migration controlled by mechanical stimulation and suggests strategies for designing artificial scaffolds that direct neuronal cell migration.

3.3 Differentiation of Pronephros

The kidney is essential for regulating waste excretion, homeostasis, and hormone secretion. Vertebrates have developed three different forms of the kidney to execute these respective functions: pronephros, mesonephros, and metanephros. All three forms consist of a basic unit called the nephron, but differ with respect to nephron number and organization. Mammals have a metanephros, which contains approximately 1 million nephrons. Waste is collected by these nephrons and excreted via a system of collecting ducts (Saxén and Sariola 1987). In contrast, the pronephros is a simple excretory organ found in *Xenopus* larvae. Unlike the metanephros, the pronephros contains open nephrons that consist of tubules, a duct, and a glomus (Fig. 3.4). The *Xenopus* pronephros is derived from the intermediate mesoderm at the late gastrula stage. Lim-1 and Pax-8 are the first molecular signals to be expressed in the pronephros anlagen in the late gastrula stage. Lim-1 is required for patterning of the anterior neural tissues (Taira et al. 1994) and is necessary for tubule development (Chan et al. 2000). From the late neurula stage, the pronephros anlagen starts to differentiate into the tubule, duct, and glomus. Changes in the shape of the tubules, duct, and glomus are regulated by the expression of Pax-2, Wnt-4, WT1, Xfz8, and other molecular signals (Wallingford et al. 1998; Saulnier et al. 2002; Satow et al. 2004). FGF signaling has also been implicated as necessary for the condensation and transition of the nephric mesenchyme during early tubulogenesis, while BMP signaling may regulate differentiation of the tubule and duct (Urban et al. 2006; Bracken et al. 2008). Patterning of the tubules and extension of the ducts coincides with the expression of all genes as described in previous sentences. As described, the molecular basis of pronephros formation has been well studied, but remains not fully characterized. We recently observed interesting phenomena with possible links to the mechanism of tubulogenesis. The next section describes this work in detail.

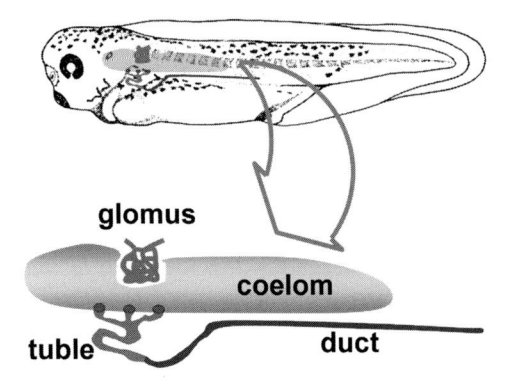

Fig. 3.4 Scheme of pronephros in *Xenopus* late tailbud stage. The pronephros contains open nephrons that consist of tubules, a duct, and a glomus

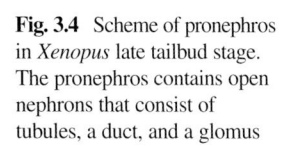

3.4 Effect of Gravity and Shear Stress on Dome Structure Formation in a Tubulogenesis Model

The A6 cell line derived from *Xenopus* renal epithelial cells (Rafferty 1969) spontaneously forms "domes". These are multicellular hemicyst structures (Valentich et al. 1979; Moberly and Fanestil 1988) that are functionally equivalent to differentiated epithelium and show trans-epithelial solute transport (Misfeldt et al. 1976; Cereijido et al. 1978; Shlyonsky et al. 2005). The dome structures form when liquid accumulates between the cell layer and the underlying support and the cellular adherence ability of the monolayer diminishes as a result (Fig. 3.5). Moreover, A6 cell structures have characteristics of tight epithelia in culture, i.e., similar to the kidney distal tubule and collecting duct (Perkins and Handler 1981). Therefore, A6 cells provide an excellent system for studies of pronephros tubulogenesis.

The three-dimensional (3D) clinostat is a valuable device for simulating a microgravity G-force that is close to zero. Previously, we described a long-term culturing system using the 3D clinostat to compare A6 cells cultured under simulated μG and static (1G) conditions. The simulated μG condition inhibited dome formation (Ichigi and Asashima 2001). In addition, the global gene expression profile of A6 cells cultured under simulated μG was determined using *Xenopus* microarrays, and genes that showed up- or downregulation of expression were investigated (Kyuno et al. 2003; Kitamoto et al. 2005; Ikuzawa et al. 2007). We identified 283 genes (153 upregulated and 130 downregulated) on day 15 of culture under simulated μG conditions; ENaC and Na$^+$/K$^+$-ATPase were downregulated. Epithelial sodium channel ENaC and Na$^+$/K$^+$-ATPases are also found at the apical and basolateral membranes in A6 cells, respectively. Based on these findings, we proposed that sodium ions in the A6 cell culture medium are incorporated into the cells by ENaC in the apical membrane, and pumped out into the extracellular space by Na$^+$/K$^+$-ATPases in the basolateral membrane (Ikuzawa et al. 2007). Therefore, the mechanism of dome formation in A6 cells is common to those reported in other epithelial cells (Misfeldt et al. 1976; Cereijido et al. 1978; Rabito et al. 1980; Hanwell et al.

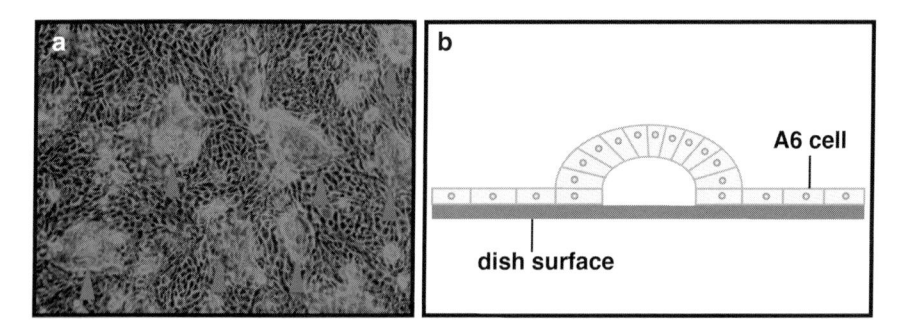

Fig. 3.5 Dome formation by A6 cells. (**a**) The A6 epithelial cell line forms a dome structure in culture. The *arrowheads* indicate the dome structure. (**b**) Schematic representation of the dome structure, showing how the cells detach from the surface of the culture dish

2002; Warth 2003; Orlowski and Grinstein 2004). Thus, the existence of gravity might be required for normal tubulogenesis in pronephros.

Polycystins are also involved in dome formation. The name comes from the identification of mutations in the genes encoding polycystin-1 (TRPP1) and -2 (TRPP2) in autosomal-dominant polycystic kidney disease (ADPKD). Upregulation of TRPP1 mRNA and protein was found in domes formed by the pancreatic ductal epithelial cell line SU.86.86 (Bukanov et al. 2002). *TRPP1* mRNA levels in A6 cells were also upregulated with dome maturation. TRPP1 is an integral membrane glycoprotein with 11 transmembrane domains and an extensive amino-terminal extracellular domain involved in cell–cell and cell–matrix interactions and signaling pathways (Wilson 2004). TRPP1 interacts with TRPP2, a 6-transmembrane domain protein of the TRP ion channel family (Giamarchi et al. 2006). TRPP1 and TRPP2 colocalize at the membrane of the primary cilia in renal epithelial cells where they are proposed to transduce luminal shear stress into calcium signaling (Nauli et al. 2003, 2008). TRPP1 might also be a mechanical sensor regulating the opening of the associated calcium-permeable channel TRPP2. However, how polycystins contribute to cellular mechanosensitivity remains obscure. A recent study implicated the TRPP1/TRPP2 ratio in pressure sensing (Sharif-Naeini et al. 2009) by regulating stretch-activated ion channel (SAC) mechanosensitivity via the filamin A (FLNa)/F-actin cytoskeletal network, and thus influencing the conversion of intraluminal pressure to local bilayer tension. This finding is important for understanding the physiological and disease states associated with pressure sensing. The precise roles of polycystins in dome formation remain to be demonstrated, and although further investigation is needed, microgravity and shear stress/pressure are candidate mechanisms.

References

Bracken CM, Mizeracka K, McLaughlin KA (2008) Patterning the embryonic kidney: BMP signaling mediates the differentiation of the pronephric tubules and duct in *Xenopus laevis*. Dev Dyn 237 : 132–144

Brock A, Chang E, Ho CC, Leduc P, Jiang X, Whitesides GM, Ingber DE (2003) Geometric determinants of directional cell motility revealed using microcontact printing. Langmuir 19:1611–1617

Bronner-Fraser M (1986) An antibody to a receptor for fibronectin and laminin perturbs cranial neural crest development in vivo. Dev Biol 117:528–536

Bukanov NO, Husson H, Dackowski WR, Lawrence BD, Clow PA, Roberts BL, Klinger KW, Ibraghimov-Beskrovnaya O (2002) Functional polycystin-1 expression is developmentally regulated during epithelial morphogenesis *in vitro*: downregulation and loss of membrane localization during cystogenesis. Hum Mol Genet 11:923–936

Carmona-Fontaine C, Matthews HK, Kuriyama S, Moreno M, Dunn GA, Parsons M, Stern CD, Mayor R (2008) Contact inhibition of locomotion *in vivo* controls neural crest directional migration. Nature 456:957–961

Cereijido M, Robbins ES, Dolan WJ, Rotunno CA, Sabatini DD (1978) Polarized monolayers formed by epithelial cells on a permeable and translucent support. J Cell Biol 77:853–880

Chan TC, Takahashi S, Asashima M (2000) A role for Xlim-1 in pronephros development in *Xenopus laevis*. Dev Biol 228:256–269

Duband JL, Rocher S, Chen WT, Yamada KM, Thiery JP (1986) Cell adhesion and migration in the early vertebrate embryo: location and possible role of the putative fibronectin receptor complex. J Cell Biol 102:160–178

Giamarchi A, Padilla F, Coste B, Raoux M, Crest M, Honoré E, Delmas P (2006) The versatile nature of the calcium-permeable cation channel TRPP2. EMBO Rep 7:787–793

Goodman SL, Newgreen D (1985) Do cells show an inverse locomotory response to fibronectin and laminin substrates? EMBO J 4:2769–2771

Greene LA, Burstein DE, Black MM (1982) The role of transcription-dependent priming in nerve growth factor promoted neurite outgrowth. Dev Biol 91:305–316

Hanwell D, Ishikawa T, Saleki R, Rotin D (2002) Trafficking and cell surface stability of the epithelial Na+ channel expressed in epithelial Madin-Darby Canine Kidney cells. J Biol Chem 277:9772–9779

Hatten ME (2002) New directions in neuronal migration. Science 297:1660–1663

He Z, Tessier-Lavigne M (1997) Neuropilin is a receptor for the axonal chemorepellent Semaphorin III. Cell 90:739–751

Ichigi J, Asashima M (2001) Dome formation and tubule morphogenesis by *Xenopus* kidney A6 cell cultures exposed to microgravity. In Vitro Cell Dev Biol Anim 37:31–44

Ikuzawa M, Akiduki S, Asashima M (2007) Gene expression profile of *Xenopus* A6 cells cultured under random positioning machine shows downregulation of ion transporter genes and inhibition of dome formation. Adv Space Res 40:1694–1702

Ingber DE (2003) Tensegrity I. Cell structure and hierarchical systems biology. J Cell Sci 116:1157–1173

Jiang X, Bruzewicz DA, Wong AP, Piel M, Whitesides GM (2005) Directing cell migration with asymmetric micropatterns. Proc Natl Acad Sci USA 102:975–978

Kane RS, Takayama S, Ostuni E, Ingber DE, Whitesides GM (1999) Patterning proteins and cells using soft lithography. Biomaterials 20:2363–2376

Kitamoto J, Fukui A, Asashima M (2005) Temporal regulation of global gene expression and cellular morphology in *Xenopus* kidney cells in response to clinorotation. Adv Space Res 35:1654–1661

Kolodkin AL, Levengood DV, Rowe EG, Tai YT, Giger RJ, Ginty DD (1997) Neuropilin is a semaphorin III receptor. Cell 90:753–762

Krull CE, Lansford R, Gale NW, Collazo A, Marcelle C, Yancopoulos GD, Fraser SE, Bronner-Fraser M (1997) Interactions of Eph-related receptors and ligands confer rostrocaudal pattern to trunk neural crest migration. Curr Biol 7:571–580

Kyuno J, Fukui A, Michiue T, Asashima M (2003) Characterization of a gene respondent to clinorotation in *Xenopus* A6 cells. Biol Sci Space 17:171–172

Mahmud G, Campbell CJ, Bishop KJM, Komarova YA, Chaga O, Soh S, Huda S, Kanderegrzybowska K, Grzybowski BA (2009) Directing cell motions on micropatterned ratchets. Nat Phys 5:606–612

Meulemans D, Bronner-Fraser M (2004) Gene-regulatory interactions in neural crest evolution and development. Dev Cell 7:291–299

Misfeldt DS, Hamamoto ST, Pitelka DR (1976) Transepithelial transport in cell culture. Proc Natl Acad Sci USA 73:1212–1216

Moberly JB, Fanestil DD (1988) A monoclonal antibody that recognizes a basolateral membrane protein in A6 epithelial cells. J Cell Physiol 135:63–70

Nauli SM, Alenghat FJ, Luo Y, Williams E, Vassilev P, Li X, Elia AE, Lu W, Brown EM, Quinn SJ (2003) Polycystins 1 and 2 mediate mechanosensation in the primary cilium of kidney cells. Nat Genet 33:129–137

Nauli SM, Kawanabe Y, Kaminski JJ, Pearce WJ, Ingber DE, Zhou J (2008) Endothelial cilia are fluid shear sensors that regulate calcium signaling and nitric oxide production through polycystin-1. Circulation 11:1161–1171

Ohnuma K, Hayashi Y, Furue M, Kaneko K, Asashima M (2006) Serum-free culture conditions for serial subculture of undifferentiated PC12 cells. J Neurosci Methods 151:250–261

Orlowski J, Grinstein S (2004) Diversity of the mammalian sodium/proton exchanger SLC9 gene family. Pflugers Arch 447:549–565

Perkins F, Handler J (1981) Transport properties of toad kidney epithelia in culture. Am J Physiol Cell Physiol 241:C154–C159

Rabito CA, Tchao R, Valentich J, Leighton J (1980) Effect of cell-substratum interaction on hemicyst formation by MDCK cells. In Vitro 16:461–468

Rafferty KA Jr (1969) Mass culture of amphibian cells: methods and observations concerning stability of cell type. In: Mizell M (ed) Biology of amphibian tumors. Springer, New York, pp 52–81

Recknor JB, Sakaguchi DS, Mallapragada SK (2006) Directed growth and selective differentiation of neural progenitor cells on micropatterned polymer substrates. Biomaterials 27:4098–4108

Satow R, Chan TC, Asashima M (2004) The role of *Xenopus* frizzled-8 in pronephric development. Biochem Biophys Res Commun 321:487–494

Saulnier DM, Ghanbari H, Brandli AW (2002) Essential function of Wnt-4 for tubulogenesis in the *Xenopus* pronephric kidney. Dev Biol 248:13–28

Saxén L, Sariola H (1987) Early organogenesis of the kidney. Pediatr Nephrol 1:385–392

Sharif-Naeini R, Folgering JHA, Bichet D, Fabrice Duprat F, Lauritzen I, Arhatte M, Jodar M, Dedman A, Chatelain FC, Uwe Schulte U, Retailleau K, Loufrani L, Patel A, Sachs F, Delmas P, Peters DJM, Honoré E (2009) Polycystin-1 and -2 dosage regulates pressure sensing. Cell 139:587–596

Shlyonsky V, Goolaerts A, Van Beneden R, Sariban-Sohraby S (2005) Differentiation of epithelial Na⁺ channel function an *in vitro* model. J Biol Chem 280:24181–24187

Taira M, Otani H, Saint-Jeannet JP, Dawid IB (1994) Role of the LIM class homeodomain protein Xlim-1 in neural and muscle induction by the Spemann organizer in *Xenopus*. Nature 372:677–679

Urban AE, Zhou X, Ungos JM, Raible DW, Altmann CR, Vize PD (2006) FGF is essential for both condensation and mesenchymal-epithelial transition stages of pronephric kidney tubule development. Dev Biol 297:103–117

Valentich JD, Tchao R, Leighton J (1979) Hemicyst formation stimulated by cyclic AMP in dog kidney cell line MDCK. J Cell Physiol 100:291–304

Wallingford JB, Carroll TJ, Vize PD (1998) Precocious expression of the Wilms' tumor gene xWT1 inhibits embryonic kidney development in *Xenopus laevis*. Dev Biol 202:103–112

Warth R (2003) Potassium channels in epithelial transport. Pflugers Arch 446:505–513

Wilson PD (2004) Polycystic kidney disease. N Engl J Med 350:151–164

Part II
Tissue and Gravity

Chapter 4
Mechanobiology in Skeletal Muscle: Conversion of Mechanical Information into Molecular Signal

Yuko Miyagoe-Suzuki and Shin'ichi Takeda

4.1 Introduction

Overload leads to muscle hypertrophy. In the process, several events occur inside and outside the myofibers, including increased protein synthesis, change in gene expression, fiber-type transition, satellite cell activation, and angiogenesis (Bassel-Duby and Olson 2006; Blaauw et al. 2009). Interestingly, at the early phase of muscle hypertrophy, protein synthesis significantly increases (Baar et al. 2006), and later the transcription of growth-related genes follows (Carson 1997). Satellite cell activation is generally thought to be a critical component for increase in muscle mass; however, it is still a debated issue whether satellite cell incorporation into hypertrophying muscle fibers is required for muscle hypertrophy (O'Connor and Pavlath 2007; McCarthy and Esser 2007). A recent paper demonstrated that the rapid incorporation of BrdU seen in overloaded muscle reflects angiogenesis but not proliferation of satellite cells (Blaauw et al. 2009). Thus, muscle hypertrophy is a complicated and dynamic process, influenced by many factors (nutrients, blood flow, hormones, energy status, or oxidative status), making it difficult to elucidate the mechanism by which mechanical information is sensed by myofibers. IGF-1, anabolic steroids, or blockage of myostatin signaling increases muscle mass without mechanical stimulation via distinct mechanisms from mechanotransduction. For myostatin or steroids, please refer to other comprehensive reviews (Joulia-Ekaza and Cabello 2007; Kadi 2008).

The muscle atrophy process is characterized by suppressed protein synthesis, and increased rate of degradation of muscle proteins (Jackman and Kandarian 2004; Ventadour and Attaix 2006). Decreased muscle activity (i.e., denervation, prolonged bed rest, space flight, immobilization, etc.) and diseases (cancer, AIDS, muscular dystrophy, sepsis, chronic heart failure, diabetes, etc.), malnutrition, or drugs (glucocorticoids) lead to muscle atrophy. In chronic diseases, elevated levels

Y. Miyagoe-Suzuki and S. Takeda (✉)
Department of Molecular Therapy, National Institute of Neuroscience, National Center of Neurology and Psychiatry, 4-1-1 Ogawa-Higashi, Kodaira, Tokyo 187-8502, Japan
e-mail: takeda@ncnp.go.jp

of proinflammatory cytokines, glucocorticosteroids, tumor-derived factors, and endotoxins are responsible for the induction of muscle atrophy, whereas the upstream mediators which trigger catabolic signal cascades in unloaded muscle are largely unknown.

4.2   Mechanosensor in Skeletal Muscle

4.2.1   *Experimental Models for Mechanotransduction*

Experiments to study mechanotransduction in skeletal muscle employ in vivo and in vitro models (Table 4.1). As discussed below, experimental designs greatly influence the responses of the muscles to the mechanical stress.

Although it is not yet a consensus, it is reported that eccentric contraction more effectively activates p70S6K than centric contraction (Eliasson et al. 2006; Nader and Esser 2001). This difference might be because eccentric contraction accompanies stretching stimuli, and thereby strongly stimulates stretch-activated channels (SACs) or other stretch-sensitive mechanoreceptors. But this hypothesis needs more investigation.

In addition to the type of mechanical stimulation, the intensity and duration of stimulation must be taken into account on interpretation. Atherton et al. reported that aerobic exercise stimuli promoted specific adaptive responses toward mito-chondrial biogenesis besides slow phenotypes, without activating mTOR and p70S6K. Resistance strength training stimuli, however, strongly activated mTOR and p70S6K (Atherton et al. 2005). But anther group did not observe such a difference (Nader and Esser 2001). As Zanchi and Lancha discuss in their review article (Zanchi and Lancha 2008), the difference might be due to the difference in the design of the experiment, i.e., in vivo versus in vitro, or in the duration of mechanical stimulation.

In overloaded muscle, the protein synthesis is initiated by mechanical stimuli, but later modulated by additional anabolic stimuli. IGF-1 is generally accepted to strongly

Table 4.1   Models to study mechanotransduction in skeletal muscle

Modes of mechanical load	
Overload	In vivo model
	Synergist ablation (functional overload)
	Electrical stimulation
	In vitro model
	Electrical stimulation in vitro
	Cell culture (stretch)
Unload	In vivo models
	Limb immobilization
	Hindlimb suspension
	Denervation

promote protein synthesis through the PI3K/Akt/mTOR pathway (Rommel et al. 2001; Bodine et al. 2001). Nonetheless, IGF-1-mediated PI3K/Akt/mTOR activation seems to be part of a late component of the hypertrophy process (Hameed et al. 2003; Haddad and Adams 2002). Furthermore, acute stimulation of skeletal muscle seems not to be always dependent on PI3K/Akt as a means to activate mTOR (Hornberger et al. 2004, 2006). Thus, acute phase might be more suitable than chronic phase to investigate the mechanotransduction system in skeletal muscle.

Lastly, skeletal muscle is composed of myofibers with different contractile and metabolic properties. For example, the soleus, which is tonic and postural, is a mixture of predominant oxidative slow-twitch (type I) fibers and glycolytic fast-twitch fibers (type II). The extensor digitorum longus (EDL) muscle, which is phasic, is composed of fast-twitch glycolytic fibers. Importantly, these two muscles respond to mechanical stimuli differently: the soleus shows a lower susceptibility to mechanical load compared with EDL (Widrick et al. 2002). The findings suggest either that different mechanosensing apparatus exists in these two muscles, or that different signal transduction pathways operate downstream of the mechanosensor in fast and slow muscles.

4.2.2 Mechanosensors in Skeletal Muscle

So far, many molecules have been proposed as a mechanosensing molecule in skeletal muscle, including SACs (Spangenburg and McBride 2006), the dystrophin-glycoprotein complex (Acharyya et al. 2005; Suzuki et al. 2007), integrins (discussed in Zanchi and Lancha 2008; Spangenburg 2009) or sarcomere structure (Gautel 2008) (Fig. 4.1), but none of them is definitive. It is quite plausible that skeletal muscle has multiple mechanosensors and integrates the mechanical information from all these sensors into anabolic or catabolic responses.

4.2.3 Stretch-Activated Channels

SACs were initially described in skeletal muscle (Franco and Lansman 1990a, b). SACs in skeletal muscle are permeable to both Na+ and Ca2+ and increase their open probability in response to stretch of the membrane. A previous report, using Ga3+ and streptomycin as specific blockers of SACs, suggested that SACs are activated during lengthening or stretch-induced contraction, activate the Akt/mTOR pathway, and induce muscle hypertrophy (Spangenburg and McBride 2006; Butterfield and Best 2009). However, since neither Ga3+ nor streptomycin completely abolished Akt/mTOR activities, the authors speculated that SACs is one of the molecules which sense mechanical load and promote protein synthesis. Currently, SACs are identified based on patch-clamping experiments, and channel activity is blocked with specific inhibitors. Cloning of genes encoding SACs and

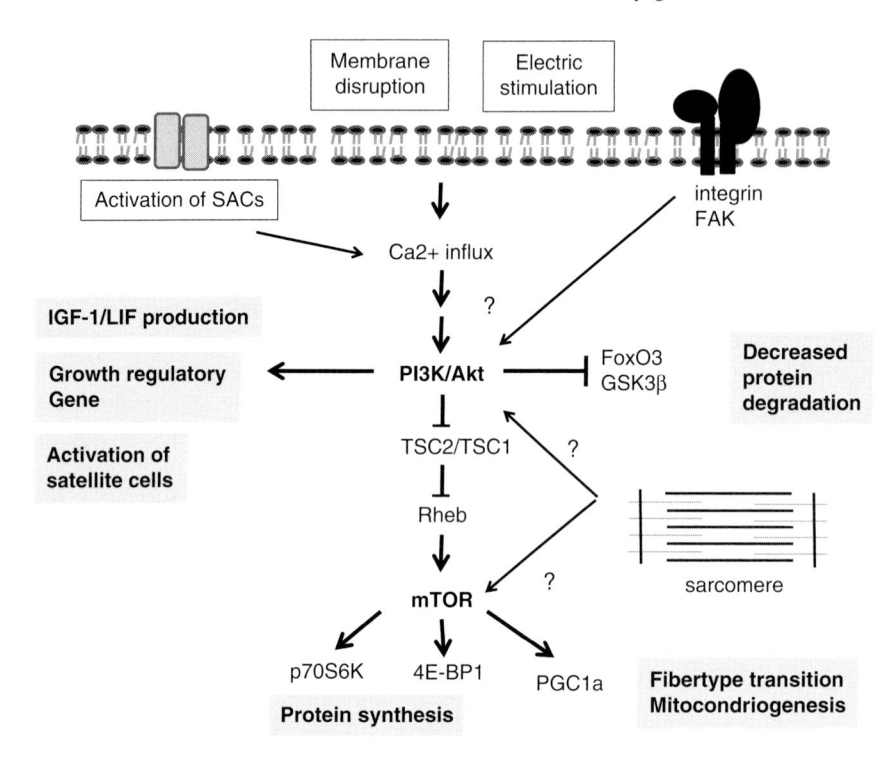

Fig. 4.1 Mechanical load and muscle hypertrophy. Stretching of myofibers activates sarcolemmal stretch-activated channels (SACs), disrupts the cytoplasmic membrane, starts a cascade of anabolic signal transduction, and through activation of mTOR increases its mass and contractile force. Active muscle contraction also activates the Akt/mTOR pathway. The sarcomere structure might work one of the sensors of muscle contraction, and convert the mechanical information into biochemical signals. Several studies suggest that a variety of mechanical information converge on mTOR, which in turn activates downstream targets (S6K, 4E-BP1, or PGC-1alph) to adapt the mechanical load. Note that some parts are not fully supported by experimental evidence

ablation of the gene in skeletal muscle would greatly accelerate the elucidation of the role of SACs in mechanotransduction in skeletal muscle, and would help develop pharmacological intervention of disuse-related muscle atrophy.

4.2.4   *Neuronal Nitric Oxide Synthase and Mechanotransduction*

Neuronal nitric oxide synthase (nNOS) is a peripheral member of the dystrophin–glycoprotein complex at the sarcolemma (Brenman et al. 1996), and especially concentrated at myotendinous junctions, which are specialized for force transduction across the muscle cell membrane (Chang et al. 1996).The expression and activity of nNOS are upregulated by exercise (Roberts et al. 1999), mechanical loading, electrical stimulation of muscle, and passive stretching (Reiser et al. 1997; Tidball et al. 1998). Criswell and co-workers reported that inhibition of NOS by L-NAME (a wide inhibitor of NOS) during functional overload attenuated the increase of muscle mass and fiber-type transition (Smith et al. 2002; Soltow et al. 2006). The same group further showed that NO upregulated contractile protein mRNAs including α-actin and type 1 myosin heavy chain (Sellman et al. 2006). Steensberg et al. reported that L-NAME treatment inhibited exercise-induced mRNA expression of IL-6, HO-1 and PDK4 (Steensberg et al. 2007). Tidball et al. showed that talin and vinculin mRNA expressions induced by mechanical stimulation are mediated by NO (Tidball et al. 1999). Koh and Tidball showed that NO positively modulates sarcomere addition in immobilized–remobilized muscles (Koh and Tidball 1999). Together, these data strongly suggest that nNOS plays an important role in mechanotransduction in skeletal muscle.

Nitric oxide (NO) gas is reported to be released from myofibers when myofibers are stretched in vitro, and activate satellite cells via release of HGF (Tatsumi et al. 2002). Furthermore, Tatsumi et al. demonstrated that stretching of myofibers in vivo indeed activates satellite cells via production of NO (Tatsumi et al. 2006). The nitric gas is supposed to be generated by nNOS. Together, these findings suggest that the mechanosensor in skeletal muscle is directly or indirectly linked with nNOS and activate nNOS when the fibers are stretched. Alternatively, nNOS itself might sense the change in mechanical stimuli and be activated.

4.2.5 nNOS and Unloading

Suzuki et al. recently reported that nNOS acts as an upstream regulator of forkhead box O (FoxO) transcription factors in unloading-induced muscle atrophy (Suzuki et al. 2007) (Fig. 4.2). The study showed that during hind limb suspension, nNOS was lost from the sarcolemma, and cytoplasmic nNOS produced high levels of NO and activated FoxO3a, which in turn upregulated MuRF-1 and atrogin-1/MAFbx, two major muscle-specific E3 ligases. Although the data suggest that nNOS is a mediator of mechanotransduction in skeletal muscle, the mechanism by which nNOS is activated remains unknown. It is also still to be determined how NO gas activates FoxO transcription factors. One possibility is that NO controls FoxO transcription factors via S-nitrosylation. NO reacts with thiol of protein cysteine residues to form S-nitrosothiol. This posttranslational modification affects the functions of a wide range of proteins, and regulates protein–protein interactions and DNA-binding activity of transcription factors (Hess et al. 2005). Recent studies show that FoxO transcription factors are necessary and sufficient for induction of autophagy in skeletal muscle (Mammucari et al. 2007; Zhao et al. 2007). Whether nNOS activates the autophagy–lysosome system through activation of FoxO3a in unloaded muscle is to be determined.

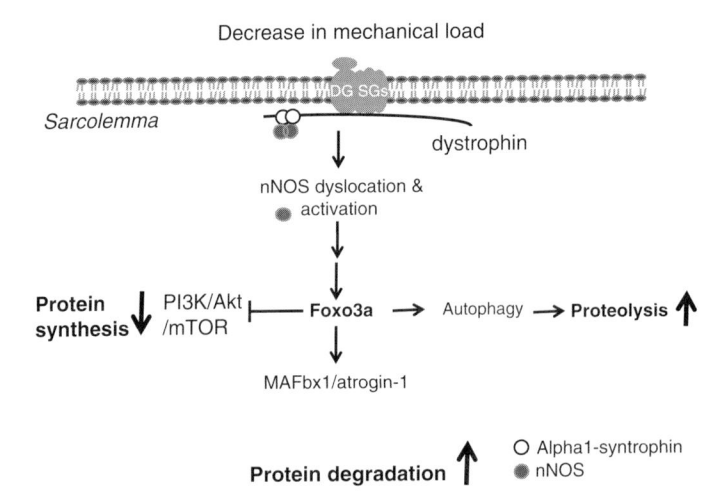

Fig. 4.2 A mechanism by which an unloaded signal activates nNOS and promotes muscle atrophy (hypothetical). In skeletal muscle, a major fraction of nNOS exists as a member of the dystrophin–glycoprotein complex at the sarcolemma. nNOS interacts directly with α1-syntrophin through its N-terminal PDZ domain. In unloading condition, nNOS is lost from the sarcolemma, and cytoplasmic nNOS is activated, produces nitric oxide and activates FoxO3a transcription factor (Suzuki et al. 2007). FoxO3a activates the ubiquitin-proteasome-mediated proteolysis and autophagy. How nNOS is activated and how it activates FoxO3a transcription factor remain to be determined in a future study

4.2.6 Phospholipase D and Phosphatidic Acid

Hornberger et al. showed that mechanical stimulation causes phospholipase D (PLD)-dependent increase in phosphatidic acid (PA), which binds to mTOR in the FRB domain and activates mTOR signaling (Hornberger et al. 2006). Although the observation is exciting, it remains to be determined whether elevation of PA concentration is necessary and sufficient for mechanical load-induced muscle hypertrophy.

4.2.7 Integrins/FAK Signaling and Mechanotransduction

The mechanical signal may be partly transmitted by focal adhesion complexes (FACs) in skeletal muscle. FACs connect cytoskeletal proteins to the extracellular matrix and therefore transmit force across the cell membrane. Beta-1 integrin in FACs is bound by focal adhesion kinase (FAK) and paxillin (Carson and Wei 2000; Flück et al. 1999). Mechanical loading on skeletal muscle results in phosphorylation of FAX or paxillin (Gordon et al. 2001). Furthermore, FAK is shown to affect activation of Akt/mTOR independent of PI3K activity in fibroblasts (Xia et al. 2004). At present, however, there is no direct evidence that integrin/FAK signaling activates protein synthesis in mechanically stimulated skeletal muscle.

4.2.8 Sarcomere and Mechanotransduction

The sarcomere is a candidate mediator of mechanical transduction. The M-band is proposed to be a hub mainly for protein kinase-regulated ubiquitin signaling and protein turnover, and the I-band and Z-disk contain stretch-sensitive pathways involving transcriptional modifiers (Gautel 2008). It is highly likely that signals from the sarcomere modify mass and contractile properties of overloaded muscle.

4.2.9 Growth Factors and Overload

It has been suggested that a number of cytokines, including LIF (Spangenburg and Booth 2006), IGF-1 (Rommel et al. 2001; Ohanna et al. 2005), or a variant of IGF-1 termed mechano-growth factor (MGF) (Goldspink et al. 2008), are produced locally in overloaded muscle, and contribute to muscle hypertrophy. Since these cytokines activate the PI3K/Akt/mTOR/p70S6K pathway, mechanical loading is thought to induce the muscle hypertrophy, through production of growth-promoting autocrine/ paracrine factors. For example, IGF-1 is shown to be a potent activator of the Akt/ mTOR signaling pathway (Adams 2002; Glass 2005), and forced overexpression of IGF-1 in skeletal muscle increased muscle mass (Musarò et al. 2001; Barton 2006). If so, is IGF-1 production a trigger of overload-induced muscle hypertrophy? Upregulation of IGF-1 is, however, observed between 24 and 72 h after the onset of the mechanical load on muscle, whereas Akt/mTOR is activated a few hours after the onset of mechanical load. Furthermore, a recent study using transgenic mice expressing a dominant negative IGF-I receptor, suggested that IGF-1 signaling is dispensable in mechanical load-induced skeletal muscle hypertrophy (Spangenburg et al. 2008). Therefore, although IGF-1 would augment the hypertrophy by accelerating protein synthesis, it is not likely the trigger of increased protein synthesis. In contrast, the loss of leukemia inhibitory factor (LIF) is reported to result in a failed hypertrophic response after mechanical loading (Spangenburg and Booth 2006). The mechanisms by which LIF promotes muscle hypertrophy process remain to be determined.

4.3 Signaling Pathways in Mechanotransduction

Although how mechanical stimuli are converted into biochemical events in skeletal muscle remains to be determined, several downstream signaling pathways are shown to be involved in the mechanotransduction.

4.3.1 mTOR Is a Key Signaling Molecule for Mechanical Overload-Induced Muscle Hypertrophy

How mechanical loading on skeletal muscle activates the Akt/mTOR signaling pathway remains largely unknown. Nonetheless, the importance of mTOR in mechanical load-induced hypertrophy was generally accepted due to the findings that rapamycin administration (a potent and selective inhibitor of mTOR) completely abolished the hypertrophy under overload conditions (Bodine et al. 2001; Hornberger et al. 2003; Bodine 2006).

mTOR activates p70S6K (ribosomal protein kinase S6) and 4E-BP1. p70S6K phosphorylated by mTOR, increases the translation of mRNAs with a 5′-tract of pyrimidine, which is contained in all known ribosomal proteins. 4E-BP1 regulates the initiation of protein synthesis by sequestering eIF4E. When phosphorylated, 4E-BP1 releases eIF4E, which in turn complexes with eIF4G and eIF3 and initiates translation (Rennie et al. 2004). Once again, we emphasize that we have a very limited understanding of how physical loading of the muscle fibers activates the mTOR signaling pathway.

4.3.2 Unload (Inactivity or Disuse) and the Catabolic Signaling Pathway

When the activity of skeletal muscle is reduced by spaceflight (microgravity), prolonged bed rest, immobilization, denervation, or hindlimb suspension, skeletal muscle loses its weight and contractile force. Two major signaling pathways are reported to play major roles in unload-induced skeletal muscle atrophy: nuclear factor-kappa B (NF-κB) and FoxO pathways.

Genetically modified animal models confirmed that NF-κB is one of most important signaling pathways linked to the loss of skeletal muscle in both physiological and pathophysiological conditions. Hunter et al. reported that skeletal muscle inactivity leads to increased NF-κB transcriptional activity (Hunter et al. 2002). Later, it was reported that a specific NF-κB pathway (p50 and Bcl3), but not a classical p50–p65 NF-κB dimer, is required for disuse muscle atrophy (Hunter and Kandarian 2004; Judge et al. 2007). NF-κB activates the transcription of many genes, including skeletal muscle-specific E3 ligases, MuRF1. But how these target molecules induce muscle atrophy remains poorly defined.

Under atrophic conditions, it is proposed that the inhibition of FoxO activity by phosphorylation (possibly by Akt) is relieved, and the dephophorylated FoxO transcription factors accumulate in the nucleus, activate an atrophy-related ubiquitin E3 ligase, atrogin-1/MAFbx-1 and induce skeletal muscle wasting (Sandri et al. 2004; Stitt et al. 2004).

Although the upstream events of NF-κB or FoxO activation are not clarified, candidate molecules which link disuse and activation of NF-κB or FoxO are reactive oxygen species (ROS) (Powers et al. 2007; Dodd et al. 2010). But, how ROS activate NF-κB or FoxO transcription factors is not clearly defined.

4.3.2.1 Downstream Targets of NF-κB and FoxO Transcription Factors

Downstream of these signaling pathways, three major proteolytic systems, (1) lysosomal proteases (cathepsins), (2) Ca^{2+}-activated proteases (i.e., calpains), and (3) the ubiquitin-proteasome system, are thought to contribute to the degradation of muscle proteins. Unloading skeletal muscle also causes caspase-3 activation and myonuclear apoptosis, maintaining a constant ratio of cytoplasm to nuclei (myonuclear domain) (reviewed in Powers et al. 2007).

NF-κB is reported to activate muscle RING finger protein 1 (MuRF1) E3 ligase (Cai et al. 2004; Mourkioti et al. 2006). FoxO3 is shown to activate atrogin-1/MAFbx to activate the ubiquitin-proteasome pathway (Sandri et al. 2004). Increased FoxO3 function also leads to decreased myocyte cell size through induction of autophagy pathway genes, including *LC3*, *Gabarapl1*, and *Atg12* (Zhao et al. 2007; Mammucari et al. 2007).

4.4 Conclusion

Skeletal muscle is sensitive to mechanical load, and adapts the demand by increasing or decreasing its mass and contractile force. Muscle hypertrophy and wasting are accompanied by changes in fiber compositions and metabolic properties. Thanks to advances in gene manipulation techniques in mice, major players in signal transduction regulating muscle hypertrophy and atrophy have been identified. But how myofibers sense change in mechanical load, and how the information is translated into biochemical signaling pathways, remain to be determined.

Acknowledgment The authors wish to thank members of the laboratory for discussion and critical readings.

References

Acharyya S, Butchbach ME, Sahenk Z et al (2005) Dystrophin glycoprotein complex dysfunction: a regulatory link between muscular dystrophy and cancer cachexia. Cancer Cell 8:421–432

Adams GR (2002) Invited review: autocrine/paracrine IGF-I and skeletal muscle adaptation. J Appl Physiol 93:1159–1167

Atherton PJ, Babraj J, Smith K et al (2005) Selective activation of AMPK-PGC-1alpha or PKB-TSC2-mTOR signaling can explain specific adaptive responses to endurance or resistance training-like electrical muscle stimulation. FASEB J 19:786–788

Baar K, Nader G, Bodine S (2006) Resistance exercise, muscle loading/unloading and the control of muscle mass. Essays Biochem 42:61–74

Barton ER (2006) Viral expression of insulin-like growth factor-I isoforms promotes different responses in skeletal muscle. J Appl Physiol 100:1778–1784

Bassel-Duby R, Olson EN (2006) Signaling pathways in skeletal muscle remodeling. Annu Rev Biochem 75:19–37

Blaauw B, Canato M, Agatea L et al (2009) Inducible activation of Akt increases skeletal muscle mass and force without satellite cell activation. FASEB J 23:3896–3905

Bodine SC (2006) mTOR signaling and the molecular adaptation to resistance exercise. Med Sci Sports Exerc 38:1950–1957

Bodine SC, Stitt TN, Gonzalez M et al (2001) Akt/mTOR pathway is a crucial regulator of skeletal muscle hypertrophy and can prevent muscle atrophy in vivo. Nat Cell Biol 3:1014–1019

Brenman JE, Chao DS, Gee SH et al (1996) Interaction of nitric oxide synthase with the postsynaptic density protein PSD-95 and alpha1-syntrophin mediated by PDZ domains. Cell 84:757–767

Butterfield TA, Best TM (2009) Stretch-activated ion channel blockade attenuates adaptations to eccentric exercise. Med Sci Sports Exerc 41:351–356

Cai D, Frantz JD, Tawa NE Jr et al (2004) IKKbeta/NF-kappaB activation causes severe muscle wasting in mice. Cell 119:285–298

Carson JA (1997) The regulation of gene expression in hypertrophying skeletal muscle. Exerc Sport Sci Rev 25:301–320

Carson JA, Wei L (2000) Integrin signaling's potential for mediating gene expression in hypertrophying skeletal muscle. J Appl Physiol 88:337–343

Chang WJ, Iannaccone ST, Lau KS et al (1996) Neuronal nitric oxide synthase and dystrophin-deficient muscular dystrophy. Proc Natl Acad Sci USA 93:9142–9147

Dodd SL, Gagnon BJ, Senf SM et al (2010) Ros-mediated activation of NF-kappaB and FoxO during muscle disuse. Muscle Nerve 41:110–113

Eliasson J, Elfegoun T, Nilsson J et al (2006) Maximal lengthening contractions increase p70 S6 kinase phosphorylation in human skeletal muscle in the absence of nutritional supply. Am J Physiol Endocrinol Metab 291:E1197–E1205

Flück M, Carson JA, Gordon SE et al (1999) Focal adhesion proteins FAK and paxillin increase in hypertrophied skeletal muscle. Am J Physiol 277:C152–C162

Franco A Jr, Lansman JB (1990a) Calcium entry through stretch-inactivated ion channels in mdx myotubes. Nature 344:670–673

Franco A Jr, Lansman JB (1990b) Stretch-sensitive channels in developing muscle cells from a mouse cell line. J Physiol 427:361–380

Gautel M (2008) The sarcomere and the nucleus: functional links to hypertrophy, atrophy and sarcopenia. Adv Exp Med Biol 642:176–191

Glass DJ (2005) Skeletal muscle hypertrophy and atrophy signaling pathways. Int J Biochem Cell Biol 37:1974–1984

Goldspink G, Wessner B, Bachl N (2008) Growth factors, muscle function and doping. Curr Opin Pharmacol 8:352–357

Gordon SE, Flück M, Booth FW (2001) Selected contribution: skeletal muscle focal adhesion kinase, paxillin, and serum response factor are loading dependent. J Appl Physiol 90:1174–1183

Haddad F, Adams GR (2002) Selected contribution: acute cellular and molecular responses to resistance exercise. J Appl Physiol 93:394–403

Hameed M, Orrell RW, Cobbold M et al (2003) Expression of IGF-I splice variants in young and old human skeletal muscle after high resistance exercise. J Physiol 547:247–254

Hess DT, Matsumoto A, Kim SO et al (2005) Protein S-nitrosylation: purview and parameters. Nat Rev Mol Cell Biol 6:150–166

Hornberger TA, McLoughlin TJ, Leszczynski JK et al (2003) Selenoprotein-deficient transgenic mice exhibit enhanced exercise-induced muscle growth. J Nutr 133:3091–3097

Hornberger TA, Stuppard R, Conley KE et al (2004) Mechanical stimuli regulate rapamycin-sensitive signalling by a phosphoinositide 3-kinase-, protein kinase B- and growth factor-independent mechanism. Biochem J 380:795–804

Hornberger TA, Chu WK, Mak YW et al (2006) The role of phospholipase D and phosphatidic acid in the mechanical activation of mTOR signaling in skeletal muscle. Proc Natl Acad Sci USA 103:4741–4746

Hunter RB, Kandarian SC (2004) Disruption of either the Nfkb1 or the Bcl3 gene inhibits skeletal muscle atrophy. J Clin Invest 114:1504–1511

Hunter RB, Stevenson E, Koncarevic A et al (2002) Activation of an alternative NF-kappaB pathway in skeletal muscle during disuse atrophy. FASEB J 16:529–538

Jackman RW, Kandarian SC (2004) The molecular basis of skeletal muscle atrophy. Am J Physiol Cell Physiol 287:C834–C843

Joulia-Ekaza D, Cabello G (2007) The myostatin gene: physiology and pharmacological relevance. Curr Opin Pharmacol 7:310–315

Judge AR, Koncarevic A, Hunter RB, Liou HC, Jackman RW, Kandarian SC (2007) Role for IkappaBalpha, but not c-Rel, in skeletal muscle atrophy. Am J Physiol Cell Physiol 292:C372–C382

Kadi F (2008) Cellular and molecular mechanisms responsible for the action of testosterone on human skeletal muscle. A basis for illegal performance enhancement. Br J Pharmacol 154:522–528

Koh TJ, Tidball JG (1999) Nitric oxide synthase inhibitors reduce sarcomere addition in rat skeletal muscle. J Physiol 519:189–196

Mammucari C, Milan G, Romanello V et al (2007) FoxO3 controls autophagy in skeletal muscle in vivo. Cell Metab 6:458–471

McCarthy JJ, Esser KA (2007) Counterpoint: satellite cell addition is not obligatory for skeletal muscle hypertrophy. J Appl Physiol 103:1100–1102

Mourkioti F, Kratsios P, Luedde T et al (2006) Targeted ablation of IKK2 improves skeletal muscle strength, maintains mass, and promotes regeneration. J Clin Invest 116:2945–2954

Musarò A, McCullagh K, Paul A et al (2001) Localized Igf-1 transgene expression sustains hypertrophy and regeneration in senescent skeletal muscle. Nat Genet 27:195–200

Nader GA, Esser KA (2001) Intracellular signaling specificity in skeletal muscle in response to different modes of exercise. J Appl Physiol 90:1936–1942

O'Connor RS, Pavlath GK (2007) Point: counterpoint: satellite cell addition is/is not obligatory for skeletal muscle hypertrophy. J Appl Physiol 103:1099–1100

Ohanna M, Sobering AK, Lapointe T et al (2005) Atrophy of S6K1(−/−) skeletal muscle cells reveals distinct mTOR effectors for cell cycle and size control. Nat Cell Biol 7:286–294

Powers SK, Kavazis AN, McClung JM (2007) Oxidative stress and disuse muscle atrophy. J Appl Physiol 102:2389–2397

Reiser PJ, Kline WO, Vaghy PL (1997) Induction of neuronal type nitric oxide synthase in skeletal muscle by chronic electrical stimulation in vivo. J Appl Physiol 82:1250–1255

Rennie MJ, Wackerhage H, Spangenburg EE, Booth FW (2004) Control of the size of the human muscle mass. Annu Rev Physiol 66:799–828

Roberts CK, Barnard RJ, Jasman A, Balon TW (1999) Acute exercise increases nitric oxide synthase activity in skeletal muscle. Am J Physiol 277:E390–E394

Rommel C, Bodine SC, Clarke BA, Rossman R et al (2001) Mediation of IGF-1-induced skeletal myotube hypertrophy by PI(3)K/Akt/mTOR and PI(3)K/Akt/GSK3 pathways. Nat Cell Biol 3:1009–1013

Sandri M, Sandri C, Gilbert A et al (2004) FoxO transcription factors induce the atrophy-related ubiquitin ligase atrogin-1 and cause skeletal muscle atrophy. Cell 117:399–412

Sellman JE, DeRuisseau KC, Betters JL et al (2006) In vivo inhibition of nitric oxide synthase impairs upregulation of contractile protein mRNA in overloaded plantaris muscle. J Appl Physiol 100:258–265

Smith LW, Smith JD, Criswell DS (2002) Involvement of nitric oxide synthase in skeletal muscle adaptation to chronic overload. J Appl Physiol 92:2005–2011

Soltow QA, Betters JL, Sellman JE et al (2006) Ibuprofen inhibits skeletal muscle hypertrophy in rats. Med Sci Sports Exerc 38:840–846

Spangenburg EE (2009) Changes in muscle mass with mechanical load: possible cellular mechanisms. Nutr Metab 34:328–335

Spangenburg EE, Booth FW (2006) Leukemia inhibitory factor restores the hypertrophic response to increased loading in the LIF(−/−) mouse. Cytokine 34:125–130

Spangenburg EE, McBride TA (2006) Inhibition of stretch-activated channels during eccentric muscle contraction attenuates p70S6K activation. Appl Physiol 100:129–135

Spangenburg EE, Le Roith D, Ward CW, Bodine SC (2008) A functional insulin-like growth factor receptor is not necessary for load-induced skeletal muscle hypertrophy. J Physiol 586:283–291

Steensberg A, Keller C, Hillig T et al (2007) Nitric oxide production is a proximal signaling event controlling exercise-induced mRNA expression in human skeletal muscle. FASEB J 21:2683–2694

Stitt TN, Drujan D, Clarke BA et al (2004) The IGF-1/PI3K/Akt pathway prevents expression of muscle atrophy-induced ubiquitin ligases by inhibiting FOXO transcription factors. Mol Cell 14:395–403

Suzuki N, Motohashi N, Uezumi A et al (2007) NO production results in suspension-induced muscle atrophy through dislocation of neuronal NOS. J Clin Invest 117:2468–2476

Tatsumi R, Hattori A, Ikeuchi Y et al (2002) Release of hepatocyte growth factor from mechanically stretched skeletal muscle satellite cells and role of pH and nitric oxide. Mol Biol Cell 13:2909–2918

Tatsumi R, Liu X, Pulido A et al (2006) Satellite cell activation in stretched skeletal muscle and the role of nitric oxide and hepatocyte growth factor. Am J Physiol Cell Physiol 290:C1487–C1494

Tidball JG, Lavergne E, Lau KS et al (1998) Mechanical loading regulates NOS expression and activity in developing and adult skeletal muscle. Am J Physiol 275:C260–C266

Tidball JG, Spencer MJ, Wehling M, Lavergne E (1999) Nitric-oxide synthase is a mechanical signal transducer that modulates talin and vinculin expression. J Biol Chem 274:33155–33160

Ventadour S, Attaix D (2006) Mechanisms of skeletal muscle atrophy. Curr Opin Rheumatol 18:631–635

Widrick JJ, Stelzer JE, Shoepe TC, Garner DP (2002) Functional properties of human muscle fibers after short-term resistance exercise training. Am J Physiol Regul Integr Comp Physiol 283:R408–R416

Xia H, Nho RS, Kahm J, Kleidon J, Henke CA (2004) Focal adhesion kinase is upstream of phosphatidylinositol 3-kinase/Akt in regulating fibroblast survival in response to contraction of type I collagen matrices via a beta 1 integrin viability signaling pathway. J Biol Chem 279:33024–33034

Zanchi NE, Lancha AH Jr (2008) Mechanical stimuli of skeletal muscle: implications on mTOR/p70s6k and protein synthesis. Eur J Appl Physiol 102:253–263

Zhao J, Brault JJ, Schild A et al (2007) FoxO3 coordinately activates protein degradation by the autophagic/lysosomal and proteasomal pathways in atrophying muscle cells. Cell Metab 6:472–483

Chapter 5
Mechanobiology in Space

Yuushi Okumura and Takeshi Nikawa

5.1 Introduction

During spaceflight, we are exposed to the long-term unloading environment. Although a number of physiological adaptations occurred in our body, these are not necessarily favorable. The loss of skeletal muscle mass is one of the most obvious phenomenons. After only 5 days in space, an 11–24% loss in muscle fiber cross-section is observed (Booth and Criswell 1997). Likewise, muscle disuse through the limb immobilization and prolonged bed rest also represent the large degree of muscle loss (Bloomfield 1997; Droppert 1993). Indeed, 2 weeks of leg immobilization has been shown to result in an almost 5% reduction in quadriceps lean mass and 27% fall in isometric strength (Jones et al. 2004). A 20% reduction in type I muscle fiber cross-sectional area and a 30% reduction in type II fiber area has been observed following 60 days of bed rest (Salanova et al. 2008). However, the mechanisms by which devices mainly transduce these changes are not well understood.

On a molecular level, skeletal muscle protein synthesis decreases and protein degradation increases (Baldwin 1996; Booth and Criswell 1997), resulting in marked skeletal muscle atrophy. During atrophy, the myofibrillar apparatus, which comprises at least 60% of muscle proteins, rapidly decreases in mass to a greater extent than the soluble components (Munoz et al. 1993). Our previous studies demonstrated that unloading condition produced by spaceflight as well as tail suspension in animals, stimulated ubiquitination of various proteins, including myosin heavy chain (MHC), and accumulation of MHC degradation fragments in atrophied rat gastrocnemius muscle (Ikemoto et al. 2001). Other studies also reported that two ubiquitin–protein ligases (E3s), atrogin-1/MAFbx and MuRF-1, are induced early in the atrophy process (Bodine et al. 2001; Gomes et al. 2001), and that the rise in atrogin-1 and MuRF-1 expression precedes the loss of muscle weight (Dehoux et al. 2003; Lecker et al. 2004; Wray et al. 2003). Indeed, animals lacking atrogin-1/MAFbx and MuRF-1

Y. Okumura and T. Nikawa (✉)

Department of Nutritional Physiology, Institute of Health Biosciences, University of Tokushima Graduate School, 3-18-15 Kuramoto-cho, Tokushima 770-8503, Japan

e-mail: nikawa@nutr.med.tokushima-u.ac.jp

M. Noda (ed.), *Mechanosensing Biology*,
DOI 10.1007/978-4-431-89757-6_5, © Springer 2011

genes are resistant to muscle atrophy caused by denervation (Bodine et al. 2001). Alternatively, the upregulation of another ubiquitin ligase, Cbl-b, was demonstrated by spaceflight (Nikawa et al. 2004). Like atrogin-1- and MuRF-1-deficient mice, Cbl-b-deficient mice are also resistant to muscle atrophy and dysfunction induced during unloading by tail suspension. In addition, the induction of Cbl-b in vivo-induced significant atrophy in rat tibialis anterior muscle (Nakao et al. 2009). In molecular basis analysis, under atrophic condition, IRS-1 was ubiquitinated by Cbl-b and thereby deactivated the skeletal muscle growth factor signaling (i.e., IGF-1/PI3K/Akt pathway). Cbl-b activation appears to account for the overall decrease in protein synthesis observed in skeletal muscle atrophy.

The initial trigger molecules of mechanosensing, which facilitate ubiquitin ligase function during unloading conditions, are still poorly understood. As the focus of this review, however, we concentrate on the alterations observed in skeletal muscle and discuss the protein degradation system underlying the cellular and molecular basis of muscle atrophy.

5.2 Unloading and Protein Degradation

Skeletal muscle atrophy caused by unloading is characterized by a combination of decreased protein synthesis and increased protein degradation. In protein degradation, three known proteolytic systems lysosomal degradation, calpain proteolysis and the ubiquitin–proteasome system, are thought to be implicated in muscle atrophy (Taillandier et al. 1996). Lysosomal enzymes, cathepsins, express in adult muscle and cell lines although their expression levels are relatively low. In some experimental studies, cathepsins increased their activities in skeletal muscle treated with hindlimb suspension (Taillandier et al. 1996) and cleaved the myofibrillar components in vitro (Bechet et al. 2005). However, Whitaker et al. (1983) suggested a possible role, their involvement in the clearance of damaged protein during muscle regeneration and repair. On the other hand, several groups have reported that the ubiquitous Ca^{2+}-activated calpain (Tidball and Spencer 2002) or caspase (Du et al. 2004) may initially cleave the myofibrillar components, thereby accelerating their disassembly and degradation by the ubiquitin–proteasome system. However, direct proof of this model's functioning in atrophying muscles seems to be lacking. Alternatively, under unloading condition such as spaceflight and tail suspension in animals, various proteins including MHC are ubiquitinated (Ikemoto et al. 2001). Other groups have also reported the accumulation of ubiquitinated proteins in atrophic situations (Berthon et al. 2007; Krawiec et al. 2005). In addition to the enzyme function, upregulation of the ubiquitin ligases gene was reported in space-flown rats (Nikawa et al. 2004). Fluck et al. (2005) also detected the induction of gene expression of proteasomal factors under unloading conditions. These findings also supported that the ubiquitin–proteasome system acts mainly on the promotion of the myofibril components disassembly and degradation during atrophy.

5.3 Muscle Specific Ubiquitin Ligases

It is well recognized that the ATP-dependent ubiquitin–proteasome pathway is associated with the degradation of most muscle protein during atrophy (Solomon and Goldberg 1996; Jagoe and Goldberg 2001; Lecker et al. 2006). For targeting of proteins to the proteasome, ubiquitin moieties must be conjugated by ubiquitin–protein ligases (E3s), which are rate-limiting enzymes in this pathway. However, to date, it is unclear whether E3s ubiquitinate many muscle components or only key regulatory ones, whose destruction promotes the breakdown of other proteins. The apparent evidence suggested that the expression of the ubiquitin ligases such as MuRF-1 and Cbl-b was specifically induced by unloading conditions. MuRF-1 is a monomeric E3 belonging to the tripartite motif family of ligases which contain a tripartite RING:B-box:coiled–coil domain that is essential for ubiquitin conjugation. Two close homologues of MuRF-1, MuRF-2 and MuRF-3, are also present in normal muscle and appear to be important for maintenance of normal contractile function (Fielitz et al. 2007; Witt et al. 2008). MuRF-1, on the other hand, is thought to be required for rapid atrophy but not for normal muscle growth. To clarify the role of its ligase activity in skeletal muscle atrophy, mice lacking MuRF-1 have been challenged in an atrophy model and were shown to be resistant to atrophy (Bodine et al. 2001). Although two substrates have been reported for MuRF-1, cardiac Troponin-I (Kedar et al. 2004) and MHC (Clarke et al. 2007), the role of MuRF-1 in degrading specific myofibrillar or soluble proteins in skeletal muscle is unknown. Specifically, it is unclear whether MuRF-1 or any E3 ligase can act on the proteins within the myofibril, which is a highly organized rigid structure, or whether mechanisms exist to release proteins from the sarcomeres before ubiquitin-dependent degradation. Model studies in muscle extracts have demonstrated rapid hydrolysis of purified actin and myosin by the ubiquitin–proteasome pathway (Solomon and Goldberg 1996). However, these proteins were much more stable when associated with each other in the actomyosin complex or intact myofibrils. These findings raised the possibility that the ubiquitin–proteasome system may not by itself be able to degrade components of intact myofibrils, and that constituent proteins have to be released by some mechanism to be substrates for degradation (Solomon and Goldberg 1996). In other cases, unloading stress resulted in skeletal muscle atrophy through the significant induction and activation of the ubiquitin ligase Cbl-b. Cbl-b is another RING-type ubiquitin ligase that attenuates tyrosine kinase signaling of growth factors by degrading downstream of signaling molecules (Ogawa et al. 2006). For example, upon induction, Cbl-b interacted with and degraded the IGF-1 signaling intermediate IRS-1. In turn, loss of IRS-1 activated the FOXO3-dependent induction of atrogin-1/MAFbx, a dominant mediator of proteolysis in atrophic muscle (Fig. 5.1). Thus, the Cbl-b-dependent destruction of IRS-1 is a critical mediator of unloading-induced muscle atrophy. Cbl-b-deficient mice were also resistant to unloading-induced atrophy and loss of muscle function.

Other experimental studies reported that muscle atrophy caused by immobilization or unloading is associated with oxidative stress and the generation of reactive oxygen

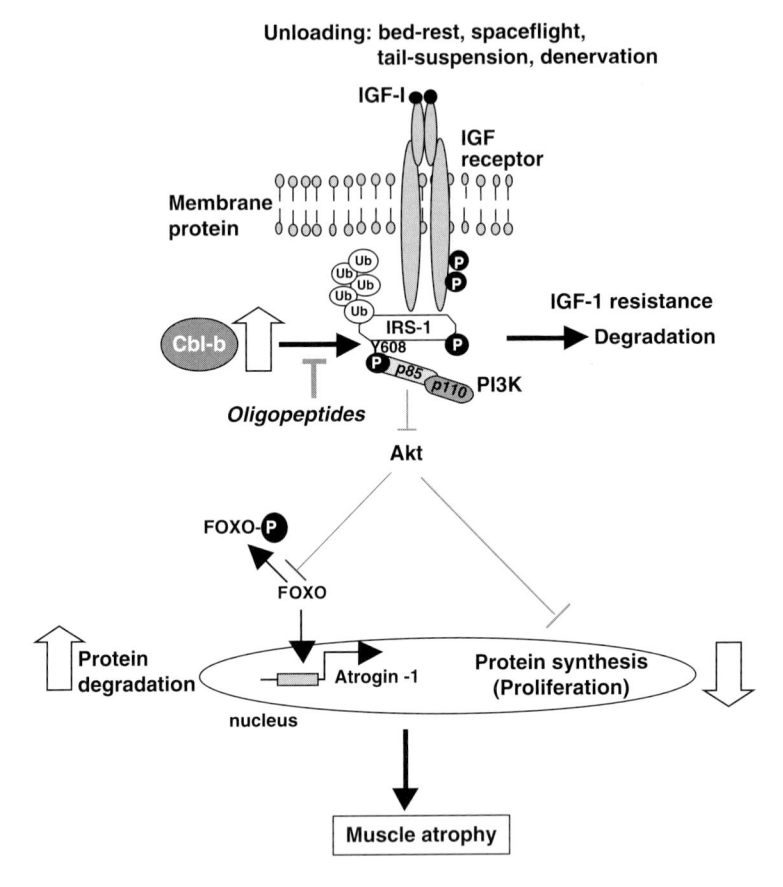

Fig. 5.1 Mechanistic model of unloading-mediated muscle atrophy. Unloading induces ubiquitin ligase Cbl-b in myocytes. Cbl-b stimulates ubiquitination and degradation of IRS-1, an important intermediate in IGF-1 signaling pathway, resulting in IGF-1 resistance in myocytes during unloading. IGF-1 resistance induces impaired protein synthesis and enhances protein degradation in muscle, leading to muscle atrophy. Cbl-b and PI3K may interact with phospho-tyrosine[608] of IRS-1. Inhibition of Cbl-b and IRS-1 interaction by oligopeptides may restore this impairment of IGF-1 signaling. GSK3, glycogen synthase kinase 3; mTOR, mammalian target of rapamycin; S6K, p70 S6 kinase; Ub, ubiquitin

species (ROS) (Bar-Shai et al. 2008). In addition, exogenous ROS significantly upregulated the expression of muscle atrophy-related ubiquitin ligase, Cbl-b (Nikawa et al. 2004). Moreover, Cbl-b expression induced by unloading conditions was regulated by the activation of Egr through MAPK(ERK)-dependent pathway following oxidative stress. Thus, oxidative stress may serve as an important trigger of signaling pathways leading to muscle atrophy during prolonged periods of disuse. In fact, supplementation of cysteine, an antioxidative nutrient, prevented unweighting-induced ubiquitination in association with redox regulation in rat skeletal muscle (Ikemoto et al. 2002). However, to date, antioxidative nutrients have not been reported to prevent activation of muscle atrophy-related ubiquitin ligase.

5.4 Deactivation of IGF-1/PI3K/Akt Pathway

Intracellular signaling molecules also provide mechanisms that underlie adaptability to unloading stress. Although skeletal muscle atrophy is not simply the converse of hypertrophy, recent studies have shown important connections between the deactivation of the IGF-1/PI3K/Akt pathway and proteolytic activity/expression of proteolytic genes. In fact, the stimulation of protein synthesis in skeletal muscle through activation of PI3K, Akt, IGF-1 and insulin also reduced the expression of atrogene-1/MAFbx (Sacheck et al. 2004). In addition, the upregulation of atrogene-1/MAFbx and MuRF-1 induced by dexamethasone-treated myotubes was antagonized by simultaneous treatment with IGF-1, acting through the PI3K/Akt pathway. Moreover, in cultured myotubes undergoing atrophy, the activity of the PI3K/Akt pathway decreased (Sandri et al. 2004). Thus, in controlling muscle mass, the IGF-1/PI3K/Akt pathway not only increases overall synthesis but also suppresses proteolysis and the expression of atrophy-related ubiquitin ligases. Therefore, inhibition of Cbl-b-mediated IRS-1 ubiquitination may be a new therapeutic strategy for unloading-induced muscle atrophy. Recent studies demonstrated that a pentapeptide mimetic of tyrosine[608]-phosphorylated IRS-1 inhibited Cbl-b-mediated IRS-1 ubiquitination and strongly decreased Cbl-b-mediated induction of atrogin-1/MAFbx (Nakao et al. 2009). The use of foods containing low molecular mass peptide-based inhibitors can be extended to develop novel functional space menus against microgravity-induced muscle atrophy.

5.5 Myostatin

Myostatin, a member of transforming growth factor (TGF)-β superfamily, plays a dominant role in the genetic control of muscle mass. It regulates muscle formation during embryogenesis and postnatal muscle development as an endogenous inhibitor of muscle mass (McPherron et al. 1997). Inactivating mutation of the myostatin gene in cattle is associated with generalized increase in skeletal muscle (McPherron and Lee 1997), Myostatin is synthesized as a precursor protein, which generates the N-terminal propeptide and the C-terminal mature myostatin peptide by a posttranslational cleavage event. Activation of myostatin requires a two-step proteolysis by different kinds of proteases. In the first step, myostatin is processed at the Arg-Ser-Arg-Arg cleavage site by proteases, which selectively recognize this site (Anderson et al. 2008). As the candidate processing proteases, we proposed proprotein convertase (PC) family member such as furin, and the type II transmembrane serine protease, mosaic serine protease large form (MSPL). After conversion of myostatin into its active form, active myostatin was folded by its prodomain. In the second step, to break down the folding prodomain, myostatin is cleaved by proteases, which are members of the bone morphogenetic protein-1/tollois (BMP-1/TLD) family of metalloproteinase (Wolfman et al. 2003; Lee 2008). Interestingly, C2C12 myoblast subjected to three-dimensional (3D)-clinorotation increased the myostatin activation as well as Egr and Cbl-b expression. This suggests that myostatin activation induced by unloading stress may increase Cbl-b expression accompanied by muscle atrophy. Additionally, the expression of the candidate processing proteases for myostatin was not affected under unloading conditions. This suggests that the activation of myostatin by unloading stress might depend on the enhanced activity of processing proteases. Although the activation and inhibitory mechanisms of these proteases are currently being investigated, they might be trigger molecules for mechanosensing.

5.6   Summary and Perspective

In this review, we have summarized some of the experimental evidence that supports a model of muscle atrophy in response to unloading condition. To understand the molecular events underlying unloading-induced muscle atrophy within a subset of mechanotransducer including ubiquitin ligase, Cbl-b will provide the precise targets to control them. Since there are no effective treatments to reverse the progression of atrophy, inhibition of ubiquitin ligase-mediated protein ubiquitination may present us with a new therapeutic strategy for unloading-mediated muscle atrophy. Further, effective countermeasures for the space environment may then be applicable for the older or immobilized person.

Acknowledgments   This study was carried out as part of "Ground Research Announcement for Space Utilization" promoted by Japan Aerospace Exploration Agency (JAXA) and Japan Space Forum, and "Promotion of Basic Research Activities for Innovative Biosciences" from Bio-oriented Technology Research Advancement Institution, Japan.

References

Anderson SB, Goldberg AL, Whitman M (2008) Identification of a novel pool of extracellular pro-myostatin in skeletal muscle. J Biol Chem 283:7027–7035
Baldwin KM (1996) Effect of spaceflight on the functional, biochemical, and metabolic properties of skeletal muscle. Med Sci Sports Exerc 28(8):983–987
Bar-Shai M, Carmeli E, Ljubuncic P, Reznic AZ (2008) Exercise and immobilization in aging animals: the involvement of oxidative stress and NF-kappaB activation. Free Radic Biol Med 44:202–214
Bechet D, Tassa A, Taillandier D, Combaret L, Attaix D (2005) Lysosomal proteolysis in skeletal muscle. Int J Biochem Cell Biol 37:2098–2114
Berthon P, Duguez S, Favier FB, Amirouche A, Feasson L, Vico L, Denis C, Freyssenet D (2007) Regulation of ubiquitin-proteasome system, caspase enzyme activities, and extracellular proteinases in rat soleus muscle in response to unloading. Pflugers Arch 454:625–633
Bloomfield SA (1997) Changes in musculoskeletal structure and function with prolonged bed rest. Med Sci Sports Exerc 29:197–206
Bodine SC, Latres E, Baumhueter S, Lai VK, Nunez L, Clarke BA, Poueymirou WT, Panaro FJ, Na E, Dharmarajan K, Pan ZQ, Valenzuela DM, DeChiara TM, Stitt TN, Yancopoulos GD,

Glass DJ (2001) Identification of ubiquitin ligases required for skeletal muscle atrophy. Science 294:1704–1708

Booth FW, Criswell DS (1997) Molecular events underlying skeletal muscle atrophy and the development of effective countermeasures. Int J Sports Med 18(Suppl 4):S265–S269

Clarke BA, Drujan D, Willis MS, Murphy LO, Corpina RA, Burova E, Rakhilin SV, Stitt TN, Patterson C, Latres E, Glass DJ (2007) The E3 ligase MuRF1 degrades myosin heavy chain protein in dexamethasone-treated skeletal muscle. Cell Metab 6:376–385

Dehoux MJ, van Beneden RP, Fernández-Celemín L, Lause PL, Thissen JP (2003) Induction of MafBx and Murf ubiquitin ligase mRNAs in rat skeletal muscle after LPS injection. FEBS Lett 544:214–217

Droppert PM (1993) A review of muscle atrophy in microgravity and during prolonged bed rest. J Br Interplanet Soc 46:83–86

Du J, Wang X, Miereles C, Bailey JL, Debigare R, Zheng B, Price SR, Mitch WE (2004) Activation of caspase-3 is an initial step triggering accelerated muscle proteolysis in catabolic conditions. J Clin Invest 113:115–123

Fielitz J, Kim MS, Shelton JM, Latif S, Spencer JA, Glass DJ, Richardson JA, Bassel-Duby R, Olson EN (2007) Myosin accumulation and striated muscle myopathy result from the loss of muscle RING finger 1 and 3. J Clin Invest 117:2486–2495

Fluck M, Schmutz S, Wittwer M, Hoppeler H, Desplanches D (2005) Transcriptional reprogramming during reloading of atrophied rat soleus muscle. Am J Physiol Regul Integr Comp Physiol 289:R4–R14

Gomes MD, Lecker SH, Jagoe RT, Navon A, Goldberg AL (2001) Atrogin-1, a muscle-specific F-box protein highly expressed during muscle atrophy. Proc Natl Acad Sci USA 98:14440–14445

Ikemoto M, Nikawa T, Takeda S, Watanabe C, Kitano T, Baldwin KM, Izumi R, Nonaka I, Towatari T, Teshima S, Rokutan K, Kishi K (2001) Space shuttle flight (STS-90) enhances degradation of rat myosin heavy chain in association with activation of ubiquitin-proteasome pathway. FASEB J 15:1279–1281

Ikemoto M, Nikawa T, Kano M, Hirasaka K, Kitano T, Watanabe C, Tanaka R, Yamamoto T, Kamada M, Kishi K (2002) Cysteine supplementation prevents unweighting-induced ubiquitination in association with redox regulation in rat skeletal muscle. Biol Chem 383:715–721

Jagoe RT, Goldberg AL (2001) What do we really know about the ubiquitin-proteasome pathway in muscle atrophy? Curr Opin Clin Nutr Metab Care 4:183–190

Jones SW, Hill RJ, Krasney PA, O'Conner B, Peirce N, Greenhaff PL (2004) Disuse atrophy and exercise rehabilitation in humans profoundly affects the expression of genes associated with the regulation of skeletal muscle mass. FASEB J 18:1025–1027

Kedar V, McDonough H, Arya R, Li HH, Rockman HA, Patterson C (2004) Muscle-specific RING finger 1 is a bona fide ubiquitin ligase that degrades cardiac troponin I. Proc Natl Acad Sci USA 101:18135–18140

Krawiec BJ, Frost RA, Vary TC, Jefferson LS, Lang CH (2005) Hindlimb casting decreases muscle mass in part by proteasome-dependent proteolysis but independent of protein synthesis. Am J Physiol Endocrinol Metab 289:E969–E980

Lecker SH, Jagoe RT, Gilbert A, Gomes M, Baracos V, Bailey J, Price SR, Mitch WE, Goldberg AL (2004) Multiple types of skeletal muscle atrophy involve a common program of changes in gene expression. FASEB J 18:39–51

Lecker SH, Goldberg AL, Mitch WE (2006) Protein degradation by the ubiquitin-proteasome pathway in normal and disease states. J Am Soc Nephrol 17:1807–1819

Lee S-J (2008) Genetic analysis of the role of proteolysis in the activation of latent myostatin. PLoS One 3(2):e1628

McPherron AC, Lee SJ (1997) Double muscling in cattle due to mutations in the myostatin gene. Proc Natl Acad Sci USA 94:12457–12461

McPherron AC, Lawler AM, Lee SJ (1997) Regulation of skeletal muscle mass in mice by a new TGF-b superfamily member. Nature 387:83–90

Munoz KA, Satarug S, Tischler ME (1993) Time course of the response of myofibrillar and sarcoplasmic protein metabolism to unweighting of the soleus muscle. Metabolism 42:1006–1012

Nakao R, Hirasaka K, Goto J, Ishidoh K, Yamada C, Ohno A, Okumura Y, Nonaka I, Yasutomo K, Baldwin KM, Kominami E, Higashibata A, Nagano K, Tanaka K, Yasui N, Mills EM, Takeda S, Nikawa T (2009) Ubiquitin ligase Cbl-b is a negative regulator for IGF-1 signaling during muscle atrophy caused by unloading. Mol Cell Biol 29:4798–4811

Nikawa T, Ishidoh K, Hirasaka K, Ishihara I, Ikemoto M, Kano M, Kominami E, Nonaka I, Ogawa T, Adams GR, Baldwin KM, Yasui N, Kishi K, Takeda S (2004) Skeletal muscle gene expression in space-flown rats. FASEB J 18:522–524

Ogawa T, Furochi H, Mameoka M, Hirasaka K, Onishi Y, Suzue N, Oarada M, Akamatsu M, Akima H, Fukunaga T, Kishi K, Yasui N, Ishidoh K, Fukuoka H, Nikawa T (2006) Ubiquitin ligase gene expression in healthy volunteers with 20-day bedrest. Muscle Nerve 34:463–469

Sacheck JM, Ohtsuka A, McLary SC, Goldberg AL (2004) IGF-I stimulates muscle growth by suppressing protein breakdown and expression of atrophy-related ubiquitin ligases, atrogin-1 and MuRF1. Am J Physiol Endocrinol Metab 287:E591–E601

Salanova M, Schiffl G, Püttmann B, Schoser BG, Blottner D (2008) Molecular biomarkers monitoring human skeletal muscle fibres and microvasculature following long-term bed rest with and without countermeasures. J Anat 212:306–318

Sandri M, Sandri C, Gilbert A, Skurk C, Calabria E, Picard A, Walsh K, Schiaffino S, Lecker SH, Goldberg AL (2004) Foxo transcription factors induce the atrophy-related ubiquitin ligase atrogin-1 and cause skeletal muscle atrophy. Cell 117:399–412

Solomon V, Goldberg AL (1996) Importance of the ATP-ubiquitin-proteasome pathway in the degradation of soluble and myofibrillar proteins in rabbit muscle extracts. J Biol Chem 271:26690–26697

Taillandier D, Aurousseau E, Meynial-Denis D, Bechet D, Ferrara M, Cottin P, Ducastaing A, Bigard X, Guezennec CY, Schmid HP et al (1996) Coordinate activation of lysosomal, Ca 2+-activated and ATP-ubiquitin-dependent proteinases in the unweighted rat soleus muscle. Biochem J 316(Pt 1):65–72

Tidball JG, Spencer MJ (2002) Expression of a calpastatin transgene slows muscle wasting and obviates changes in myosin isoform expression during murine muscle disuse. J Physiol 545 (Pt 3):819–828

Whitaker JN, Bertorini TE, Mendell JR (1983) Immunocytochemical studies of cathepsin D in human skeletal muscle. Ann Neurol 13:133–142

Witt CC, Witt SH, Lerche S, Labeit D, Back W, Labeit S (2008) Cooperative control of striated muscle mass and metabolism by MuRF1 and MuRF2. EMBO J 27:350–360

Wolfman NM, McPherron AC, Pappano WN, Davies MV, Song K, Tomkinson KN, Wright JF, Zhao L, Sebald SM, Greenspan DS, Lee SJ (2003) Activation of latent myostatin by the BMP-1/tolloid family of metalloproteinases. Proc Natl Acad Sci USA 100:15842–15846

Wray CJ, Mammen JM, Hershko DD, Hasselgren PO (2003) Sepsis upregulates the gene expression of multiple ubiquitin ligases in skeletal muscle. Int J Biochem Cell Biol 35:698–705

Chapter 6
Mechanical Stress and Bone

**Masaki Noda, Tadayoshi Hayata, Tetsuya Nakamoto,
Takuya Notomi, and Yoichi Ezura**

6.1 Introduction

Bone has been known to adapt to mechanical stress (Amin 2010; Beier and Loeser 2010; Currey 2010; Temiyasathit and Jacobs 2010). The presence of mechanical stress increases bone mass and the absence of mechanical stress reduces bone mass. Bending of weight-bearing long bone increases pressure on the concave side and decreases it on the convex side. Under such circumstances, bone is accumulated on the concave side and is reduced on the convex side. This can be seen after angular deformity due to malunion of fractures in children where minor angular deformity could be corrected during the growth of the children. This is considered to be due to the response of the bone adjusting itself to the axis of the weight-bearing line. It is seen in tennis players that the thickness of their arm bones is not symmetrical as the more frequently used side accumulates more bone mass than the other side (Kontulainen et al. 2002). When the whole body is subjected to unloading, systemic bone resorption is seen, as in the cases of bedridden patients or astronauts in space flight (Goodship et al. 1998). Unloading-induced bone loss can also be observed in animal models. Tail suspension of rodents is an established model of unloading condition for the hind limbs, and bone loss is observed within several weeks. Unloading-induced bone loss is due to suppression of bone formation as well as

M. Noda (✉), T. Hayata, T. Nakamoto, T. Notomi, and Y. Ezura
Department of Molecular Pharmacology, Division of Advanced Molecular Medicine,
Medical Research Institute, Tokyo Medical and Dental University, 1-5-45 Yushima,
Bunkyo-ku, Tokyo 113-8510, Japan

M. Noda, T. Nakamoto, and T. Notomi
Global Center of Excellence Program "International Research Center for Tooth
and Bone Diseases", Tokyo Medical and Dental University, 1-5-45 Yushima,
Bunkyo-ku, Tokyo 113-8510 Japan

M. Noda
Hard Tissue Genome Research Center, Tokyo Medical and Dental University;
Department of Orthopedic Surgery, Tokyo Medical and Dental University,
1-5-45 Yushima, Bunkyo-ku, Tokyo 113-8510, Japan
e-mail: noda.mph@mri.tmd.ac.jp

M. Noda (ed.), *Mechanosensing Biology*,
DOI 10.1007/978-4-431-89757-6_6, © Springer 2011

increase in bone resorption (Sakai et al. 2002). These features are considered to be reminiscent to the human situations in the cases of unloading or disuse-induced osteopenia. These observations indicate that bone mass is controlled by loading or unloading and/or mechanical stress. However, mechanisms underlying such phenomena have not been well understood.

6.2 Nervous System Is Involved in Unloading-Induced Bone Loss

Both local and systemic controls have been suggested to be involved in determination of bone mass. Systemic bone mass is regulated not only by the endocrine system but also the nervous system (Elefteriou et al. 2005). Therefore, the involvement of sympathetic nervous tone in bone loss in the case of unloading due to tail suspension was examined. When mice were subjected to tail suspension, the mass of the hind limb bones was reduced in 2–4 weeks. In bone, osteoblasts express beta 2 adrenergic receptors (β2AR). Injection of blockers for β2ARs was conducted to examine their effects on bone loss due to tail suspension. Propranolol is an inhibitor of β2AR, and treatment with propranolol reduced the loss of bones induced by tail suspension (Kondo et al. 2005).

In order to see whether this reduction of unloading-induced bone loss is due to the effects on the bone formation side, the bone formation rate (BFR), the mineral apposition rate (MAR) (which indicates the activity of individual osteoblasts) and the mineralized surface (roughly corresponding to the number of osteoblasts on the surface of the bone) were examined. These three parameters of bone formation were all suppressed by hind limb loading. However, propranolol treatment reduced such suppression in the bone formation activities of all three parameters. The bone resorption side in histomorphometry was also examined. The levels of osteoclast number as well as osteoclast surface in bone were increased by unloading due to tail suspension. However, propranolol treatment reduced such an unloading-induced increase in the osteoclastic parameters. These observations indicated that elevation of bone resorption induced by unloading is suppressed by blocking the beta 2 adrenergical receptor in vivo.

To test whether such beta blocker effects on bone formation and bone resorption could be based on a cell-autonomous regulation, bone marrow cells were obtained after 2 weeks of hind limb unloading and were cultured under osteogenic or osteoclastogenic conditions. In the presence of beta-glycerophosphate and ascorbic acid (osteogenic condition), bone marrow cells form calcified nodules after 3 weeks in culture. The levels of nodule formation were reduced in the cells obtained from mice subjected to unloading. However, when these mice were treated with propranolol during the period of unloading in vivo, a reduction in nodule formation was no longer observed, indicating that unloading effects on bone formation were at least in part cell-autonomous. When the bone marrow cells were cultured in the presence of RANKL and M-CSF, that stimulate cells to form osteoclasts in culture, unloading in vivo enhanced the number of osteoclasts developed in culture within one week. Treatment with propranolol during unloading in vivo, however, suppressed such increase in osteoclast number in the culture of bone marrow cells, indicating that the propranolol affected cells that would give rise to osteoclasts. Therefore, the beta blocker propanolol reduced unloading-induced enhancement of osteoclastogenesis in a cell-autonomous manner.

To find out whether such effects of propranolol on bone formation and bone resorption in vivo as well as osteoblasts and osteoclasts in vitro are due to its action on beta adrenergic signaling, another beta blocker, guanethidine, that depletes adrenaline at the presynaptic ending was tested. Guanethidine acts on presynaptic nerve endings to block β adrenergic signaling. Guanethidine treatment reduced unloading-induced bone loss (BV/TV). Analysis of bone formation parameters revealed that guanethidine treatment reduced the levels of unloading-induced suppression of bone formation parameters, including BFR, MAR and the mineralizing surface. The bone resorption side was also examined, and guanethidine treatment reduced the levels of unloading-induced increase in the osteoclast number and osteoclast surface. In vitro experiments indicated that guanethidine treatment in vivo reduced unloading-induced decrease in mineralized nodule formation in vitro as well as the increase in osteoclast development in the culture of bone marrow cells taken from the mice (Kondo et al. 2005).

With respect to the effects of sympathetic tone activation, isoproterenol treatment was conducted. Isoproterenol treatment reduced bone mass in the mice (control, loaded) to the levels similar to the low levels of the bone mass observed after unloading. If these two signaling pathways share a certain extent of the bases, the simultaneous presence of these should not further suppress the levels of the bone mass. These isoproterenol effects on bone mass were due to the suppression of bone formation as well as enhancement of bone resorption. These isoproterenol effects on bone formation and bone resorption were similar to unloading-induced effects on bone formation and enhancement of bone resorption. The simultaneous presence of both hind limb unloading and isoproterenol treatment did not further suppress bone formation nor did it enhance bone resorption parameters compared to the situation of either one of the two alone. Mineralized nodule formation assay as well as osteoclast development in the cultures were conducted using bone marrow cells taken from mice either subjected to unloading alone or isoproterenol treatment alone or subjected to both unloading and isoproterenol treatment in vivo. Combination resulted in suppression of mineralized nodule formation and enhancement of osteoclast development in culture and the levels were comparable to those observed under the condition where unloading or isoproterenol treatment was conducted alone. Therefore, these observations indicated that increase in the sympathetic tone mimics the effects of unloading on bone mass as well as bone formation and bone resorption in vivo. Sympathetic tone also mimics unloading effects on the cultures of bone marrow cells taken from the animals. Moreover, unloading and isoproterenol treatment are suggested to have common pathway to regulate bone mass.

Since we have examined the effects of drugs (agonists and blockers) for the beta adrenergic receptor, one cannot exclude the possibility that our observations could be influenced by the systemic effects of the drugs on organs other than bone. Therefore, we further tested whether genetic manipulation of the sympathetic nervous tone could influence unloading-induced bone loss. For this purpose, mice deficient in dopamine beta hydroxylase (DBH) gene were used. These mice cannot produce an enzyme which is important to produce transmitters used in the synapses of adrenergic system. In the case of DBH-deficient mice, unloading-induced reduction in nodule formation was suppressed compared to the suppression seen in the cells of wild-type mice. Total bone mass in control wild-type was reduced by unloading. However, unloading did not reduce bone mass in DBH(+/−) mice as much as that in wild-type. These genetic data in combination with pharmacological data further supported the notion that sympathetic nervous tone plays an important role in the unloading-induced bone mass reduction (Kondo et al. 2005).

6.3 Central Control of Bone Mass Under Unloading Condition

Sympathetic nervous tone is under the control of central system. Within the hypothalamic system, ventromedial hypothalamic (VMH) neurons are responsible for sympathetic nervous tone. Therefore, involvement of VMH neurons in the regulation of bone mass due to unloading was tested. To do this, mice were treated with gold-thioglucose (GTG) to destroy VMH neurons. Then, these mice were subjected to unloading to induce bone loss. Based on micro-CT analysis, VMH neuron deficiency blocked unloading-induced bone loss (Lorenzo et al. 2008). BFR, MAR and the mineralizing surface were all reduced by unloading in control mice while these three parameters of bone formation were unchanged in the case of GTG-treated mice even after unloading. Cell-autonomous effects were tested using bone marrow cell cultures. With respect to gene expression, unloading reduced expression of osteocalcin, alkaline phosphatase and Runx2 (Cbfa1) gene in the control mice. GTG treatment per se did not affect basal expression levels of these genes. However, unloading-induced reduction was totally blocked by the treatment with GTG. Interestingly, GTG treatment enhanced the expression of the genes related to fat cell differentiation, such as PPARγ2, CEBP α and CEBP β. The expression levels of these adipogenic genes were enhanced by unloading in wild-type but not further enhanced by unloading in GTG-treated mice. In fact, fat levels of bone marrow were increased by unloading in control mice. In contrast, GTG-treated mice exhibited high levels of adipose tissue mass under these conditions. Unloading did not enhance the mass of adipose tissue in the bone marrow of GTG-treated mice. These data further indicated that unloading induces bone loss through the sympathetic nervous tones involving the central nervous system, and that this would at least be through the signals using both in peripheral sympathetic neuronal transmitters and brain nuclei.

6.4 Calcium Channel Involvement in Unloading-Induced Bone Loss

As mentioned above, the central and peripheral nervous systems are required for the unloading-induced bone loss in mouse models. The question would be how these signaling events occur in both the nervous system and the bone system. One of the possible mechanisms involved in the signaling of these events which are occurring during unloading would be via cation channels (Lorenzo et al. 2008; Thodeti et al. 2009; Son et al. 2009; Mendoza et al. 2010). Therefore, the effects of unloading on bones of the mice that were deficient in candidate cation channels were examined. One of these candidate cation channels is TRPV4. This is an ion channel identified as being responsive to mechanical and physical stimuli (Mizuno et al. 2003; Suzuki et al. 2003a–c; Sokabe et al. 2010; Guilak et al. 2010; Masuyama et al. 2008). For instance, TRPV4 knockout mice are unable to respond to stimuli such as physical pressure, temperature or osmotic pressure. TRPV4 is a member of transient receptor potential vanylloid, and is one of the subgroups where several types of TRP channels belong. TRPV4 is a six times membrane-spanning receptor, with both N and C terminals within the cytoprasmic region. TRPV4 forms multimers (tetramer) and lets the calcium and other cationic ions go through the membrane from the outside to the inside. In addition to the physical stimuli, phorbol ester, anadamide, or arachidonic acid, and epoxyeicosatrienoic acid also stimulate the calcium in flux. The temperature also affects the calcium entry through TRPV4. Furthermore, a hearing problem was observed in the knockout mice. Such hearing is a sensation-created physical stimulus that is related to this cation channel. Upon hind limb unloading, wild-type bone mass levels were suppressed. In contrast, resistance to bone loss induced by unloading was observed in TRPV4-deficient mice (Mizoguchi et al. 2008). This phenotype was observed in both two-dimensional as well as three-dimensional bone quantification. Micro-CT analysis indicated that the number of trabecular bone was reduced by unloading in the wild-type but this reduction was not observed in TRPV4 knockout mice.

The unloading-induced bone loss was particularly observed in the primary trabecular bone. The primary trabecular bone loss due to unloading was not observed in the TRPV4 mice. Unloading reduced BFR and MAR in wild-type but not in TRPV4 knockout mice. Unloading enhanced osteoclast number as well as anadamide, arachidonic acid, osteoclast surface within the secondary trabecular region in wild-types. However, such an increase in osteoclasts due to unloading was not observed in the absence of TRPV4 channel. TRPV4 mRNA expression was observed in nervous tissues as well as in the RAW246.7 macrophage cell line in addition to the expression in vivo in cortical bone and cancellous bone. These data indicated that TRPV4 is present in both bone and nervous systems and is required for unloading-induced bone loss. Therefore, this molecule could be one of the possible signaling entry points for unloading-induced events that lead to bone loss which are controlled by the central nervous system.

6.5   Tooth Movement Model and Extracellular Matrix Protein Action in Mechanical Stimulation

The involvement of bone in mediating signal of mechanical stress suggest that bone matrix proteins could work to transmit signal from bone to cells. As such, the bone matrix molecule, osteopontin (OPN), was examined in a tooth movement model. Mechanical stress-induced events during tooth movement includes interactions between the tooth and surrounding bone in the jaw, and this is regulated at the interface between the cells and matrix (Rios et al. 2008; Yu et al. 2009; Ziegler et al. 2010). A traction system was set up on the first molar tooth of mice to pull it in a direction towards the front side (incisor). This system was using a shape-memory metal spring coil to constantly pull the tooth by applying 10 g in one direction (Chung et al. 2008). Micro-CT examination of the first molar with respect to the morphology of the tooth root indicated the presence of root resorption activity on the pressure side in wild-type. When the force was applied in the case of OPN knockout mice, odontoclasts on the surface of the tooth that are indicative of root resorbing activity were not observed (Chung et al. 2008). Accordingly, the surface of the root was relatively smooth. Quantification of this roughness of the surface indicated that, when the force was not applied, the roughness values of tooth and bone were similar between wild-type and OPN knockout mice. In contrast, the roughness was increased after force application in wild-type, and this was significantly lower in the case of OPN knockout mice.

To examine the cellular events which give rise to the roughness on the surface of the tooth, histological sections were prepared and then stained for tartrate-resistant acid phosphatase (TRAP). In wild-type mice, there were a few osteoclasts present on the bone surface of the alveolar sockets. When the force was applied towards the front end (mesial direction), TRAP positive osteoclasts appeared on the surface of the trabecular bone of the pressure side in the sockets as well as on the surface of the root of the tooth in the wild-type mice. In OPN knockout mice, before the force was applied, there were a few osteoclasts on the distal wall of the alveolar bone within the sockets due to physiological posterior drift. However, the number of the osteoclasts was similar to the bone observed in the wild-type, which were not subjected to the force application. When directional force was applied to the tooth in OPN knockout mice, the osteoclast number was increased on the surface of the pressure side alveolar bone within the socket but not on the tooth root surface. The mineral content was also examined; however, the levels of the calcium and phosphate were similar in both alveolar bone and tooth regardless of the genotype. Therefore, knockout phenotypes were not explained by the mineral content.

As tooth movement has been known to be controlled by the local cytokine release, the levels of cytokines were examined. TNFα has been considered to be present in the microenvironment of the tooth during movement. OPN-deficiency effects on TNFα-induced release of the calcium 45, which was used to label bone, was examined in vitro. In wild-type bone, TNFα treatment increased the release of the calcium 45 into the medium indicating the enhancement of the bone resorption by osteoclasts present in the cultured long bone. In contrast, OPN-deficient long bone did not respond to TNFα with respect to the calcium 45 release into the medium, indicating the reduction in the bone resorption activities. These observations revealed that OPN deficiency suppressed the response of the alveolar bone upon the force application to the tooth, and this may in part be due to the resistance of the bone to cytokines present in the microenvironment of the tooth root.

6.6 Transcription Factor Modulates Unloading-Induced Bone Loss

Since bone matrix accumulation is under the control of the osteoblastic activities, determinants of bone cell activity may also be involved in the response of the bone to the mechanical stress. Therefore, the effects of the haploinsufficiency of Runx2 were examined. Runx2 is a transcription factor which is necessary for the differentiation of the osteoblasts. In terms of Runx2 knockout, animals can form a cartilaginous skeletal pattern but no bone is formed (Komori 2010). Therefore, Runx2 knockout mice die just after birth because of the lack of bone in their ribs to support their respiration. Although the Runx2 homozygous knockout is lethal, Runx2 heterozygous mutants can survive to become adult. This is similar to the human mutation in the Runx2 gene which is known as cleidocranial dysplasia (Pineda et al. 2010; Lou et al. 2009). In the case of the heterozygous knockout mice, the skeletal patterning and also the bone shape were similar to those in wild-type mice. When the wild-type mice were subjected to tail suspension, they lost bone with respect to cancellous bone volume as well as BMD. Heterozygous Runx2 knockout mice responded to unloading by exhibiting more reduction in bone volume and bone mineral density than those in wild-type (Salingcarnboriboon et al. 2006). Such Runx2-deficiency effects on the enhancement of bone loss was not only observed in cancellous bones but also in the cortical bone. In fact, the cortical bone parameters such as cortical thickness, cortical area, indicated that unloading induced the reduction in wild-type while more reduction was observed in the case of Runx2 hetero-knockout mice. To see how such enhancement in bone loss occurs in these Runx2 hetero-knockout mice, dynamic histomorphometric parameters were examined. Calcein-labeling-based measurement of the BFR as well as MAR was carried out. Unloading reduced endosteal MAR in wild-type and such reduction was further enhanced in of Runx2 hetero-knockout mice. The base line levels of MAR in control (loaded) wild-type and the control (loaded) Runx2 knockout mice were similar. Also, endosteal BFR indicated reduction by unloading in wild-type, and unloading of knockout mice further reduced endosteal BFR. Endosteal MAR also exhibited a similar trend.

With respect to a trabecular MAR, Runx2 heterozygous-deficient mice in the control (loaded) group indicated reduced levels compared to the control (loaded) wild type. Unloading reduced MAR in wild-type. Unloading in Runx2 hetero-knockout mice further reduced the levels of MAR in cancellous bone. A similar trend was observed with respect to BFR in trabecular bone. Regarding osteoclast surface and osteoclast number as well as urinary excretion of deoxypyridinoline, the control (loaded) wild-type group and the control (loaded) Runx2 heterozygous knockout mice group were similar and unloading did not significantly increase the levels of bone resorption parameters in wild-type or in Runx2 knockout mice. These data indicated that the effects of lack of Runx2 were significant in terms of the levels of the decrease due to tail suspension.

Molecular analysis also indicated that the levels of osteocalcin and osterix gene expression in bone were suppressed by unloading in wild-type bone. With respect to osterix, absence of a half gene dosage of Runx2 completely suppressed its expression upon unloading. Interestingly, OPN in heterozygous Runx2 mice was increased in the unloaded group compared to the wild-type control (loaded) group. In wild-type, OPN was not reduced by unloading while in Runx2 hetero-knockout mice, OPN expression was reduced by unloading. Unloading increased RANK gene expression in wild-type while no such increase was observed in Runx2 heterozygous knockout mice. These data indicated that Runx2 is a target of unloading and the prerequisite molecule for the recruitment of osteoblasts, at least in part. Although most skeletal patterning can be preserved even in the absence of a half gene dosage of Rux2, unloading severely reduced bone formation in the Runx2 hetero-knockout mice. This is indicative of the fact that, upon the suppressive condition against bone formation, both alleles of the Runx2 gene are fully driven to overcome the unloading-induced suppression of the bone formation. In contrast, during the normal levels of bone maintenance, a half gene dosage of Runx 2 did not result in the reduction of bone mass suggesting that the steady state levels of recruitment of new osteoblasts are relatively large and may not require activation of the double Runx2 gene dosage in their adulthood. Upon the suppressive condition for osteoblastic recruitment, Runx2 may be under the control of the feedback mechanism to compensate bone formation. How these events are controlled for the Runx2 gene expression is still to be elucidated. However, if such signaling events are potentiated in the case of bedridden patients, it may be beneficial for prevention of disuse osteoporosis in the future.

6.7 Unloading-Induced Bone Loss Requires Nucleocytoplasmic Shuttling Protein

Mechanical stress affects bone cells possibly through the geometric changes within the microenvironment of bone to let the bone cells to perceive minor bending of substrate. This is induced by the signal through the attachment proteins on the matrix surface (Paszek et al. 2009). The cell body attaches to extracellular matrix via adhesion plaques which consist of a number of molecules including p130Cas, FAK and other signaling molecules (Kirsch et al. 2002). Cas-interacting zinc finger protein (CIZ) is one of the components of such adhesion plaque. This molecule was first cloned by Nakamoto and Hirai as a protein that binds to the SH2 domain of p130Cas (Nakamoto et al. 2000). CIZ is shuttling between the cell adhesion plaque and nuclei of the cells. Within the nuclei, CIZ binds to the DNA consensus sequence, C/GAAAAA. This particular consensus sequence has been identified in the promoter region of genes encoding MMP7 and other matrix metalloproteinase. CIZ has been also identified as a protein that binds to the regulatory region of collagen gene. As CIZ has been named for its binding to Cas, this protein has been identified to be a component of the adhesion plaque. However, it is also a nuclear matrix protein that binds to the part of the genome and/or regulates its expression. In addition, CIZ has a transcription factor function (Nakamoto et al. 2000). A unique property of this molecule is a shuttling protein linking the cell adhesion plaque, nuclei and regulation of gene expression. CIZ may be involved in the regulation of cellular events regulated by mechanical stress.

Therefore, this hypothesis was tested by using an unloading model. As previously, known, wild-type mice show reduction is the bone volume (BV/TV) after unloading. CIZ-deficient mice exhibited resistance to reduction of bone mass due to unloading (Hino et al. 2006). Base line BV/TV values in CIZ knockout mice were higher than those of wild-type mice. After the mice were subjected to unloading, BV/TV levels in the knockout CIZ mice were still higher than those for wild-type mice subjected to unloading. Such resistance to bone loss was not only in trabecular bone but also in cortical bone that was less sensitive than trabecular bone regarding the reduction upon unloading in wild-type mice. Resistance against unloading-induced bone loss in CIZ knockout mice was seen with respect to both cortical bone area and cortical thickness.

As for dynamic bone formation parameters, MAR was reduced by unloading in wild-type mice and base line MAR in CIZ knockoutmice was higher than that in wild-type. Under this condition, unloading did not reduce MAR in CIZ knockout mice. The levels of osteoblast number on the surface of the trabecular bone can be accessed by the mineralizing surface per bone surface (MS/BS). This MS/BS levels were suppressed in wild-type upon unloading. Knockout mice show higher MS/BS values under control (loaded) condition. Upon unloading, there was no suppression with respect to MS/BS levels in CIZ knockout mice. BFR was reduced in wild-type upon unloading. BFR was higher under in CIZ knockout mice under loaded condition compared to wild-type. Upon unloading, BFR was not in suppressed in CIZ knockoutmice. Osteoclast surface (Oc.S/BS) was unchanged at the end of 2 weeks time period of unloading in wild-type and the levels were similar in both wild-type and CIZ knockoutmice. There was no change in CIZ knockoutmice after unloading with respect to osteoclast surface. Similarly, the number of osteoclast (Oc.N/BS, N/mm) was not changed after unloading in wild-type and CIZ knockoutmice. These data indicated that CIZ is involved in unloading-induced events at least in part via osteoblasts.

Cell autonomous changes were examined. Bone marrow cells were obtained from the mice after they were subjected to unloading or loaded conditions and cultured. In wild-type mice, bone marrow cells of mice after unloading revealed a reduction in the mineralized nodule formation. Nodule formation in CIZ knockoutmice from the loaded group showed higher levels compared to loaded wild-type mice. In contrast to wild-type, unloading did not change the levels of mineralized nodule formation in CIZ knockoutmice. Therefore, these data indicated that cell-autonomous changes in bone marrow environment for the cells in osteoblast lineages were the targets of the unloading condition and that CIZ is required for such reduction in the cells of osteoblastic activity present in bone marrow. With respect to osteoclastic development in the cultures of bone marrow cells subjected to the culture in the presence of M-CSF and RANKL, 2 weeks unloading protocol in wild-type bone did not show any difference after unloading in terms of the development of osteoclasts in bone marrow cells. Unloading did not affect the osteoclast development in CIZ knockout. These data further supported the notion that unloading conditions affect osteoblastic activities and that CIZ is important for such reduction of osteoblastic activities due to unloading.

6.8 Interaction of PTH Signaling and Unloading

Unloading condition affects bone through local and systemic control systems. Some of the systemic controls which are known to consist of multiple layers are sympathetic tone, serotonin, oxytocin and calciotropic endocrine systems. Among these, a unique endocrine molecule anabolic for osteoblastic activities is parathyroid hormone. Especially, the signaling of parathyroid hormone in the osteoblastic cell lineages is considered to be a major route to promote anabolic action of this hormone. Osteoblast-specific expression of a constitutively active mutant form of parathyroid/parathyroid hormone-related peptide receptor (caPPR) shows in high bone mass phenotype (Calvi et al. 2001). To test whether the unloading condition would affect the signaling system within osteoblastic cells, the transgenic mice harboring Jansen-type mutation (H223R) of PTH/PTHrP receptor (caPPR) were subjected to unloading condition. Bone mass (BV/TV) in wild-type of the littermate mice was reduced upon unloading. Littermate PPR-Tg (transgenic) mice indicated higher bone mass in the loaded group compared to the loaded wild-type. Upon unloading, however, PPR-Tg mice did not show reduction of bone mass (Ono et al. 2007). This was observed in cancellous bone in both femur and spine. CaPPR is expressed in osteoblastic cells and signaling alteration would be within osteoblasts. Therefore, it was surprising to see that, regardless of the resistance to the loss of bone mass levels upon unloading, all bone formation parameter levels, including those of BFR, MAR and the mineralizing surface, were suppressed upon unloading in PPR-Tg mice. Moreover, the level of this suppression was comparable to that of the suppression by unloading in wild-type mice. RT-PCR analyses to look into the expression levels of genes such as Runx2 and ostelix revealed that unloading suppressed expression of these genes in wild-type, and similarly suppressed the expression of osteoblastic genes in caPPR-Tg mice with respect to the messenger of RNAs within hind limb bones. To address systemic changes, bone formation parameters such as osteocalcin levels in the blood were examined. Osteocalcin levels in systemic circulation were reduced in wild-type upon unloading. CaPPR-Tg mice showed suppression of osteocalcin levels upon unloading. The cell autonomous activity of osteoblasts was examined by using mineralized nodule formation assay in culture. Unloading suppressed in vitro mineralization in the cells obtained from wild-type. The base line levels in PPR-Tg mice were higher than those of wild-type but suppression was also observed. Therefore, unloading suppresses bone formation in both PPR-Tg mice and wild-type.

The bone resorption side was then analyzed with respect to the resistance of caPPR-Tg mice against unloading-induced bone loss. The number of osteoclasts and the osteoclast surface of bone were increased upon unloading in wild-type mice as known before. Surprisingly, unloading suppressed the number of osteoclasts (MS.Oc/BS) and the osteoclast surface of bone (OcS/BS). These observations may be limited to particular bones subjected unloading and, therefore, systemic parameters were examined. Urinary excretion of deoxypyridinoline (Dpyd) (nM/mM.Cr) level is a biochemical parameter to evaluate systemic bone resorption. This was increased by unloading in wild-type. Control (loaded) PPR-Tg mice showed higher base line levels than those of wild-type under loaded condition. Unloading reduced urinary deoxypyridinoline levels in PPR-Tg mice confirming histomorphometrical data.

To examine cellular events, we obtained the bone marrow cells of PPR-Tg mice or wild-type mice either subjected to loading or unloading and cultured under osteoclastogenic conditions. Osteoclastogenesis in the cultures of bone marrow cells from the wild-type mice subjected to either loading or unloading conditions was similar. Bone marrow cells taken from PPR-Tg mice showed similar levels of osteoclastogenesis compared to the wild-type mice when the cells were taken from the group of the mice subjected to the loading condition. However, bone marrow cells taken from PPR-Tg mice subjected to the unloading condition revealed significantly lower levels of osteoclast development compared to the loading group of PPR-Tg mice or compared to the wild-type mice.

PPR-Tg signaling should be operating specifically in osteoblasts as the mutant PPR was expressed under the control of 2.3 kb of type I collagen promoter. The caPPR signaling should be involved in regulation in osteoblasts first and then may indirectly affect osteoclastogenesis and bone resorption. Therefore, RANKL expression in bone was examined. RANKL expression was enhanced by unloading in wild-type. The base line levels of the RANKL mRNA expression in transgenic mice were higher than those of the wild-type mice subjected to loading condition. However, upon unloading, RANK mRNA expression was not enhanced by unloading in PPR-Tg mice. Furthermore, M-CSF expression was higher in loaded PPR-Tg mice than wild-type and this level was suppressed by unloading in the PPR-Tg mice. These data suggest that although PPR signaling would be operating specifically in osteoblasts, it would indirectly activate cells in osteoclast lineage.

How these would be occurring was addressed by examining the gene expression in vivo. RANKL expression or OPG expression were not changed in PPR-Tg mice and, therefore, does not explain the reduction of bone resorption and osteoclastic activities upon unloading in PPR-Tg mice. Our examination revealed two candidate genes, M-CSF and MCP-1, which are related to osteoclastogenesis by unloading in PPR-Tg mice. Unloading did not suppress expression of M-CSF and MCP-1 gene in wild-type. However, unloading suppressed M-CSF and MCP-1 in association with the suppression of bone resorption in PPR-Tg mice upon unloading. How this reversal of the direction of bone resorption would occur in osteoclastic cytokine expression is still to be determined. Our data indicated that PTH-dependent modulation in the osteoblastic activity would change the response of bone resorption by the unloading condition. PTH treatment has been clinically proven to be an effective treatment for severe osteoporosis. If our observations in animal experiments could be extrapolated to humans, PTH signaling may also be beneficial to reverse the direction of the bone resorption upon disuse osteoporosis. These analyses on PTH PTHrP receptor transgenic mice have identified a new type of signaling which may exist in unloading conditions.

6.9 Role of Noncollagenous Matrix Protein, OPN, in Bone in Unloading-Induced Bone Loss

Mechanical stress affects bone tissue by slightly vending the bone per se. This will affect the interface between bone matrix and osteoblasts or osteoclasts. The proteins present in the interface are matrix proteins, such as collagen and noncollagenous proteins in bone and integrins expressed on the surface of the cells. Noncollagenous proteins have been studied with respect to the response of the cells to unloading. OPN is of interest because mechanical stress enhances expression of this protein in bone cells. OPN knockout mice were examined with respect to their response to unloading. Unloading reduces hind limb bone mass of mice 2–4 weeks after the experimental unloading in wild-type animals. In contrast, OPN deficiency suppressed such reduction in bone mass even after several weeks of unloading (Ishijima et al. 2001, 2002, 2006, 2007). Dynamic bone histomorphometry indicated that unloading suppressed the levels of a bone formation parameters including MAR that represents the activity of a individual osteoblasts, and mineralizing surface (MS/BS) and eventually BFR in wild-type mice. In contrast, these three parameters of bone formation were not suppressed in OPN-deficient mice even after tail suspension.

Bone resorption parameters including the osteoclast surface as well as osteoclast number were increased by unloading in wild-type but not in OPN-deficient mice. With these changes in osteoclast parameters of bone resorption in long bone, deoxypyridinoline levels in urine were increased by unloading in wild-type mice. However, no such increase was observed after hind limb unloading in OPN-deficient mice. These observations indicated that OPN plays a key role in bone loss induced by unloading.

6.10 Role of Noncollagenous Matrix Protein, OPN in Mechanical Force Dependent Bone Formation

As mentioned previously, OPN is known to be expressed in response to mechanical stress in bone. To further test whether positive stimuli-induced bone formation is also under the influence of OPN, a model of a traction-induced bone formation in the calvarial suture was designed (Morinobu et al. 2003). Sagittal suture of adult mice was expanded by a small orthodontic device used for the application of forces to the tooth in human. This small dental coil spring could apply approximately 10 g of force to the right and left parietal bone so that forces could gradually open the sagittal suture. The cells in the sutures and on the contacting surfaces for each of the parietal bones facing sagittal suture were stimulated to form bone that eventually reduced the gap of the expanded sagittal suture. During this bone formation period, cells within the suture and those on the inner edges of the parietal bones were continuously exposed to the traction forces. After several weeks of the application of traction force, the level of bone formation was measured in wild-type and OPN-deficient mice. In wild-type mice, OPN expression in the cells around the suture and within the inner edges of the bone was increased during the initial period of the force application in osteocytes and osteoblasts. At the same time, cells covering the inner edge of the bone facing the sagittal suture were also showing an increase in OPN expression. Such increase in the expression of OPN was not observed in both osteocytes within the inner edge of the bone or in the cells covering the inner surface of the bone facing the sagittal bone in the mice where the dead spring (no force) was applied. Therefore, this system provided an opportunity to observe the role of ostepontin in response to the traction force to form bone in vivo. In this model, OPN deficiency suppressed bone formation by traction force applied to the bone to expand the sagittal suture in adult mice. These experiments indicated that the presence of OPN is required for the response not only to the unloading but also to externally applied force to form bone.

6.11 Intracellular Mechanism of Sensing Mechanical Stress

Although cells are basically the first step responder to the mechanical response of the tissue in vivo, how cells are sensing and responding to the mechanical forces is largely unknown. The molecular phenomena have been analyzed by using partial fragments of proteins present in the interface between substrate and cells. Mechanical force would be sensed at the interface between the substrate and the cells. The cells use their attachment machinery facing the scaffold to sense the stress. When an externally applied force is present, the interface at the substrate and the cells would give signals to the attachment complex in the cells facing the extra cellular matrix proteins. The intracellular machinery such as adhesion plaques would be a complex of proteins that are linked to cytoskeletons. Within such cell adhesion plaques, integrins are mediating the extra cellular signals into the cells via extra and intracellular domains. One of the intracellular signaling proteins within the adhesion plaques is p130Cas. The knockout of p130Cas is lethal. However, the knockout embryos (p130Cas $^{-/-}$) have been used to examine the behavior of the cells, and the cells are impaired in terms of their migration capability. p130Cas is with the protein complex and plays roles in mediation of the mechanical stress inside cells (Sawada et al. 2006; Tamada et al. 2004). Unfolding of a fragment of p130Cas (substrate domain, SB) resulted in exposure of the phosphorylation site in p130Cas to modulate the activity of Src. p130Cas activation subsequently activates Rap1, a small GTPase. Interestingly, no changes are observed in kinase activity. Antibodies were raised to recognize extended p130Cas (SB) in vitro at the different locations in the different domains in the cells. In a particular location of the spread portion of the cells, where their traction forces are expected to be high, extension associated to p130Cas SB could be seen. Physiological force transduction results in expansion of p130Cas SB. The intact p130Cas phosphorylation has not been observed in the cells. Such structural alteration of molecules may explain the response of cells to external physical forces. It is still to be elucidated how p130Cas could lead to the activation of a cellular event and response to mechanical stress.

6.12 Conclusion

Mechanical stress controls bone formation and bone resorption via multiple signaling events. Comprehensive understanding of the mechanism would benefit contemplating new strategy for treatment.

References

Amin S (2010) Mechanical factors and bone health: effects of weightlessness and neurologic injury. Curr Rheumatol Rep 12:170–176

Beier F, Loeser RF (2010) Biology and pathology of Rho GTPase, PI-3 kinase-Akt, and MAP kinase signaling pathways in chondrocytes. J Cell Biochem 110:573–580

Calvi LM, Sims NA, Hunzelman JL, Knight MC, Giovannetti A, Saxton JM, Kronenberg HM, Baron R, Schipani E (2001) Activated parathyroid hormone/parathyroid hormone-related protein receptor in osteoblastic cells differentially affects cortical and trabecular bone. J Clin Invest 107:277–286

Chung CJ, Soma K, Rittling SR, Denhardt DT, Hayata T, Nakashima K, Ezura Y, Noda M (2008) OPN deficiency suppresses appearance of odontoclastic cells and resorption of the tooth root induced by experimental force application. J Cell Physiol 214:614–620

Currey JD (2010) Mechanical properties and adaptations of some less familiar bony tissues. J Mech Behav Biomed Mater 3:357–372

Elefteriou F, Ahn JD, Takeda S, Starbuck M, Yang X, Liu X, Kondo H, Richards WG, Bannon TW, Noda M, Clement K, Vaisse C, Karsenty G (2005) Sympathetic regulation of osteoclastogenesis is required for gonadectomy-induced bone loss and antagonized by CART. Nature 433:7028

Goodship AE, Cunningham JL, Oganov V, Darling J, Miles AW, Owen GW (1998) Bone loss during long term space flight is prevented by the application of a short term impulsive mechanical stimulus. Acta Astronaut 43:65–75

Guilak F, Leddy HA, Liedtke W (2010) Transient receptor potential vanilloid 4: the sixth sense of the musculoskeletal system. Ann NY Acad Sci 1192:404–409

Hino K, Nifuji A, Morinobu M, Tsuji K, Ezura Y, Nakashima K, Yamamoto H, Noda M (2006) Unloading-induced bone loss was suppressed in gold-thioglucose treated mice. J Cell Biochem 99:845–852

Ishijima M, Rittling SR, Yamashita T, Tsuji K, Kurosawa H, Nifuji A, Denhardt DT, Noda M (2001) Enhancement of osteoclastic bone resorption and suppression of osteoblastic bone formation in response to reduced mechanical stress do not occur in the absence of osteopontin. J Exp Med 193:399–404

Ishijima M, Tsuji K, Rittling SR, Yamashita T, Kurosawa H, David TD, Nifuji A, Noda M (2002) Resistance to unloading-induced three-dimensional bone loss in osteopontin-deficient mice. J Bone Miner Res 17:661–667

Ishijima M, Ezura Y, Tsuji K, Rittling SR, Kurosawa K, Denhardt DT, Emi M, Nifuji A, Noda M (2006) Osteopontin is associated with nuclear factor κB gene expression during tail-suspension-induced bone loss. Exp Cell Res 312:3075–3083

Ishijima M, Tsuji K, Rittling SR, Yamashita T, Kurosawa H, Denhardt DT, Nifuji A, Ezura Y, Noda M (2007) Osteopontin is required for mechanical stress-dependent signals to bone marrow cells. J Endocrinol 193:236–243

Kirsch K, Kensinger M, Hanafusa H, August A (2002) A p130Cas tyrosine phosphorylated substrate domain decoy disrupts v-crk signaling. BMC Cell Biol 3:18

Komori T (2010) Regulation of osteoblast differentiation by runx2. Adv Exp Med Biol 658:43–49

Kondo H, Nifuji A, Takeda S, Ezura Y, Rittling S, Denhardt DT, Nakashima K, Karsenty G, Noda M (2005) Unloading induces osteoblastic cell suppression and osteoclastic cell activation to lead to bone loss via sympathetic nervous system. J Biol Chem 280:30192–30200

Kontulainen S, Sievänen H, Kannus P, Pasanen M, Vuori I (2002) Effect of long-term impact-loading on mass, size, and estimated strength of humerus and radius of female racquet-sports players: a peripheral quantitative computed tomography study between young and old starters and controls. J Bone Miner Res 17:2281–2289

Lorenzo IM, Liedtke W, Sanderson MJ, Valverde MA (2008) TRPV4 channel participates in receptor-operated calcium entry and ciliary beat frequency regulation in mouse airway epithelial cells. Proc Natl Acad Sci USA 105:12611–12616

Lou Y, Javed A, Hussain S, Colby J, Frederick D, Pratap J, Xie R, Gaur T, van Wijnen AJ, Jones SN, Stein GS, Lian JB, Stein JL (2009) A Runx2 threshold for the cleidocranial dysplasia phenotype. Hum Mol Genet 18:556–568

Masuyama R, Vriens J, Voets T, Karashima Y, Owsianik G, Vennekens R, Lieben L, Torrekens S, Moermans K, Vanden Bosch A, Bouillon R, Nilius B, Carmeliet G (2008) TRPV4-mediated calcium influx regulates terminal differentiation of osteoclasts. Cell Metab 8:257–265

Mendoza SA, Fang J, Gutterman DD, Wilcox DA, Bubolz AH, Li R, Suzuki M, Zhang DX (2010) TRPV4-mediated endothelial Ca2+ influx and vasodilation in response to shear stress. Am J Physiol Heart Circ Physiol 298:466–476

Mizoguchi F, Mizuno A, Hayata T, Nakashima K, Heller S, Ushida T, Sokabe M, Miyasaka N, Suzuki M, Ezura Y, Noda M (2008) Transient receptor potential vanilloid 4 deficiency suppresses unloading-induced bone loss. J Cell Physiol 216:47–53

Mizuno A, Matsumoto N, Imai M, Suzuki M (2003) Impaired osmotic sensation in mice lacking TRPV4. Am J Physiol Cell Physiol 285(1):C96–C101

Morinobu M, Ishijima M, Rittling SR, Tsuji K, Yamamoto H, Nifuji A, Denhardt DT, Noda M (2003) Osteopontin expression in osteoblasts and osteocytes during bone formation under mechanical stress in the calvarial suture in vivo. J Bone Miner Res 18:1706–1715

Nakamoto T, Yamagata T, Sakai R, Ogawa S, Honda H, Ueno H, Hirano N, Yazaki Y, Hirai H (2000) CIZ, a zinc finger protein that interacts with p130(cas) and activates the expression of matrix metalloproteinases. Mol Cell Biol 20:1649–1658

Ono N, Nakashima K, Schipani E, Hayata T, Ezura Y, Soma K, Kronenberg HM, Noda M (2007) Constitutively active parathyroid hormone receptor signaling in cells in osteoblastic lineage suppresses mechanical unloading-induced bone resorption. J Biol Chem 282:25509–25516

Paszek MJ, Boettiger D, Weaver VM, Hammer DA (2009) Integrin clustering is driven by mechanical resistance from the glycocalyx and the substrate. PLoS Comput Biol 5:e1000604

Pineda B, Hermenegildo C, Laporta P, Tarín JJ, Cano A, García-Pérez MA (2010) Common polymorphisms rather than rare genetic variants of the Runx2 gene are associated with femoral neck BMD in Spanish women. J Bone Miner Metab. doi:10.1007/s00774-010-0183-2

Rios HF, Ma D, Xie Y, Giannobile WV, Bonewald LF, Conway SJ, Feng JQ (2008) Periostin is essential for the integrity and function of the periodontal ligament during occlusal loading in mice. J Periodontol 79:1480–1490

Sakai A, Sakata T, Tanaka S, Okazaki R, Kunugita N, Norimura T, Nakamura T (2002) Disruption of the p53 gene results in preserved trabecular bone mass and bone formation after mechanical unloading. J Bone Miner Res 17:119–127

Salingcarnboriboon R, Tsuji K, Komori T, Nakashima K, Ezura Y, Noda M (2006) Runx2 is a target of mechanical unloading to alter osteoblastic activity and bone formation in vivo. Endocrinology 147:2296–2305

Sawada Y, Tamada M, Dubin-Thaler BJ, Cherniavskaya O, Sakai R, Tanaka S, Sheetz MP (2006) Force sensing by mechanical extension of the Src family kinase substrate p130Cas. Cell 127:1015–1026

Sokabe T, Fukumi-Tominaga T, Yonemura S, Mizuno A, Tominaga M (2010) The TRPV4 channel contributes to intercellular junction formation in keratinocytes. J Biol Chem 285:18749–18758

Son AR, Yang YM, Hong JH, Lee SI, Shibukawa Y, Shin DM (2009) Odontoblast TRP channels and thermo/mechanical transmission. J Dent Res 88:1014–1019

Suzuki M, Hirao A, Mizuno A (2003a) Microtubule-associated [corrected] protein 7 increases the membrane expression of transient receptor potential vanilloid 4 (TRPV4). J Biol Chem 278(51):51448–51453. Epub 2003 Sep 28. Erratum in: J Biol Chem 2005: 280(27):25948

Suzuki M, Watanabe Y, Oyama Y, Mizuno A, Kusano E, Hirao A, Ookawara S (2003b) Localization of mechanosensitive channel TRPV4 in mouse skin. Neurosci Lett 353(3):189–192

Suzuki M, Mizuno A, Kodaira K, Imai M (2003c) Impaired pressure sensation in mine lacking TRPV4. J Biol Chem 278(25):22664–22668

Tamada M, Sheetz MP, Sawada Y (2004) Activation of a signaling cascade by cytoskeleton stretch. Dev Cell 7:709–718

Temiyasathit S, Jacobs CR (2010) Osteocyte primary cilium and its role in bone mechanotransduction. Ann NY Acad Sci 1192:422–428

Thodeti CK, Matthews B, Ravi A, Mammoto A, Ghosh K, Bracha AL, Ingber DE (2009) TRPV4 channels mediate cyclic strain-induced endothelial cell reorientation through integrin-to-integrin signaling. Circ Res 104:1123–1130

Yu V, Damek-Poprawa M, Nicoll SB, Akintoye SO (2009) Dynamic hydrostatic pressure promotes differentiation of human dental pulp stem cells. Biochem Biophys Res Commun 386:661–665

Ziegler N, Alonso A, Steinberg T, Woodnutt D, Kohl A, Müssig E, Schulz S, Tomakidi P (2010) Mechano-transduction in periodontal ligament cells identifies activated states of MAP-kinases p42/44 and p38-stress kinase as a mechanism for MMP-13 expression. BMC Cell Biol 11:10

Chapter 7
TRP Channels and Mechanical Signals

Makoto Suzuki and Atsuko Mizuno

7.1 Transient Receptor Potential Channels

Transient receptor potential (TRP) channels are a unique cellular mechanism converting a wide variety of signals, including mechanical stress, to cation flow. The protein was first discovered in studies that examined *Drosophila* phototransduction (Lo and Pak 1981). The photoreceptor cells of wild *Drosophila* exhibit sustained receptor potential, while a mutant showing a transient receptor potential (*trp*) in response to continuous light exposure was reported (Cosens and Manning 1969). The ionic basis for the sustained receptor potentials is an influx of Ca^{2+} from the extracellular space. The *trp* gene was cloned in 1989 (Montell and Rubin 1989) and was subsequently shown to encode a Ca^{2+}-permeable cation channel (Hardie and Minke 1992). Since then, many channels that bear sequence and structural similarities to the *Drosophila* TRP have been cloned from flies, worms, and mammals.

There are 28 TRP channels in mammals (28 mouse proteins, 27 human proteins), and these are divided into six subfamilies according to sequence homology. When the *trp* gene was discovered, seven other mammalian homologues were found in a computer-based sequence similarity search. This family of TRPC (canonical) includes a receptor-operated Ca^{2+} permeable cation channel followed by activation of the peptide hormone. TRP channels are classified into TRPC, TRPV (vanilloid), and TRPM (melastatin) subfamilies. More recent classification has expanded the TRP superfamily to include three additional but more distantly related subfamilies: TRPP (polycystin), TRPML (mucolipin), and TRPA (ankyrin) (Fig. 7.1).

Structurally, all these TRP channels have six putative transmembrane (TM) segments and N-terminal and C-terminal cytoplasmic tails that are similar to the

M. Suzuki
Edogawabashi-clinic, DSD building 4F, 348 Yamabuki-cho, Shinjuku-ku,
Tokyo 162-0801, Japan

A. Mizuno (✉)
Department of Pharmacology, Jichi Medical University, 3311-1 Yakushiji,
Shimotsuke, Tochigi 329-0498, Japan
e-mail: aaamiz@jichi.ac.jp

M. Noda (ed.), *Mechanosensing Biology*,
DOI 10.1007/978-4-431-89757-6_7, © Springer 2011

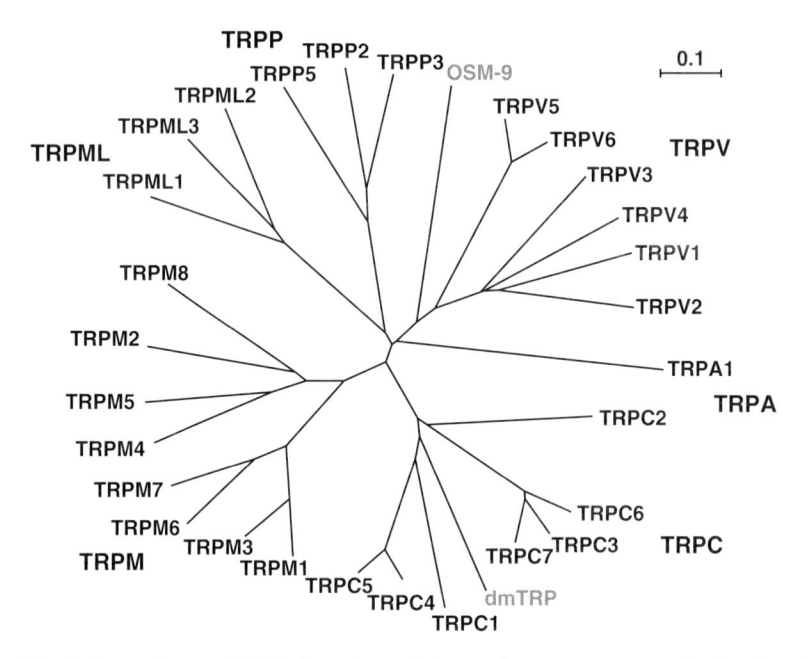

Fig. 7.1 Phylogenetic tree of TRP channels. A phylogenetic tree was generated using ClustalW by aligning the whole amino acid sequences of mice TRPs, dmTRP (*Drosophila melanogaster*), and OSM-9 (*Caenohabitis elegans*). TRPC2 in humans is a pseudogene. The *scale bar* indicates genetic distance

topologies of voltage-gated channels (Fig. 7.2). The fourth TM segment of the TRP channels lacks the complete set of positively charged residues necessary for voltage sensing in many voltage-gated channels. The six TM polypeptide subunits of TRP channels likely assemble as tetramers to form cation-permeable pores. Ankyrin binding repeats are 33-residue motifs that mediate cytoskeletal anchoring or protein–protein interaction. Some subfamilies contain a conserved stretch of 25 amino acids, known as the TRP domain.

In relation to mechanical gating, we propose a description of the following in each section below. TRPC may include a stretch-activated channel, although the results were still controversial. TRPM is a member of the melastatin family and is thought to be a stretch-activated channel. TRPA1 has long ankyrin repeats and is thought to be mechano-gated channel opened by link to cytoskeletal proteins through the domain. TRPV1 is a capsaicin receptor that was independently cloned by expression cloning, resulting in a similar structure to the TRP channel. The receptor (TRPV1) is essentially sensitive to hot temperatures suggesting that this family may be closely related to transmission of physical force to ionic flow. TRPV2 is reported to be a stretch-activated channel in the cells overexpressed the TRPV2 cDNA. TRPV4 is activated by a cell exposed to hypotonic solution. TRPV1 may include a hyperosmolarity-activated mechanism. TRPP is a family of polycystin, a protein responsible for polycystic kidney disease. TRPP2 is a Ca^{2+}-permeable channel and is a unique component of the urine flow-sensor. TRPV4 coupled with TRPP2, which is located at the base of cilia, forms

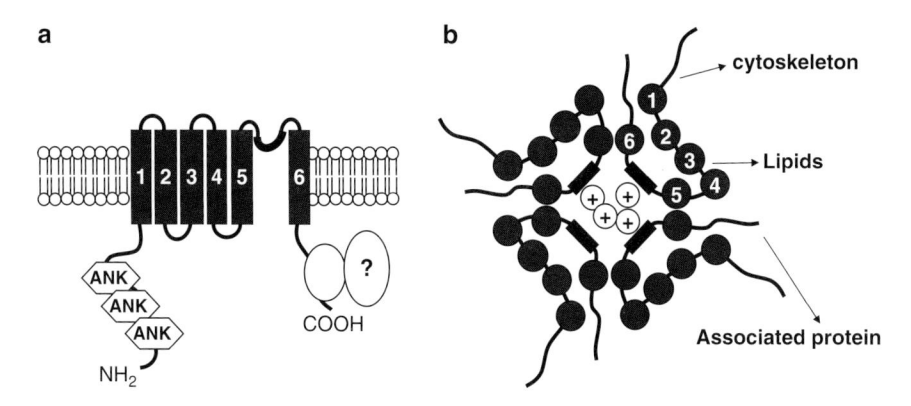

Fig. 7.2 (**a**) Proposed structure of TRP channels. There are ankyrin-like repeats in the N-terminal and the C-terminal that possibly interact with other binding proteins. (**b**) Suggestive mechanisms of the pore opening by mechanical pulling. Using an analog of a voltage-gated K channel, TRP channels have been composed by fifth and sixth TM. The stretch of the membrane deflects the cytoskeletal location that pulls the first TM through the N-terminal ankyrin–cytoskeletal interaction. The third or fourth TM is sensitive to unique lipids so that the portion may be pulled by alteration of the lipid moiety induced by mechanical stress. The C-terminal probably connects to a binding protein in the location of the neighboring pore. Because C-terminal is nearly located on the pore of channel, pulling the C-terminal appears most influential on the gating

a complex that is capable of sensing flow. Flow rate accompanies the change of shear stress, and thereby the TRPV4/TRPP2 may be sensitive to the pressure.

Before discussing mechanosensitive TRP channels, it is necessary to consider the types of mechanical stress. In the past, exposing a cell to hypotonic solutions was considered to induce membrane stretch because of swelling or volume gains and to place the same stress on the direct membrane stretch. However, hypotonicity may not change the capacitance of the whole cell in the whole-cell configuration under a patch-clamp. Hypotonicity per se activates the membrane bound phospholipase, the metabolites of which lead to channel opening. Thus, direct stretch on the membrane should be discriminated from hypo-osmotic cell volume change. In the same way, mechanical stresses observed in vivo should be differently and independently considered and tested for the effects of a gentle touch, harmful pressure, shear stress, blood pressure, sound, and gravity. A TRP channel that is suggested as mechanosensitive should, therefore, be re-examined through in vitro studies using different cell types and through in vivo studies in genetically manipulated animals.

7.2 TRPC Channels: Activation by Direct Stretch Via a Lipid-Dependent Mechanism

TRPC involves the cation channel playing a role in receptor-operated Ca^{2+} influx and is thought to be a stretch-activated channel resulting from the interaction of the membrane lipid moiety (Fig. 7.3). Any channel embedded in a lipid bilayer is exposed

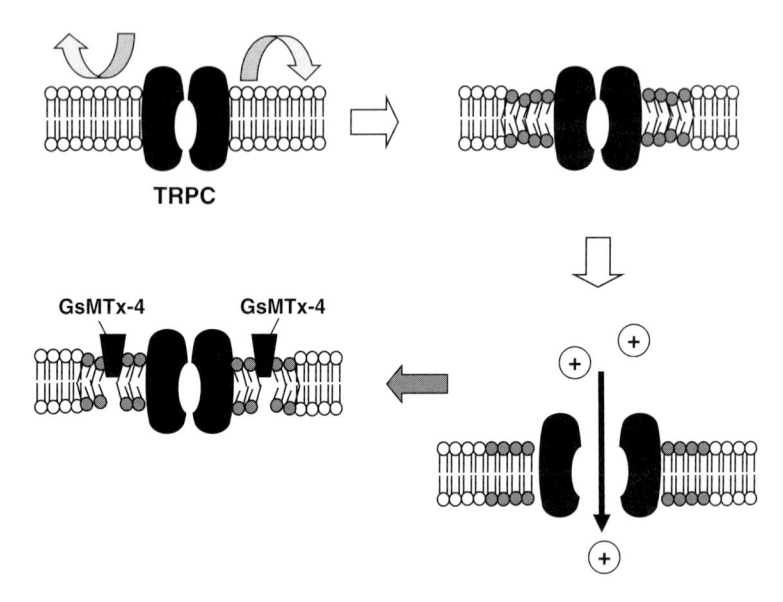

Fig. 7.3 The schema of the mechanism under lipid-dependent opening. The stretch alters the lipid moiety around the channel. The alteration of lipid moiety accompanies the deformity of the membrane which provides the channel opening energy. TRPC channels have a site regulated by polar lipids including PIP2, and suggest the hypothesis. GsMTx4 is a spider venom and is a specific blocker of stretch channels, and is also believed to play a role in blocking lipid changes

to negative and positive pressures created by the bilayer. Under equilibrium conditions, the conformational energy of the channel matches the energy profile. A change in the forces acting within the lipid membranes may alter the composition and offer the gating of channels. The two-pore potassium channel, such as MEC-2 or TREK-1 (Huber et al. 2006; Honore 2007), is activated under this model. A phospholipid, phosphatidylinositol, is altered by a membrane stretch and then opens the channels.

TRPC1 has been identified as a mechanosensitive cation channel in *Xenopus laevis* oocytes (Maroto et al. 2005). TRPC1 is also mechanosensitive in artificial liposomes. In vivo experiments have also hypothesized that TRPC1 is involved in stretch-induced muscle damage by elevating intracellular free calcium on the membrane stretch (Allen et al. 2005). TRPC6 is another receptor-activated cation channel that is activated by mechanical stretch or hypo-osmolality. It is well known that members of the TRPC3/6/7 channel subfamilies are activated directly by the lipid DAG, either added exogenously to cells or generated in response to PLC-coupled receptors (Venkatachalam et al. 2002, 2003). Like TREK-1, the alteration in lipid composition by DAG or PIP2 offers membrane deflection that induces the gating energy of TRPC6 (Spassova et al. 2006). A tarantula peptide, GsMTx-4, is a specific peptide inhibitor of mechanosensitive channels, which disturbs the lipid–channel boundary (Suchyna et al. 2000, 2004). Suchyna et al. also showed that the stretch activation of TRPC6 was blocked by this toxin. All these data suggest that TRPC1/6, and probably TRPC3/7, are stretch-activated cation channels, where the stretch of the membrane alters the local lipid composition leading to gating of the channel.

However, recent results have failed to confirm the mechanosensitivity of the TRPC1 channel in transfected cells or native arterial myocytes (Dietrich et al. 2007; Gottlieb et al. 2008). Similar to TRPC1, stretch activation of the homomultimeric TRPC6 channel could not be confirmed in transiently transfected COS cells and CHO cells (Dietrich et al. 2007; Gottlieb et al. 2008), although DAG still activated the TRPC6 channel in the inside-out patch configuration. The in vivo experiments may not support the stretch activity of the TRPC1 channel. Pressure-induced constriction, for example, was not impaired in TRPC1 knockout mice (Dietrich et al. 2007). In vitro experiments using isolated vascular muscles were also performed, which demonstrated that this channel might not play a significant role in mechanotransduction (Dietrich et al. 2007).

Thus, it remains controversial as to whether the TRPC family is actually involved in the stretch-activated cation channel, although the mechanism of lipid–protein interaction is highly interesting.

7.3   TRPM Channel: Is It Mechanosensitive?

TRPM is a melastatin family that was discovered by differential expression cloning between metastatic and nonmetastatic tumors. TRPM molecules have a long amino acid terminal that exhibits enzymatic activity. Translocation of TRPM is usually regulated by this enzyme (Launay et al. 2004). The short form of TRPM3 was shown to be activated by hypotonic cell swelling (Grimm et al. 2003). However, a mechanistic insight and further studies are required. TRPM4 is a Ca^{2+}-activated 25-pS cation channel and is voltage-dependent (Nilius et al. 2003). Expression of TRPM4B in HEK293 cells results in the appearance of stretch-activated cation channels that share properties of the native stretch-activated channels of cerebral artery myocytes (Morita et al. 2007). However, the stretch-activated channel is usually not dependent on the presence of Ca^{2+}.

The Ca^{2+}-activated large conductance K^+ channel is clearly discriminated by its conductance from other small channels. When the Ca^{2+}-activated K^+ channel is located close to the Ca^{2+}-permeable channel in a cell, swelling of the cell or even stretch of the local membrane in cell-attached configuration have been shown, as if the large conductance K^+ current is mechanosensitive. Likewise, the Ca^{2+}-activated cation channel, TRPM4, might be activated by the rise in intracellular Ca^{2+} via influx through another adjacent stretch-activated Ca^{2+} channel.

It is proposed that TRPM7 is a stretch- and swelling-activated cation channel that plays an important role in volume regulation (Numata et al. 2007). Whole-cell experiments and excised patches suggested TRPM7 as a stretch-augmented single-channel activity. The single-channel conductance and its rectification are, however, not matched with the initial findings (Runnels et al. 2002).

Taken together, the TRPM family may involve a stretch-activated channel although the results are not yet confirmed and the actual molecular mechanism, such as the lipid–protein interaction, requires further study.

7.4 TRPA1: Activation by Interaction of Cell Cytoskeleton?

TRPA1 was first considered as a candidate mechanosensor in an auditory system and was proposed as the mechanically gated hair cell transduction channel in zebrafish (Corey et al. 2004). Interestingly, TRPA1 contains a 14- to 16-amino terminal ankyrin repeat domain that has been hypothesized to act as a gating spring in mechanosensing (Fig. 7.2) (Howard and Bechstedt 2004; Sotomayor et al. 2005). However, TRPA1 knockout mice (*trpa1*$^{-/-}$) revealed a normal startle reflex to loud noise, a normal auditory brain stem response, and normal transduction currents in vestibular hair cells (Bautista et al. 2006; Kwan et al. 2006). Mutations in *trpa1*, the *Caenorhabditis elegans* orthologue of mouse TRPA1, cause defects in nose-touch responses, and *trpa1*$^{-/-}$ revealed a weakness in response to mechanical stimuli (Kwan et al. 2006), suggesting the contribution of TRPA1 to mechanosensation (Kindt et al. 2007).

However, exogenously expressed TRPA1 in HEK cells did not reveal direct mechanosensitivity but sensitivity to cationic amphipathic molecules. Interestingly, the amphipathic molecule chlorpromazine inserts in the inner leaflet of the bilayer and thus produces a negative (concave) deformation of the membrane (Patel et al. 1998). Further, the tarantula toxin, GsMTx-4, a blocker of cardiac stretch-activated channels through a bilayer-dependent mechanism, causes activation of TRPA1 channels (Hill and Schaefer 2007).

Therefore, although several lines of evidence suggest that TRPA1 is involved in mechanosensory function, the molecular mechanism of the gating is still under hypothesis and remains to be confirmed.

7.5 TRPV: Activation by Mechanical Stress Through Lipid Metabolites

TRPV1 was isolated by expression cloning using capsaicin, the hot pepper compound, as an activator (Caterina et al. 1997). TRPV1 is also temperature sensitive, making it the first reported channel in response to a physical force. TRPV1 is thought to play a role in sensing hypertonic stimuli by using TRPV1-knockout mouse (*trpv1*$^{-/-}$) (Ciura and Bourque 2006). Hypertonic conditions induce cell shrinkage that is temporally associated with an increase in ruthenium-red-sensitive cation conductance. The current is inhibited by stretch, namely, the mechanosensitive stretch-inhibitable channel. The provocation of the osmosensing neurons in the organum vasculosum laminae terminalis (OVLT) and supraoptic nucleus (SON) of *trpv1*$^{-/-}$ was not found, suggesting that TRPV1 functions as a hypertonic sensor. Many previous studies have suggested that the loci are essential in thirst sensation and release of a vasopressin. However, exogenously expressed TRPV1 does not show the current activated by hypertonic solution. An N-terminal truncated form of TRPV1 is thought to show the cation current (Sharif Naeini et al. 2006). This truncated form is a hypertonicity-activated channel but the finding is still under consideration.

TRPV2 expression was also reported to be heat-sensitive and mechanosensitive (Caterina et al. 1999). TRPV2 is located in vascular smooth muscle cells. Knockdown of TRPV2 with antisense oligonucleotide suppresses swelling acti-vated currents and is clearly observed in wild, isolated smooth muscle cells. TRPV2 expressed in CHO cells showed a stretch-activated channel in the cell-free membrane (Muraki et al. 2003). Furthermore, when the elastic silicone membrane, seeding with TRPV2-expressing cells, was stretched, the intracellular-free calcium level of the cells were elevated (Iwata et al. 2003). However, the results have not been confirmed, and the mechanism remains obscure.

TRPV4 was first reported as a mechanosensitive channel by homology cloning to TRPV1 (Liedtke et al. 2000). TRPV4 is widely located in the body, including osmo-sensing neurons, and is richly present in the lungs, endothelial, and kidneys. In our preliminary experiment, TRPV4 cDNA was detected in isolated distal renal tubules, medullary thick ascending limb (mTAL), distal connecting tubules (DCT), and corti-cal collecting duct (CCD). When intracellular-free Ca^{2+} was measured in terms of fluorescent intensity of fluo-3, hypotonic but not hypertonic solution increased intracellular-free Ca^{2+} levels. We first generated TRPV4 knockout mice ($trpv4^{-/-}$) and found that TRPV4 may play a role in sensing a range of pressures (Suzuki et al. 2003). As shown in Fig. 7.4, the swelling-induced increase in intracellular-free Ca^{2+} was completely absent in $trpv4^{-/-}$ (Suzuki M, unpublished data). Furthermore, perfusion

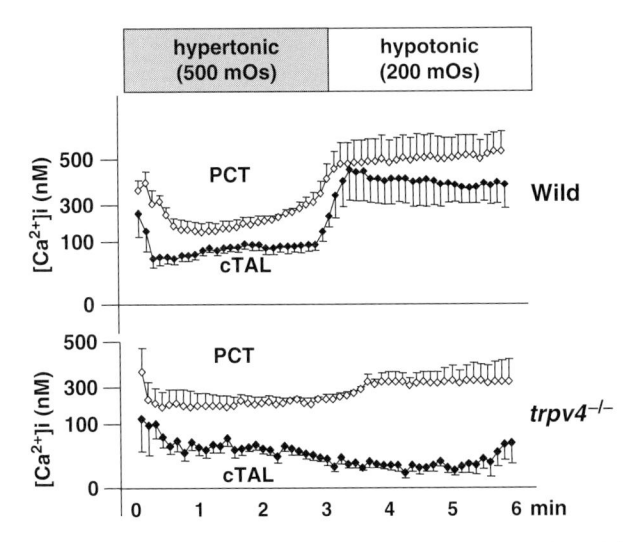

Fig. 7.4 Intracellular free Ca^{2+} ([Ca^{2+}]i) in an iso-osmotic fluid in wild and $trpv4^{-/-}$ renal tubules. The renal proximal convoluted (PCT) and cortical thick ascending tubule (cTAL) were isolated manually. The ratio of fluorescent dye, fluo-3 and fura-red, was used to calculate [Ca^{2+}]i. Hypertonic solution reduced [Ca^{2+}]i in both segments either in wild or $trpv4^{-/-}$ kidneys. Hypotonic solution increased [Ca^{2+}]i in PCT and cTAL in wild, while failed to increase [Ca^{2+}]i in cTAL in $trpv4^{-/-}$. Trpv4 is rich in cTAL rather than PCT. A hypotonic solution induced an increase in [Ca^{2+}]i that is completely absent in the cTAL of $trpv4^{-/-}$. Therefore, TRPV4 plays a role in increasing [Ca^{2+}]i through cell-swelling

of the renal tubules indicated that the response to shear stress was also absent in $trpv4^{-/-}$ (Taniguchi et al. 2007). Thus, TRPV4 is a mechanosensitive channel to hypotonic stimuli and shear stress. However, TRPV4 is, as first reported, not a stretch-activated channel (Strotmann et al. 2000). We attempted to detect the stretch channel in CHO cells in cell-attached patches and found it in approximately 5% of patches. TRPV4-expressed CHO cells revealed a similar frequency, indicating that TRPV4 itself is not a stretch-activated channel. These behaviors of TRPV4 as a mechanosensor are established in vitro and in vivo, thereby making the investigation of the molecular mechanism highly informative.

The mechanism for the gating of TRPV4 by hypotonic stimuli was discovered in 2003 (Watanabe et al. 2003). Cells exposed to hypotonic solutions activate TRPV4 through phospholipase 2 (PLA2)-dependent formation of arachidonic acid (AA) (Basavappa et al. 1998) and its subsequent metabolization to 5,6-epoxyeicosatrienoic acids (EET) through a cytochrome P450 epoxygenase-dependent pathway. Thus, the hypotonic-activated currents driven by TRPV4 are completely blocked by all PLA2 inhibitors; methylarachidonyl fluorophosphate (MAFP), arachidonyl trifluoromethyl ketone (AACOCF3), 3-[(4-octadecyl) benzoyl]acrylic acid (OBAA, 100 μM), N-(p-amylcinnamoyl) anthranilic acid (ACA, 80 μM), and p-bromophenacyl bromide (BPB, 100 μM) in TRPV4-expressing HEK cells (Watanabe et al. 2003). PLA2 is divided into two groups: cytosolic PLA2 (cPLA2) in the cell interior and secretory PLA2 (sPLA2) in plasma. The effect of inhibitors on the PLA subclasses is different: MAFP, AACOCF3, and OBAA are inhibitors of cPLA2, while BPB is a specific inhibitor of sPLA2 (Hernandez et al. 1998). AA released from PLA2 is metabolized into three pathways (via cyclooxygenase, lipoxygenase and epoxygenase). Among these, the epoxygenase endoproduct, 5,6-EET has been proven to be a direct activator of TRPV4 in a single-channel analysis. Furthermore, the inhibition of the cytochrome P-450 epoxygenase pathway by 5,8,11,13-eicosatetraynoic acid, miconazole, 17-octadecynoic acid (all 10 μM) (Vriens et al. 2004) or sulfaphenazole (for CYP2C9) (Vriens et al. 2005) blocks the hypotonic activation of TRPV4. These lipid metabolites structurally interact with TM3/4 of TRPV4 and open the channel (Fig. 7.5).

AA is derived from various fatty acids of the plasma membrane. Hydrolysis of the endocannabinoids, anandamide (AEA) and 2-arachidonoylglycerol (2-AG), which are endogenous ligands of the CB1 and CB2 metabotropic cannabinoid receptors, is one of the pathways resulting in the formation of AA. AEA and 2-AG activate TRPV4 in an AA-dependent pathway (Watanabe et al. 2003). All mechanisms under the hypotonic-activation and stimulation by an endocannabinoid have been examined in TRPV4-expressing HEK cells and primary cultured endothelial cells. In these cells, the primary target of the swelling is the activation of PLA2, although the molecular mechanism for the direct coupling of the enzyme and the physical force has not been fully clarified. Interestingly, methanandamide, which blocks the conversion of anandamide to AA, completely blocks the hypotonic reduction of AQP5 in the epithelial cells of mice lungs through the activation of TRPV4 (Sidhaye et al. 2006). Therefore, PLA2, which hydrolyzes anandamide into AA, may be a primary target of cell swelling. The lipid-dependent activation is similar

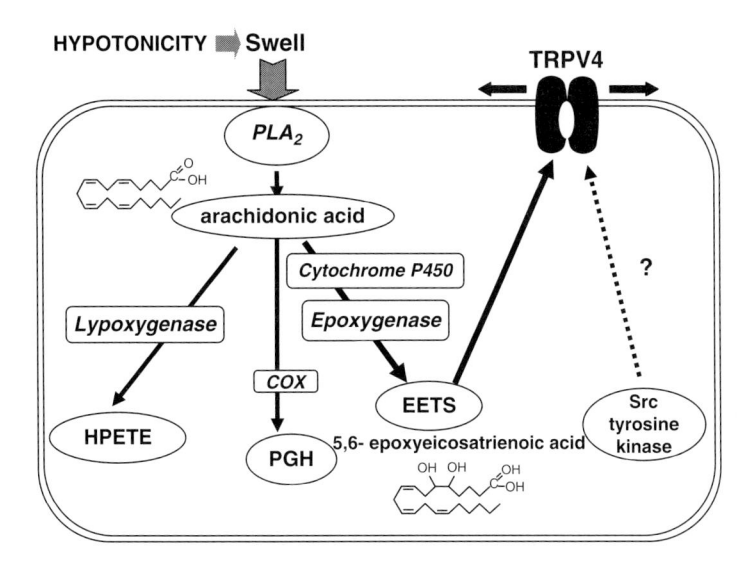

Fig. 7.5 Arachidonate cascade that activates TRPV4. Swelling of the cell activates phospholipase A2, leading to the release of arachidonate. Metabolites of cytochrome P450, but neither those of cyclooxygenase nor of lipoxygenase, activate TRPV4 by binding to TM4. Src tyrosine kinase may be activated by the swelling. The phosphorylation of TRPV4 by this kinase may activate TRPV4

to TRPV1. Capsaicin is a lipid soluble compound, directly activating TRPV1 requiring TM3/4 alignment (Gavva et al. 2004).

If the mechanosensing is restricted to the direct interaction of the membrane and the channel, the swelling of the cells is fundamentally different from this mechanism. TRPV4 is suggested to play a role in the direct sensing of mechanical stress in vivo.

Impairment of the pressure sensation in *trpv4*$^{-/-}$ mice in vivo has been indicated (Suzuki et al. 2003), although at slightly different ranges of pressure. To test the mechanosensation, von-Fray hair is a useful tool to detect the touch sensation in animals and humans. Von-Fray hair is composed of different sized plastic hairs providing variable pressure (gram). When a mouse senses the touch of the hair on its hind paw, it shakes the hair off. Liedtke and Friedman (2003) indicated, using an automated von-Fray detector, that the elevation in the threshold of avoidance was observed in *trpv4*$^{-/-}$ mice. We performed the test manually and failed to find a significant difference. However, we used our own device to measure the pressure sensation on the tail and found that sensation to mechanical stimuli is impaired in *trpv4*$^{-/-}$ mice. An additional study, in which the direct neural activity of femoral nerves evoked by pressure on the paw was measured, confirmed the result. Additionally, avoidance of touch on a nematode (*C. elegans*) nose is restored by mammalian TRPV4 expressed exogenously in the sensory neurons of worms (Liedtke et al. 2003). Thus, TRPV4 is a direct sensor of mechanical pressure in vivo.

TRPV4 is also important for sensing fluid flow. We have demonstrated that *trpv4*$^{-/-}$ mice showed an inability to sense flow in microperfused renal tubules (Taniguchi et al. 2007). Many other experiments have been performed and suggest that TRPV4 is sensitive to shear stress in the vascular system and also to flow in secreting ducts (Thodeti et al. 2009).

7.6 TRPP2: Making a Flow-Sensing Complex with TRPV4

TRPP is a family of genes that cause autosomal-dominant polycystic kidney disease (ADPKD). The polycystin family is divided structurally and functionally into two subfamilies, the polycystin 1 (PKD1)-like and polycystin 2 (PKD2)-like proteins (Igarashi and Somlo 2002; Delmas 2004, 2005; Nauli and Zhou 2004). The PKD2-like subgroup contains three homologous proteins, TRPP2, TRPP3 and TRPP5. The PKD1-like subgroup contains five homologous proteins with an 11-transmembrane topology, and is now not considered to be a member of the TRP superfamily (Fig. 7.1). The finding that TRPP2 is retained in the endoplasmic reticulum of most cell systems supported the belief that TRPP2 might function as a reticular Ca^{2+}-release channel (Koulen et al. 2002). On the other hand, TRPP2 has been shown to reside and act at the plasma membrane in Madine–Darby canine kidney (MDCK) cells derived from CCDs (Scheffers et al. 2002; Luo et al. 2003). Although the subcellular localization of TRPP2 is still obscure, TRPP2 has been shown to assemble with PKD1 to form a receptor–ion channel complex (Hanaoka et al. 2000; Delmas et al. 2004). Both proteins localize to the primary cilia of renal epithelial cells, where they are implicated in mechanosensitive transduction signals (Pazour et al. 2002; Yoder et al. 2002; Nauli et al. 2003).

The primary cilium is proposed to act as a flow sensor because it was shown to be essential for the ability of MDCK cells to detect laminar fluid flow (Praetorius and Spring 2001, 2003). Fluid shear-force bending of the cilium causes Ca^{2+} influx through mechanically sensitive channels. In microperfused renal tubules, an increase in flow rate, which was changed by the height of influx fluid, increased intracellular Ca^{2+} (Taniguchi et al. 1994). In this model, the PKD1–TRPP2 complex can be a mechanosensor, which is used to transduce energy pressure into a change in membrane Ca^{2+} permeability.

A recent study provided the first evidence for the presence of single cation channel currents from isolated primary cilia of a renal cell line (Raychowdhury et al. 2005). Future work of the molecular identity and mechanosensitivity will be important to reveal a cilia-related mechanosensitive channel.

Recently, Walz et al, also using *trpv4*$^{-/-}$, indicated that TRPV4, another flow sensor, and TRPP2 interact and colocalize in the primary cilium (Kottgen et al. 2008). TRPC1 is known to interact with TRPP2 in expression systems (Tsiokas et al. 1999) and might form functional heteromers with TRPP2 (Bai et al. 2008).

7.7 Hypothesis: Mechanosensitive Channel Complex

Based on the above findings in the TRP channels, one can assume that the TRP channel is not a direct sensor of mechanical stress but is actually involved in the mechanosensitive channel complex (Fig. 7.6). TRP channels may locate downstream of the signal transduction of the mechanical stresses. The energy of the mechanical stress on the plasma membrane is presumably converted into changes in lipid moiety, cytoskeletal filaments, or the location of other molecules leading to gating of a TRP channel.

Direct lipid–channel interaction, such as TRPC, TRPM, and TRPV2, has rarely been found and has not been reproduced in other studies. The reproduction of stretch-activated gating is found in one cell type but not in the others. Therefore, lipid composition may differ under various cell types or culture conditions. The cytoskeletal component is required in some mechanotransduction. The stretch-activated channel is blocked by reagents that disrupt cytoskeletal filaments. However, no direct evidence has been found for the mechanosensitive connection of the cytoskeleton to the TRP channel, such as the N-terminal of TRPA1 or TRPV4. Other proteins might be required for this connection, so that investigation of binding proteins of N- or C-terminal of TRP channels is important for solving the problem. A more complex protein(s) mixture is supposed in TRPP2–PKD1, TRPP2–TRPV4, TRPC6–cell adhesion molecules, where the in vivo findings provide clues for elucidating the mechanism.

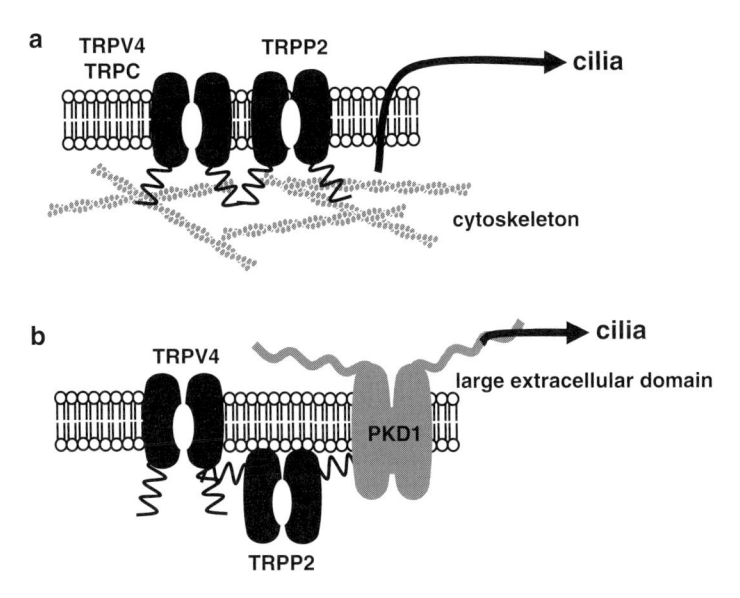

Fig. 7.6 Hypothetical TRP complexes as mechanosensors. (**a**) Two TRP channels such as TRPV4/TRPC and TRPP2 compose a mechano-gated complex and connect to each other's C-terminal. The complex is pulled, for example, by cytoskeletal filaments. (**b**) PKD1 is coupled with TRPP2 and has a large extracellular domain. TRPV4 and TRPP2 complex might be bound to PKD1, and PKD1 might be pulled by cilia or cytoskeleton through the large domain. The complex TRPs and other molecules might be a molecular unit of the mechanosensor

Acknowledgments This work was supported by grants-in-aid for scientific research (17590255, 18590899, 22590291), and grants from Japan Space Forum, JAXA.

References

Allen DG, Whitehead NP, Yeung EW (2005) Mechanisms of stretch-induced muscle damage in normal and dystrophic muscle: role of ionic changes. J Physiol 567:723–735

Bai CX, Giamarchi A, Rodat-Despoix L, Padilla F, Downs T, Tsiokas L, Delmas P (2008) Formation of a new receptor-operated channel by heteromeric assembly of TRPP2 and TRPC1 subunits. EMBO Rep 9:472–479

Basavappa S, Pedersen SF, Jorgensen NK, Ellory JC, Hoffmann EK (1998) Swelling-induced arachidonic acid release via the 85-kDa cPLA2 in human neuroblastoma cells. J Neurophysiol 79:1441–1449

Bautista DM, Jordt SE, Nikai T, Tsuruda PR, Read AJ, Poblete J, Yamoah EN, Basbaum AI, Julius D (2006) TRPA1 mediates the inflammatory actions of environmental irritants and proalgesic agents. Cell 124:1269–1282

Caterina MJ, Schumacher MA, Tominaga M, Rosen TA, Levine JD, Julius D (1997) The capsaicin receptor: a heat-activated ion channel in the pain pathway. Nature 389:816–824

Caterina MJ, Rosen TA, Tominaga M, Brake AJ, Julius D (1999) A capsaicin-receptor homologue with a high threshold for noxious heat. Nature 398:436–441

Ciura S, Bourque CW (2006) Transient receptor potential vanilloid 1 is required for intrinsic osmoreception in organum vasculosum lamina terminalis neurons and for normal thirst responses to systemic hyperosmolality. J Neurosci 26:9069–9075

Corey DP, Garcia-Anoveros J, Holt JR, Kwan KY, Lin SY, Vollrath MA, Amalfitano A, Cheung EL, Derfler BH, Duggan A, Geleoc GS, Gray PA, Hoffman MP, Rehm HL, Tamasauskas D, Zhang DS (2004) TRPA1 is a candidate for the mechanosensitive transduction channel of vertebrate hair cells. Nature 432:723–730

Cosens DJ, Manning A (1969) Abnormal electroretinogram from a *Drosophila* mutant. Nature 224:285–287

Delmas P (2004) Polycystins: from mechanosensation to gene regulation. Cell 118:145–148

Delmas P (2005) Polycystins: polymodal receptor/ion-channel cellular sensors. Pflugers Arch 451:264–276

Delmas P, Nauli SM, Li X, Coste B, Osorio N, Crest M, Brown DA, Zhou J (2004) Gating of the polycystin ion channel signaling complex in neurons and kidney cells. FASEB J 18:740–742

Dietrich A, Kalwa H, Storch U, Mederos Y, Schnitzler M, Salanova B, Pinkenburg O, Dubrovska G, Essin K, Gollasch M, Birnbaumer L, Gudermann T (2007) Pressure-induced and store-operated cation influx in vascular smooth muscle cells is independent of TRPC1. Pflugers Arch 455:465–477

Gavva NR, Klionsky L, Qu Y, Shi L, Tamir R, Edenson S, Zhang TJ, Viswanadhan VN, Toth A, Pearce LV, Vanderah TW, Porreca F, Blumberg PM, Lile J, Sun Y, Wild K, Louis JC, Treanor JJ (2004) Molecular determinants of vanilloid sensitivity in TRPV1. J Biol Chem 279:20283–20295

Gottlieb P, Folgering J, Maroto R, Raso A, Wood TG, Kurosky A, Bowman C, Bichet D, Patel A, Sachs F, Martinac B, Hamill OP, Honore E (2008) Revisiting TRPC1 and TRPC6 mechanosensitivity. Pflugers Arch 455:1097–1103

Grimm C, Kraft R, Sauerbruch S, Schultz G, Harteneck C (2003) Molecular and functional characterization of the melastatin-related cation channel TRPM3. J Biol Chem 278:21493–21501

Hanaoka K, Qian F, Boletta A, Bhunia AK, Piontek K, Tsiokas L, Sukhatme VP, Guggino WB, Germino GG (2000) Co-assembly of polycystin-1 and -2 produces unique cation-permeable currents. Nature 408:990–994

Hardie RC, Minke B (1992) The trp gene is essential for a light-activated Ca2+ channel in *Drosophila* photoreceptors. Neuron 8:643–651

Hernandez M, Burillo SL, Crespo MS, Nieto ML (1998) Secretory phospholipase A2 activates the cascade of mitogen-activated protein kinases and cytosolic phospholipase A2 in the human astrocytoma cell line 1321N1. J Biol Chem 273:606–612

Hill K, Schaefer M (2007) TRPA1 is differentially modulated by the amphipathic molecules trinitrophenol and chlorpromazine. J Biol Chem 282:7145–7153

Honore E (2007) The neuronal background K2P channels: focus on TREK1. Nat Rev Neurosci 8:251–261

Howard J, Bechstedt S (2004) Hypothesis: a helix of ankyrin repeats of the NOMPC-TRP ion channel is the gating spring of mechanoreceptors. Curr Biol 14:R224–R226

Huber TB, Schermer B, Muller RU, Hohne M, Bartram M, Calixto A, Hagmann H, Reinhardt C, Koos F, Kunzelmann K, Shirokova E, Krautwurst D, Harteneck C, Simons M, Pavenstadt H, Kerjaschki D, Thiele C, Walz G, Chalfie M, Benzing T (2006) Podocin and MEC-2 bind cholesterol to regulate the activity of associated ion channels. Proc Natl Acad Sci USA 103:17079–17086

Igarashi P, Somlo S (2002) Genetics and pathogenesis of polycystic kidney disease. J Am Soc Nephrol 13:2384–2398

Iwata Y, Katanosaka Y, Arai Y, Komamura K, Miyatake K, Shigekawa M (2003) A novel mechanism of myocyte degeneration involving the Ca2+-permeable growth factor-regulated channel. J Cell Biol 161:957–967

Kindt KS, Viswanath V, Macpherson L, Quast K, Hu H, Patapoutian A, Schafer WR (2007) *Caenorhabditis elegans* TRPA-1 functions in mechanosensation. Nat Neurosci 10:568–577

Kottgen M, Buchholz B, Garcia-Gonzalez MA, Kotsis F, Fu X, Doerken M, Boehlke C, Steffl D, Tauber R, Wegierski T, Nitschke R, Suzuki M, Kramer-Zucker A, Germino GG, Watnick T, Prenen J, Nilius B, Kuehn EW, Walz G (2008) TRPP2 and TRPV4 form a polymodal sensory channel complex. J Cell Biol 182:437–447

Koulen P, Cai Y, Geng L, Maeda Y, Nishimura S, Witzgall R, Ehrlich BE, Somlo S (2002) Polycystin-2 is an intracellular calcium release channel. Nat Cell Biol 4:191–197

Kwan KY, Allchorne AJ, Vollrath MA, Christensen AP, Zhang DS, Woolf CJ, Corey DP (2006) TRPA1 contributes to cold, mechanical, and chemical nociception but is not essential for hair-cell transduction. Neuron 50:277–289

Launay P, Cheng H, Srivatsan S, Penner R, Fleig A, Kinet JP (2004) TRPM4 regulates calcium oscillations after T cell activation. Science 306:1374–1377

Liedtke W, Friedman JM (2003) Abnormal osmotic regulation in trpv4–/– mice. Proc Natl Acad Sci USA 100:13698–13703

Liedtke W, Choe Y, Marti-Renom MA, Bell AM, Denis CS, Sali A, Hudspeth AJ, Friedman JM, Heller S (2000) Vanilloid receptor-related osmotically activated channel (VR-OAC), a candidate vertebrate osmoreceptor. Cell 103:525–535

Liedtke W, Tobin DM, Bargmann CI, Friedman JM (2003) Mammalian TRPV4 (VR-OAC) directs behavioral responses to osmotic and mechanical stimuli in *Caenorhabditis elegans*. Proc Natl Acad Sci USA 100(Suppl 2):14531–14536

Lo MV, Pak WL (1981) Light-induced pigment granule migration in the retinular cells of *Drosophila melanogaster*. Comparison of wild type with ERG-defective mutants. J Gen Physiol 77:155–175

Luo Y, Vassilev PM, Li X, Kawanabe Y, Zhou J (2003) Native polycystin 2 functions as a plasma membrane Ca2+-permeable cation channel in renal epithelia. Mol Cell Biol 23:2600–2607

Maroto R, Raso A, Wood TG, Kurosky A, Martinac B, Hamill OP (2005) TRPC1 forms the stretch-activated cation channel in vertebrate cells. Nat Cell Biol 7:179–185

Montell C, Rubin GM (1989) Molecular characterization of the *Drosophila* trp locus: a putative integral membrane protein required for phototransduction. Neuron 2:1313–1323

Morita H, Honda A, Inoue R, Ito Y, Abe K, Nelson MT, Brayden JE (2007) Membrane stretch-induced activation of a TRPM4-like nonselective cation channel in cerebral artery myocytes. J Pharmacol Sci 103:417–426

Muraki K, Iwata Y, Katanosaka Y, Ito T, Ohya S, Shigekawa M, Imaizumi Y (2003) TRPV2 is a component of osmotically sensitive cation channels in murine aortic myocytes. Circ Res 93:829–838

Nauli SM, Zhou J (2004) Polycystins and mechanosensation in renal and nodal cilia. Bioessays 26:844–856

Nauli SM, Alenghat FJ, Luo Y, Williams E, Vassilev P, Li X, Elia AE, Lu W, Brown EM, Quinn SJ, Ingber DE, Zhou J (2003) Polycystins 1 and 2 mediate mechanosensation in the primary cilium of kidney cells. Nat Genet 33:129–137

Nilius B, Prenen J, Droogmans G, Voets T, Vennekens R, Freichel M, Wissenbach U, Flockerzi V (2003) Voltage dependence of the Ca2+-activated cation channel TRPM4. J Biol Chem 278:30813–30820

Numata T, Shimizu T, Okada Y (2007) Direct mechano-stress sensitivity of TRPM7 channel. Cell Physiol Biochem 19:1–8

Patel AJ, Honore E, Maingret F, Lesage F, Fink M, Duprat F, Lazdunski M (1998) A mammalian two pore domain mechano-gated S-like K+ channel. EMBO J 17:4283–4290

Pazour GJ, San Agustin JT, Follit JA, Rosenbaum JL, Witman GB (2002) Polycystin-2 localizes to kidney cilia and the ciliary level is elevated in orpk mice with polycystic kidney disease. Curr Biol 12:R378–R380

Praetorius HA, Spring KR (2001) Bending the MDCK cell primary cilium increases intracellular calcium. J Membr Biol 184:71–79

Praetorius HA, Spring KR (2003) Removal of the MDCK cell primary cilium abolishes flow sensing. J Membr Biol 191:69–76

Raychowdhury MK, McLaughlin M, Ramos AJ, Montalbetti N, Bouley R, Ausiello DA, Cantiello HF (2005) Characterization of single channel currents from primary cilia of renal epithelial cells. J Biol Chem 280:34718–34722

Runnels LW, Yue L, Clapham DE (2002) The TRPM7 channel is inactivated by PIP(2) hydrolysis. Nat Cell Biol 4:329–336

Scheffers MS, Le H, van der Bent P, Leonhard W, Prins F, Spruit L, Breuning MH, de Heer E, Peters DJ (2002) Distinct subcellular expression of endogenous polycystin-2 in the plasma membrane and Golgi apparatus of MDCK cells. Hum Mol Genet 11:59–67

Sharif Naeini R, Witty MF, Seguela P, Bourque CW (2006) An N-terminal variant of Trpv1 channel is required for osmosensory transduction. Nat Neurosci 9:93–98

Sidhaye VK, Guler AD, Schweitzer KS, D'Alessio F, Caterina MJ, King LS (2006) Transient receptor potential vanilloid 4 regulates aquaporin-5 abundance under hypotonic conditions. Proc Natl Acad Sci USA 103:4747–4752

Sotomayor M, Corey DP, Schulten K (2005) In search of the hair-cell gating spring elastic properties of ankyrin and cadherin repeats. Structure 13:669–682

Spassova MA, Hewavitharana T, Xu W, Soboloff J, Gill DL (2006) A common mechanism underlies stretch activation and receptor activation of TRPC6 channels. Proc Natl Acad Sci USA 103:16586–16591

Strotmann R, Harteneck C, Nunnenmacher K, Schultz G, Plant TD (2000) OTRPC4, a nonselective cation channel that confers sensitivity to extracellular osmolarity. Nat Cell Biol 2:695–702

Suchyna TM, Johnson JH, Hamer K, Leykam JF, Gage DA, Clemo HF, Baumgarten CM, Sachs F (2000) Identification of a peptide toxin from *Grammostola spatulata* spider venom that blocks cation-selective stretch-activated channels. J Gen Physiol 115:583–598

Suchyna TM, Tape SE, Koeppe RE 2nd, Andersen OS, Sachs F, Gottlieb PA (2004) Bilayer-dependent inhibition of mechanosensitive channels by neuroactive peptide enantiomers. Nature 430:235–240

Suzuki M, Mizuno A, Kodaira K, Imai M (2003) Impaired pressure sensation in mice lacking TRPV4. J Biol Chem 278:22664–22668

Taniguchi J, Takeda M, Yoshitomi K, Imai M (1994) Pressure- and parathyroid-hormone-dependent Ca2+ transport in rabbit connecting tubule: role of the stretch-activated nonselective cation channel. J Membr Biol 140:123–132

Taniguchi J, Tsuruoka S, Mizuno A, Sato J, Fujimura A, Suzuki M (2007) TRPV4 as a flow sensor in flow-dependent K+ secretion from the cortical collecting duct. Am J Physiol Renal Physiol 292:F667–F673

Thodeti CK, Matthews B, Ravi A, Mammoto A, Ghosh K, Bracha AL, Ingber DE (2009) TRPV4 channels mediate cyclic strain-induced endothelial cell reorientation through integrin-to-integrin signaling. Circ Res 104:1123–1130

Tsiokas L, Arnould T, Zhu C, Kim E, Walz G, Sukhatme VP (1999) Specific association of the gene product of PKD2 with the TRPC1 channel. Proc Natl Acad Sci USA 96:3934–3939

Venkatachalam K, van Rossum DB, Patterson RL, Ma HT, Gill DL (2002) The cellular and molecular basis of store-operated calcium entry. Nat Cell Biol 4:E263–E272

Venkatachalam K, Zheng F, Gill DL (2003) Regulation of canonical transient receptor potential (TRPC) channel function by diacylglycerol and protein kinase C. J Biol Chem 278:29031–29040

Vriens J, Watanabe H, Janssens A, Droogmans G, Voets T, Nilius B (2004) Cell swelling, heat, and chemical agonists use distinct pathways for the activation of the cation channel TRPV4. Proc Natl Acad Sci USA 101:396–401

Vriens J, Owsianik G, Fisslthaler B, Suzuki M, Janssens A, Voets T, Morisseau C, Hammock BD, Fleming I, Busse R, Nilius B (2005) Modulation of the Ca2 permeable cation channel TRPV4 by cytochrome P450 epoxygenases in vascular endothelium. Circ Res 97:908–915

Watanabe H, Vriens J, Prenen J, Droogmans G, Voets T, Nilius B (2003) Anandamide and arachidonic acid use epoxyeicosatrienoic acids to activate TRPV4 channels. Nature 424:434–438

Yoder BK, Hou X, Guay-Woodford LM (2002) The polycystic kidney disease proteins, polycystin-1, polycystin-2, polaris, and cystin, are co-localized in renal cilia. J Am Soc Nephrol 13:2508–2516

Part III
Skeletal Response

Chapter 8
Osteoblast Biology and Mechanosensing

8.1 Mechanical Forces and Skeletal Integrity

It is widely accepted that the skeleton adapts in response to mechanical forces. Therefore, the maintenance of the structural integrity of the skeleton is strongly dependent on mechanical loading (Frost 2003; Turner 1992; Rubin et al. 2001). Consistently, changes in loading and unloading induced by exercise or immobilization distinctly affect bone architecture and mass. Mechanical loading increases bone density and orientation and trabecular connectivity, resulting in enhanced bone resistance and increased fracture risk (Burr et al. 2002; Robling et al. 2006; Skerry 2008; Turner 1998). In accordance with the important role of loading on the maintenance of bone structure, skeletal unloading induces marked alterations in bone mass (Wakley et al. 1992; Morey-Holton and Globus 1998), an effect resulting from uncoupling between bone resorption and formation.

It has been proposed that the changes in bone remodeling in response to loading and unloading are initiated by an internal mechanostat that would sense strain (Frost 2003). This suggests that changes in bone remodeling occur in response to decreased or increased strain in order to adjust bone mass to a level that is appropriate to maintain skeletal integrity (Turner 1998; Huiskes et al. 2000; Forwood and Turner 1995). When strain applied on the skeleton increases, bone formation is increased and bone resorption is decreased, resulting in increased bone mass to optimize bone resistance and reduce fracture risks. Conversely, decreased skeletal strain applied on the skeleton reduces bone formation and increases bone resorption, resulting in changes in bone mass that is more appropriate to the change in mechanical strength (Frost 2003). In this view, mechanosensing of external loading appears as an important physiological regulator that maintains bone mass and resistance throughout life (Robling et al. 2006; Skerry 2008).

P.J. Marie (✉)
Laboratory of Osteoblast Biology and Pathology, Inserm U606, Hopital Lariboisiere,
2 rue Ambroise Pare, 75475 Paris Cedex 10, France
and
University Paris Diderot, Paris, France
e-mail: pierre.marie@inserm.fr

M. Noda (ed.), *Mechanosensing Biology*,
DOI 10.1007/978-4-431-89757-6_8, © Springer 2011

105

The maintenance of bone mass and microarchitecture is controlled by the balance between bone resorption and formation. At the cellular level, this balance is largely dependent on the number and activity of bone forming and resorbing cells. Mechanical forces and loading were found to affect the recruitment and activity of bone cells (Duncan and Turner 1995; Vico et al. 2001). Therefore, any alteration in the number or activity of bone cells will result in an imbalance between resorption and formation, resulting in microarchitecture deterioration and altered bone mass and strength.

8.2   Effects of Mechanical Forces on Osteoblastogenesis

8.2.1   Osteoblast Biology

Osteoblastogenesis is a complex process that is dependent on the recruitment, differentiation and function of osteoblasts. The osteogenic process starts by the commitment of osteoprogenitor cells into osteoblasts under the control of transcription factors, followed by their progressive differentiation into mature osteoblasts (Aubin 2001). The early commitment of mesenchymal stem cells towards osteoblasts involves the expression of transcription factors such as Runx2 and other genes that control phenotypic osteoblast genes (Marie 2008). Differentiation of committed osteoblasts is characterized by the expression of alkaline phosphatase, an early marker of osteoblast phenotype, followed by the synthesis and deposition of type I collagen, bone matrix proteins, and increased expression of osteocalcin (OC) and bone sialoprotein at the onset of mineralization. Once the bone matrix synthesis has been deposited and calcified, most osteoblasts become flattened lining cells, about 10% of osteoblasts are embedded within the matrix and become osteocytes, and the other cells die by apoptosis. In theory, osteoblasts, lining cells and osteocytes may play a role in the transduction of mechanical forces into biological signals and skeletal adaptation to loading (Burger and Klein-Nulend 1998). However, given the fact that osteocytes are more numerous, are more in contact with the bone matrix, and can exchange multiple information via canaliculi, it is likely that osteocytes are primary mechanosensing cells and osteoblasts are effective cells (Bonewald and Johnson 2008). Nevertheless, as discussed below, both mechanical forces induce important effects at all stages of osteoblastogenesis, both in vitro and in vivo. However, cells that are exposed to physiological loading are responding more to mechanical forces than cells present in normally unloaded bones (Rawlinson et al. 1995).

8.2.2   Osteoblast Responses to Mechanical Forces

In vitro, mechanical forces induce numerous effects on cultured osteoblastic cells, although these effects are variable due to the different models used (Owan et al. 1997; Basso and Heersche 2002). Both shear and strain forces have been shown to be implicated in modulating osteoblast proliferation (Boutahar et al. 2004; Kaspar et al. 2002; Turner et al. 1998). Multiple in vitro studies have revealed that mechanical forces promote osteoblast differentiation as shown by the increased alkaline phosphatase and matrix protein expression in osteoblastic cells (Li et al. 2007; Norvell et al. 2004; Ozawa et al. 1990; Pavlin et al. 2001; Rath et al. 2008; Saunders et al. 2006; Toma et al. 1997; You et al. 2001; Ziambaras et al. 1998; Ziros et al. 2002). Consistent with these findings, mechanical stimulation promotes osteogenic mineralization in vitro (Bancroft et al. 2002; David et al. 2007; Robling et al. 2006; Sikavitsas et al. 2003). Additionally, loading may affect proteins that control signals from the cell adhesion site to the nucleus in osteoblasts. For example, deficiency of CIZ, a nucleocytoplasmic shuttling protein, prevents the decreased bone formation induced by unloading (Hino et al. 2007). Although mechanical forces induce multiple effects on osteoblast replication and differentiation in vitro, these effects depend on the magnitude and frequency of the strain applied. High strain levels increase cell proliferation and decrease osteoblast marker expression, whereas low strain affects mature osteoblasts by decreasing cell proliferation and increasing cell differentiation (Jones et al. 1995).

8.2.3 Effects of Mechanical Forces on Osteoblastogenesis In Vivo

In vivo, loading also induces major effects on bone formation (Burr et al. 2002; Hillam and Skerry 1995; Turner et al. 1998). In animal models, mechanical stimulation induces a rapid recruitment of osteoblasts (Boppart et al. 1998) followed by expression of phenotypic osteoblast genes (Miles et al. 1998). Conversely, skeletal unloading induced by hindlimb suspension (Morey-Holton and Globus 1998) causes a rapid trabecular bone loss in the long bone metaphysis resulting from reduced trabecular architecture and inhibition of endosteal bone formation (Bikle and Halloran 1999; David et al. 2006; Globus et al. 1986). Although both the number and activity of osteoblasts are decreased in the unloaded metaphyseal bone, the number of osteoblasts is more affected than their activity (Globus et al. 1986; Machwate et al. 1993). Several mechanisms were shown to underlie the effects of skeletal unloading on osteoblasts (Bikle and Halloran 1999; Marie and Zerath 2000; Marie and Kaabeche 2006) (Fig. 8.1). We (Machwate et al. 1993) and others (Barou et al. 1998; Kostenuik et al. 1997; Zhang et al. 1995) have shown that the decreased bone formation in unloaded rat bone results mainly from an impaired recruitment of osteoblast precursor cells in the bone marrow stroma and in the metaphysis. This alteration of osteoblast recruitment is associated with reduction of the expression of histone H4 and c-fos, markers of cell proliferation, in unloaded metaphyseal bone cells (Machwate et al. 1993). In addition to affecting osteoblast recruitment, skeletal unloading in this model alters the function of differentiated osteoblasts. Notably, the altered bone formation induced by skeletal unloading is associated with reduced Runx2 expression and decreased expression of bone matrix type 1 collagen (Col 1), OC and osteopontin levels (Bikle et al. 1994; Machwate et al. 1995;

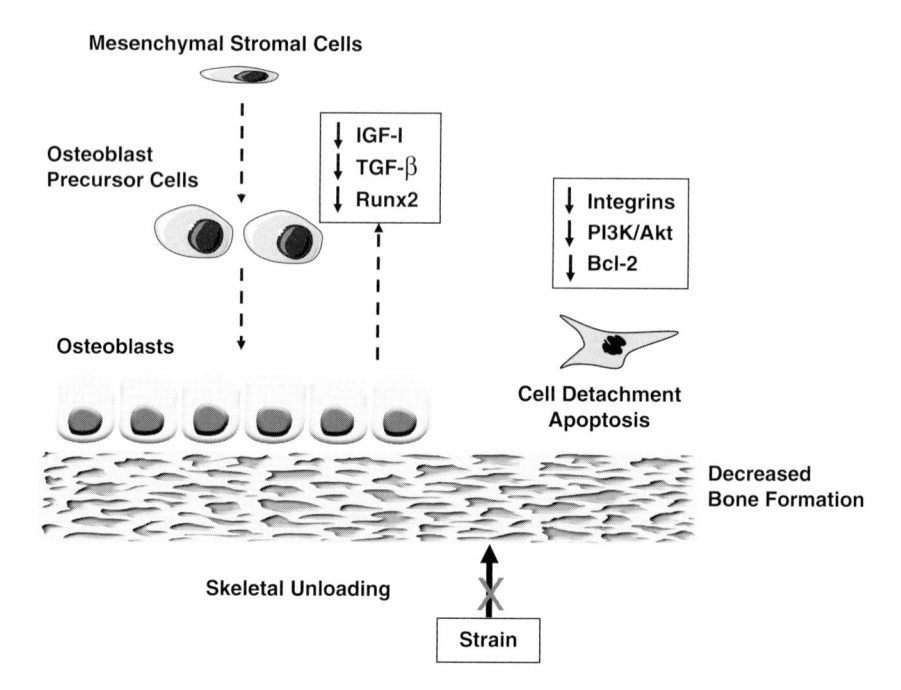

Fig. 8.1 Effects of skeletal unloading on osteoblastogenesis in vivo. The lack of strain induced by skeletal unloading results in decreased expression of growth factors and Runx2 by osteoblasts, causing decreased differentiation of mesenchymal stromal cells into osteoblast precursor cells and osteoblasts (*dotted lines*). Additionally, skeletal unloading causes reduction in integrin expression and reduced PI3K/Akt/Bcl-2 survival pathway, resulting in osteoblast detachment and apoptosis, which contributes to the decreased bone formation by osteoblasts induced by unloading

Zhang et al. 1995). We showed that Col 1 and OC mRNA levels are rapidly decreased in unloaded metaphyseal bone cells, and this effect correlates well with the decreased bone matrix synthesis measured at the tissue level by bone histomor-phometry. However, Col 1 and OC expression rise at 14 days of unloading, which is also consistent with the change in bone matrix synthesis observed at this same late time-point (Drissi et al. 1999). This partial restoration of bone formation after 2 weeks of unloading may be part of the reported adaptive response of bone cells to long-term unloading (Morey-Holton and Globus 1998). Such alterations are consistent with the effects of unloading in other rat models in which there is a reduction of the osteogenic capacity of bone marrow osteoblast precursor cells and a decreased expression of bone matrix proteins in long bones (Wakley et al. 1992; Keila et al. 1994).

In addition to reducing osteoblast differentiation, skeletal unloading or modeled microgravity increases the differentiation of bone marrow mesenchymal stromal cells towards adipocytes (Ahdjoudj et al. 2002; Zayzafoon et al. 2004), suggesting that unloading not only impairs osteoprogenitor cell differentiation into osteoblasts but also promotes adipocyte differentiation. Notably, skeletal unloading in rats increases the expression of adipogenic transcription factors C/EBPσ and PPARγ, resulting in activation of adipocyte gene expression in bone marrow stromal cells (Ahdjoudj et al. 2005) (Fig. 8.2). Thus, PPARγ and other transcription factors are involved in adipogenic conversion of bone marrow stromal cells in vivo, indicating that PPARγ is a negative regulator of bone mass in unloaded rats (Ahdjoudj et al. 2004). The exaggerated reciprocal relationship between osteoblastogenesis and adipogenesis may account for the decreased bone formation associated with the increased bone marrow adipogenesis in unloaded rats (Fig. 8.2). Consistent with this finding, stretching was found to downregulate PPARγ2 and adipocyte differentiation in mouse preadipocytes (Tanabe et al. 2004), suggesting that mechanical forces may play a dual role in the control of Runx2 and PPARγ expression in preosteoblasts. Adipocyte differentiation decreases while bone formation increases in running rats, also indicating that mechanical stimuli regulate the conversion of bone marrow stromal cells into adipocytes or osteoblasts (David et al. 2007). Thus, manipulation of PPARγ may provide a possible therapeutic intervention in the treatment of bone loss associated with immobilization (Ahdjoudj et al. 2004).

Skeletal unloading has major effects on osteoblast lifespan. Unloading induced by hindlimb suspension increases osteoblast apoptosis, independently of endogenous glucocorticoids, which may contribute to the altered osteoblastogenesis and osteopenia induced by unloading (Dufour et al. 2007; Capulli et al. 2009). Oteocyte apoptosis is also induced by weightlessness in rats (Dufour et al. 2007) and mice in which this effect precedes osteoclast recruitment and bone loss, suggesting that

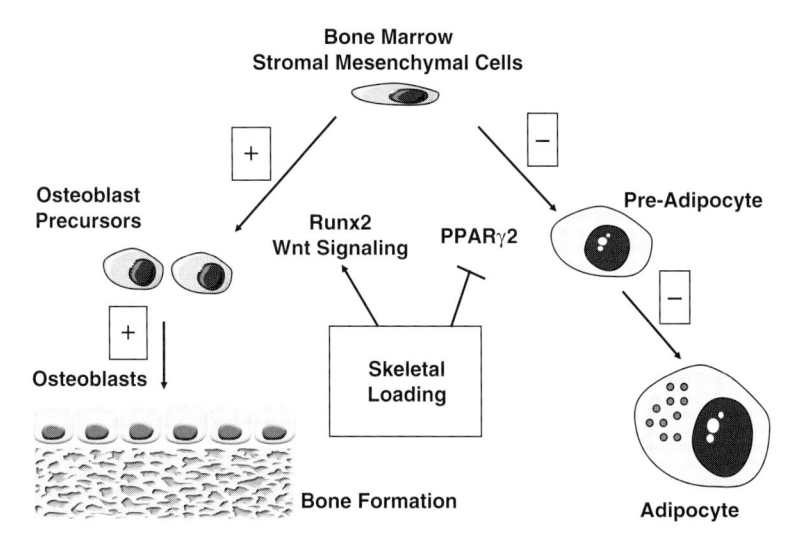

Fig. 8.2 Effects of mechanical forces on osteoblast/adipocyte differentiation in vivo. The differentiation of bone marrow stromal mesenchymal cells towards adipocytes or osteoblasts is governed by the balance between PPARγ2 and Runx2. Loading increases Runx2 and Wnt signaling and decreases PPARγ2 expression, resulting in increased osteoblastogenesis and decreased adipogenesis whereas unloading induces opposite effects

osteocyte apoptosis induced by unloading may provide messages that control remodeling of bone (Aguirre et al. 2006). Thus, removal of mechanical forces on the skeleton affects all stages of osteoblastogenesis, from the recruitment of osteoblast progenitor cells to the end of life of osteoblasts, resulting in a marked reduction in bone formation (Fig. 8.2).

8.3 Mechanosensing Mechanisms in Osteoblasts

8.3.1 Mechanical Stimuli

Given the marked effect of loading and unloading on bone formation, it is expected that bone cells may be directly or indirectly responsive to mechanical forces (Duncan and Turner 1995). Loading may generate deformation in bone tissue through pressure in the bone marrow and cortical bone. Loading may also induce transient pressure waves and shear forces through canaliculi. Loading may also provide dynamic electric fields via interstitial fluid flow on bone crystals (Rubin et al. 2006). Although strain increases bone formation, only dynamic cyclic exposure of cells to pressure or strain is anabolic, while chronic exposure to intermediary or low strain may lead to bone loss. Furthermore, the nature and amplitude of strains are essential for efficiently stimulating bone formation (Rubin et al. 2001). Osteoblastic cells were found to be responsive to mechanical stress such as hypotonic swelling, stretching or bending of the cell substratum, fluid shear stress, dynamic strain, tensile and compressive strain, altered gravity, and vibration. The response may, however, vary according to the variable forces, mechanical stretch and pressure, which drive fluid flow and stress at the cellular level, and it is not known which of these effects may reflect a physiological stress (Cowin 1998; Duncan and Turner 1995; Jones et al. 1995).

8.3.2 Mechanoreceptors

Whether and how osteoblasts can be sensitive to physiological loading remains an unanswered question (Jones et al. 1995). In theory, the transduction of mechanical signals in bone cells may involve several membrane mechanoreceptors leading to activation of signal transduction within bone cells. Possible candidate mechanoreceptors include the cytoskeleton, integrins, connexins, ion channels, and other membrane receptors (Davidson et al. 1990; Donahue 2000; Katsumi et al. 2004; Rubin et al. 2006; Ryder and Duncan 2000; Yellowley et al. 2000) (Fig. 8.3). Some of these receptors may act on signaling proteins that alter gene activity (Pavalko et al. 2003a). The extracellular-matrix-integrins-cytoskeletal axis may be a major pathway involved in the signal transduction of mechanical strain in cultured bone

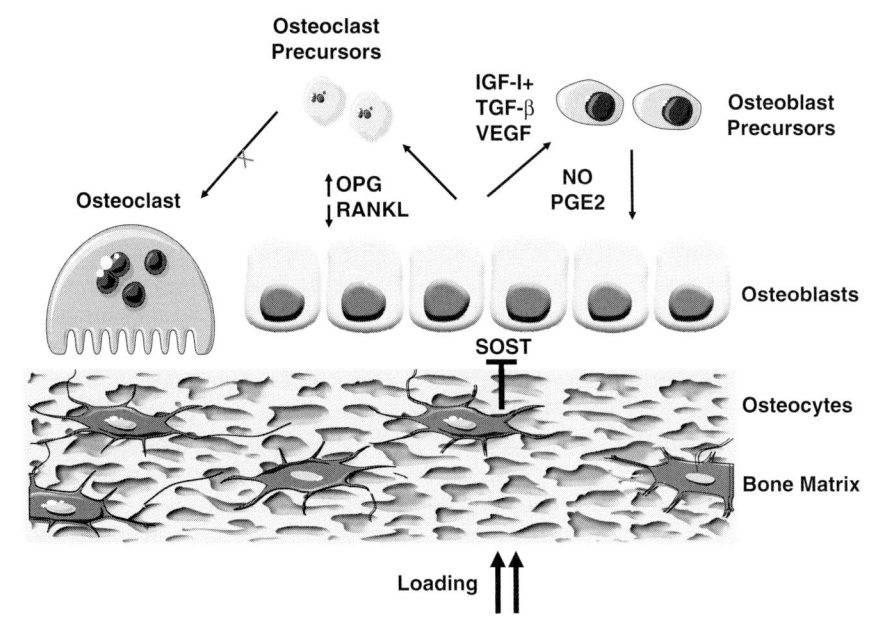

Fig. 8.3 Mechanisms involved in the response of osteoblasts to mechanical forces. Loading or mechanical forces induce increased expression of mediators (NO, PGE2) and growth factors (IGF-1, TGF-β, VEGF) by osteoblasts which in turn activate osteoblast replication, differentiation and function to promote bone formation. Mechanical forces also block the expression of the bone forming inhibitor sclerostin (SOST) by osteocytes, thus contributing to increase bone formation. Osteoblasts also respond to mechanical forces by decreasing RANKL and increased OPG expression, resulting in decreased osteoclastogenesis

cells. The organization of the cytoskeleton is an important link in the response to low-mechanical stimulation (Li et al. 2007; Jackson et al. 2008). Both microtubules and actin filaments may be involved in the cellular response to strain and are required for transduction of mechanical signals. Mechanical forces induce shape alterations in osteoblasts which are linked to signaling systems that regulate osteo-blast differentiation (Rubin et al. 2006). Mechanical stimulation in osteoblasts alters focal contacts and the cytoskeleton and induces tyrosine phosphorylation of several proteins linked to the cytoskeleton (Guignandon et al. 1997; Malone et al. 2007a). Consistently, disruption of the cytoskeleton abolishes the response to stress, suggesting that the cytoskeleton is involved in cellular mechanotransduction. Additionally, mechanical forces can control the switch between osteoblast/adipocyte differentiation since changes in cell shape or modulation of the cytoskeletal-related GTPase RhoA induce stem cell adipogenic or osteoblast differentiation (McBeath et al. 2004; Meyers et al. 2005). This effect may be related to MAPK-ERK activation downstream of the RhoA-ROCK signaling pathway (Khatiwala et al. 2009). Additionally, RhoA activation induced by mechanical signals may act by promoting Wnt signaling and Runx2 expression (Arnsdorf et al. 2009).

Integrins are transmembrane proteins that link extracellular matrix proteins to the cytoskeleton and control focal adhesion and cell adherence to the matrix. In bone, integrin–matrix interactions are important modulators of osteoblast differentiation induced by loading (Bikle 2008; Chen et al. 1999; Katsumi et al. 2004; Shyy and Chien 1997). Activation of integrins by stretch and shear forces in osteoblasts activates focal adhesion kinase (FAK) which leads to MAPK-ERK activation (Carvalho et al. 1998, 2002; Pavalko et al. 1998; Pommerenke et al. 2002; Salter et al. 1997) via interaction with c-src, Grb2 and Ras (Kapur et al. 2003; Weyts et al. 2002). In addition to ERK, strain induces activation of JNK, phospholipase C and PKC and intracellular calcium mobilization (Geng et al. 2001; You et al. 2001). Conversely, modeled microgravity results in disruption of collagen 1/integrin signaling during osteoblast differentiation (Meyers et al. 2004). The abnormal integrin-mediated signaling also results in disuse-induced osteoblast apoptosis whereas fluid shear stress inhibits osteoblast apoptosis through phosphatidyl-inositol-3 kinase (PI3K) activation in vitro (Pavalko et al. 2003b). Consistently, skeletal unloading decreases $\alpha 5\beta 1$ integrin expression and reduces PI3K activity in metaphyseal long bone, resulting in decreased levels of the survival protein Bcl-2 and osteoblast apoptosis in vivo (Dufour et al. 2008).

Besides the integrin–cytoskeletal system, several membrane proteins may be mechanosensitive in osteoblasts (Davidson et al. 1990). For example, stretch-sensitive channels are upregulated by chronic intermittent strain in osteoblasts, suggesting that these channels may act as a signal transducer for mechanical strain on osteoblastic cells. Ion channels are other possible mechanosensitive channels (Rawlinson et al. 1996; Rubin et al. 2006). Crosstalks exist between integrins and ion channels since FAK activation increases the activity of PLCγ1 which is involved in Ca^{2+}-fluxes (Rubin et al. 2006). Lipid rafts may also be involved in mechanosensing as these molecules can translocate signaling molecules induced by mechanical forces, resulting in activation of MAPKs (Kawamura et al. 2003). Additionally, glutamate receptors may be involved in the effects of strain in bone cells (Mason 2004). Mechanical loading modulates glutamate receptor subunit expression in bone (Szczesniak et al. 2005), and glutamate receptors are therefore likely to play a role in mechanotransduction in osteoblasts (Skerry and Genever 2001). The estrogen receptor may interact with mechanical strain and modulate the response to microgravity (Aguirre et al. 2007; Jagger et al. 1996; Jessop et al. 2001; Zaman et al. 2000) as discussed elsewhere in this book. Connexins are other molecules that are responsive to mechanical forces. Shear stress increases connexin expression in osteoblasts, resulting in increased communication between the cells (Rubin et al. 2006; Ziambaras et al. 1998). Connexins are required for mechanotransduction in osteoblasts since conditional ablation of the connexin 43 gene attenuates the response to mechanical loading in mice (Grimston et al. 2006, 2008). Transmission of soluble molecules via connexins–gap junctions may therefore be an important mechanism by which osteoblasts can synchronously respond to mechanical forces (Grimston et al. 2006). Finally, ciliated structures that are expressed by osteoblasts and osteocytes and that control osteoblast differentiation (Xiao et al. 2006) may be involved in the response of osteoblasts to mechanical forces (Malone et al. 2007b).

8.4 Role of Wnt Signaling in Mechanotransduction in Osteoblasts

There is accumulating evidence that mechanical loading upregulates the Wnt signaling pathway. It was first discovered that Wnt co-receptor LRP5 is essential for skeletal mechanotransduction (Sawakami et al. 2006). Consistently, Wnt signaling and β-catenin translocation to the nucleus are activated by mechanical stimulation in vitro and in vivo (Armstrong et al. 2007; Hens et al. 2005; Norvell et al. 2004), indicating that the Wnt/β-catenin signaling is a normal physiological response to mechanical loading in bone (Robinson et al. 2006). Additionally, mechanical strain activates Akt and inactivates GSK3β, resulting in β-catenin nuclear translocation (Case et al. 2008) and prevention of adipogenic differentiation in mesenchymal stem cells (Sen et al. 2008). Thus, β-catenin appears to be an important signal involved in the increased osteogenic response to mechanical stimulation. This pathway may be modulated by Wnt antagonists in response to loading. One important Wnt signaling modulator called sclerostin, the protein product of the Sost gene, is produced by osteocytes and is modulated by mechanical strain. Loading reduces Sost whereas unloading increases Sost expression by osteocytes in vivo (Robling et al. 2008). Since sclerostin is a potent inhibitor of bone formation by antagonizing Lrp5 receptor signaling, this provides a mechanism by which mechanical forces control osteoblastogenesis by modulating Wnt signaling. Recent studies indicate that the noncanonical Wnt signaling is also responsive to mechanical stimuli. Fluid flow upregulates Wnt5a and Ror2 that are required in RhoA-mediated Runx2 expression, and alters the association of N-cadherin with β-catenin, indicating that noncanonical Wnt signaling and N-cadherin related beta-catenin signaling play a role in mechanically induced osteogenic cell fate (Arnsdorf et al. 2009). Although these studies clearly reveal an important role of Wnt signaling in response to mechanical forces, it is likely that crosstalks between this and other pathways are involved in the control of osteoblastogenesis by the mechanical environment (Fig. 8.4).

8.5 Mechanoresponsive Genes in Osteoblasts

Mechanical forces exert multiple effects on signaling pathways in osteoblastic cells, resulting in activation of numerous target genes, including transcription factors, soluble molecules, growth factors and matrix proteins (Marie et al. 2000; Turner et al. 2009).

8.5.1 Transcription Factors

Several transcription factors are activated following mechanical activation in osteoblasts. Activation of MAPKs by mechanical forces leads to increased expression of c-Fos and c-Jun which are components of the activator protein (AP-1) transcription

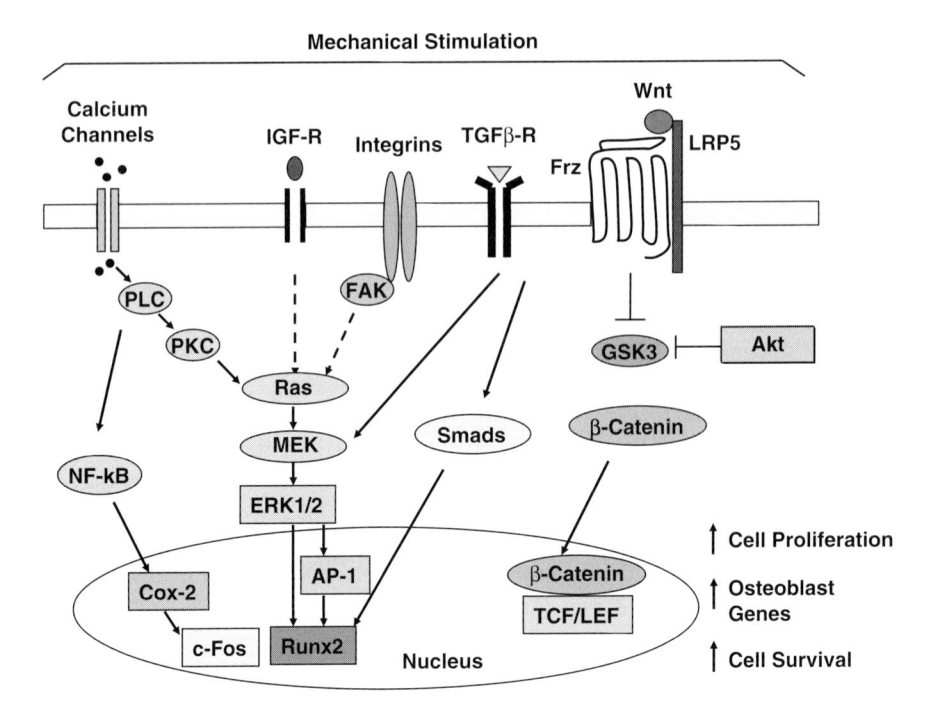

Fig. 8.4 Mechanoreceptors and signaling mechanisms involved in mechanotransduction in osteoblasts. Osteoblasts may respond to mechanical stimulation via multiple receptors that transduce the response via activation of several kinases, resulting in activation of transcription factors regulating osteoblast proliferation, differentiation and survival (see text for details)

factor (Granet et al. 2001; Kletsas et al. 2002; Peake et al. 2000). Other transcription factors such as egr-1, NFkB c-jun, junB, junD, Fra-1 and Fra-2 are activated by strain (Chen et al. 2003; Granet et al. 2002; Inoue et al. 2004; Nomura and Takano-Yamamoto 2000). Mechanical forces also cause ERK-mediated Runx2 phosphorylation (Ziros et al. 2002), and Runx2 is required for the normal response to loading in vivo (Salingcarnboriboon et al. 2006), indicating that Runx2 is a critical gene targeted by loading and unloading. A role for p53 in osteoblast apoptosis induced by unloading is provided by the observation that skeletal unloading does not induce bone loss in p53-deficient mice (Sakai et al. 2002). Thus, the changes in the expression and activity of these transcription factors may contribute to the changes in osteoblast proliferation, differentiation and survival induced by mechanical forces, resulting in bone formation in response to loading (Fig. 8.3).

8.5.2 Soluble Mediators

The activation of mechanoreceptors results in the release of several soluble molecules that mediate the cell response to strain and mechanical forces

(Burger and Klein-Nulend 1998; McAllister and Frangos 1999; Rubin et al. 2006; Smalt et al. 1997; Zaman et al. 1999) although the response depends on the force applied (Mullender et al. 2006). The response of bone cells to fluid flow or strain includes expression of inducible cycloxygenase COX-2 in osteoblasts (Wadhwa et al. 2002). This effect is dependent on cytoskeleton–integrin interactions (Pavalko et al. 1998) and occurs via an ERK-signaling pathway. This results in prostaglandin (PG) synthesis (Forwood and Turner 1995; Genetos et al. 2005; El Haj et al. 1990; Reich and Frangos 1991; Yeh and Rodan 1984) and is associated with DNA synthesis and matrix synthesis (Ajubi et al. 1999; Klein-Nulend et al. 1997; Rawlinson et al. 1991; Zaman et al. 1997). In vivo, loading also induces PG production, suggesting that PGs are involved in strain-induced osteogenesis (Rawlinson et al. 1991). The PGE_2-mediated cAMP increase leads to enhancement of connexin 43 expression (Carvalho et al. 1994; Cherian et al. 2005; Jalali et al. 2001; Li et al. 1999; Ziambaras et al. 1998). Inhibition of COX-2, the key enzyme in the formation of PGs, prevents mechanically induced bone formation in vivo, suggesting a major role of COX-2 and PGs in maintaining skeletal integrity. Strain also increases intracellular levels of inositol triphosphate (IP3), an effect that is partly dependent on PG synthesis (Reich and Frangos 1991). Nitric oxide (NO) is also produced by osteoblasts in response to mechanical stimulation (Klein-Nulend et al. 1995; Fan et al. 2004; Rubin et al. 2006) as a result of activation of endothelial nitric oxide synthase (ecNOS). The sequence of events in response to mechanical strain in osteoblasts appears to involve a rise in calcium levels which precedes activation of PKA, PKC, increased IP3 and c-fos and COX-2 transcription resulting in the production of PGE2, intracellular cAMP levels and downstream target molecules in osteoblasts (Jones et al. 1995) (Fig. 8.3). However, the actual intracellular signaling pathways that are activated by mechanical loading in physiological strain conditions in vivo remain to be identified.

Mechanical forces applied on osteoblastic cells release soluble factors that control osteoclastogenesis. Hindlimb suspension causes increased IL-6 secretion which may result in enhanced osteoclastogenesis (Grano et al. 2002). Mechanical forces may also modulate osteoclastogenesis via changes in the production of RANKL and OPG, two important molecules that are essential for osteoclast differentiation and function (Rubin et al. 2002, 2006). Modeled microgravity stimulates osteoclastogenesis by increasing RANKL/OPG (Rucci et al. 2007). Conversely, mechanical loading inhibits the expression of RANKL in marrow stromal cells by osteoblasts, causing reduction in osteoclast formation (Fan et al. 2004; Kanematsu et al. 2002; Rubin et al. 2000). However, mature osteoclasts per se seem to respond to increased mechano-stimulation by enhancing their bone-resorbing potential (Rubin et al. 2000) perhaps linked to increased RANKL/OPG ratio by more mature osteoblasts. In vivo, however, the administration of OPG in tail-suspended mice reduces bone loss by inhibiting bone resorption (Bateman et al. 2000). The modulation of OPG/RANKL ratio may thus contribute to the alterations of osteoclast formation and bone resorption induced by loading and unloading (Fig. 8.2).

8.5.3   Growth Factors

Several growth factors produced by osteoblasts were found to be modulated by microgravity and mechanical forces. Mechanical stimulation of cultured osteoblasts increases the expression of IGFs (Cheng et al. 2002; Rawlinson et al. 1993; Zaman et al. 1997) and TGFβ in vitro (Klein-Nulend et al. 1995). These factors may thus act as paracrine/autocrine factors on bone cells in response to loading or unloading. Skeletal unloading causes decreased expression of IGF-I in bone (Drissi et al. 1999; Zhang et al. 1995), and this effect correlates well with the alterations of osteoblast markers and osteoblast activity in unloaded metaphyseal bone (Drissi et al. 1999). Unloading also induces a transient decrease in IGF-I receptor expression (Drissi et al. 1999), indicating that the IGF-I/IGF-IR signaling plays an important role in the temporal control of bone formation during skeletal unloading in rats (Marie and Zerath 2000). Furthermore, skeletal unloading was found to alter TGF-β2 expression in rats (Zhang et al. 1995). In addition to this effect, we found that skeletal unloading induces a transient decrease in TGF-β type II receptor expression in bone, suggesting that defective TGF-β signaling mediates some effects of unloading on osteoblasts and bone formation in vivo (Drissi et al. 1999). Additional evidence for an important role of growth factors in the alterations of bone formation induced by skeletal unloading comes from our findings that the administration of IGF-I at therapeutic doses can prevent the defective osteoblast activity and number and the altered metaphyseal bone formation in unloaded rats (Machwate et al. 1994, 1995). IGF-I supplementation increases both the recruitment and differentiation of osteoblastic cells derived from unloaded bone marrow stroma (Machwate et al. 1994; Yamaguchi and Kishi 1994). Interestingly, IGF-I supplementation had no effect in normally loaded bone, because IGF-I expression by metaphyseal osteoblasts is initially reduced by unloading (Drissi et al. 1999). The administration of TGF-β2 can also prevent the defective bone formation and bone loss induced by unloading (Machwate et al. 1995). Exogenous TGF-β2 increases bone formation by increasing osteoblast recruitment, Runx2 expression and collagen synthesis in unloaded bone (Machwate et al. 1995). In contrast, systemic administration of BMP-2 in unloaded rats only slightly increased bone formation. However, this effect was not sufficient to prevent the trabecular bone loss induced by unloading (Zerath et al. 1998). In addition to promoting osteoblastogenesis in unloaded bone, TGF-β2 administration can increase PPARγ phosphorylation and inhibit adipocyte differentiation of bone marrow stromal cells through MAPK phosphorylation (Ahdjoudj et al. 2002, 2005). Exogenous TGF-β2 can also protect osteoblast apoptosis induced by skeletal unloading by targeting α5β1/PI3K/Akt signaling cascade and downstream Bcl-2 and phospho-Bad survival proteins (Dufour et al. 2008). Thus, restoration of the defective local TGF-β2 concentrations in unloaded bone can restore the imbalance between adipogenesis and osteoblastogenesis and protect against apoptosis in unloaded osteoblasts, which results in prevention of the altered bone formation induced by skeletal unloading (Fig. 8.2). It is, however, likely that a combination of growth factors and signaling pathways contribute to the anabolic response to mechanical forces in vivo (Lau et al. 2006).

There is also evidence that mechanical forces can affect VEGF production by osteoblasts, an important factor involved in angiogenesis. Mechanical signals modulate vascular endothelial growth factor expression in osteoblastic cells (Faure et al. 2008; Motokawa et al. 2005). Mechanically induced selection of soluble or matrix-bound VEGF production may modify osteoblast and endothelial cell cross-talks crucial during osteogenesis and fracture healing. This may have important implications in vivo since the increased angiogenesis and bone mass in response to exercise depends on VEGF expression (Yao et al. 2004).

8.5.4 Matrix Proteins

Several matrix proteins are affected by mechanical forces in osteoblasts (Harter et al. 1995). Mechanical forces induces type I collagen expression by osteoblasts by acting via activation of transcription factors or indirectly through the effects of growth factors (Nomura and Takano-Yamamoto 2000). Accordingly, skeletal unloading reduces type I collagen expression and OC synthesis in rats (Machwate et al. 1994; Drissi et al. 1999). Osteopontin is also targeted by mechanical forces. Mechanical stimuli increase osteopontin expression in cultured osteoblasts and in vivo (Terai et al. 1999; Zhang et al. 1995). The increased osteopontin expression is mediated by activation of integrins (Carvalho et al. 2002) or intracellular calcium, resulting in activation of ERK and p38 MAPK in osteoblasts (You et al. 2001). Consistently, skeletal unloading reduces osteopontin expression in vivo (Toma et al. 1997). Osteopontin is required for the effect of mechanical strain in vivo since osteopontin-deficient mice do not show loose bone in response to unloading. This may be due to the important role of osteopontin in osteoclastic bone resorption and osteoblast adhesion (Ishijima et al. 2001).

8.6 Conclusions

There is accumulating evidence suggesting that mechanical forces may directly or indirectly affect osteoblast proliferation, differentiation and survival by activating multiple mechanosensing receptors. Among the signaling pathways that may mediate the response to loading in osteoblasts, the Wnt signaling pathway plays a major role in controlling osteoblast fate, number and function in response to loading. Despite a better understanding of the cellular and molecular mechanisms involved in mechanotransduction in osteoblasts, the actual intracellular signaling pathways that are activated by mechanical load have not been definitely identified in physiological strain conditions in vivo. A major future challenge is therefore to determine the physiological mechanisms that are influenced by mechanical forces in osteoblasts in vivo, which may lead to the development of novel therapeutic strategies for the defective bone formation and bone loss in disuse osteoporosis.

Acknowledgments Due to space limitations, only a selected number of references on the subject could be quoted in this chapter. The reader is invited to read the indicated reviews for a larger selection of papers related to the subject. The author's work on skeletal unloading was in part supported by the French National Spatial Agency (CNES, Paris, France).

References

Aguirre JI, Plotkin LI, Stewart SA, Weinstein RS, Parfitt AM, Manolagas SC, Bellido T (2006) Osteocyte apoptosis is induced by weightlessness in mice and precedes osteoclast recruitment and bone loss. J Bone Miner Res 21(4):605–615

Aguirre JI, Plotkin LI, Gortazar AR, Millan MM, O'Brien CA, Manolagas SC, Bellido T (2007) A novel ligand-independent function of the estrogen receptor is essential for osteocyte and osteoblast mechanotransduction. J Biol Chem 282(35):25501–25508

Ahdjoudj S, Lasmoles F, Holy X, Zérath E, Marie PJ (2002) Transforming growth factor inhibits adipocyte differentiation induced by skeletal unloading in rat bone marrow stroma. J Bone Miner Res 17(4):668–677

Ahdjoudj S, Fromigué O, Marie PJ (2004) Plasticity and regulation of human bone marrow stromal cells:potential implication in the treatment of age-related bone loss. Histol Histopathol 19:151–157

Ahdjoudj S, Kaabeche K, Holy X, Fromigué O, Modrowski D, Zérath E, Marie PJ (2005) Transforming growth factor-beta inhibits CCAAT/enhancer-binding protein expression and PPARgamma activity in unloaded bone marrow stromal cells. Exp Cell Res 303(1):138–147

Ajubi NE, Klein-Nulend J, Alblas MJ, Burger EH, Nijweide PJ (1999) Signal transduction pathways involved in fluid flow-induced PGE2 production by cultured osteocytes. Am J Physiol 276(1 Pt 1):E171–E178

Armstrong VJ, Muzylak M, Sunters A, Zaman G, Saxon LK, Price JS, Lanyon LE (2007) Wnt/beta-catenin signaling is a component of osteoblastic bone cell early responses to load-bearing and requires estrogen receptor alpha. J Biol Chem 282(28):20715–20727

Arnsdorf EJ, Tummala P, Jacobs CR (2009) Non-canonical Wnt signaling and N-cadherin related beta-catenin signaling play a role in mechanically induced osteogenic cell fate. PLoS One 4(4):e5388

Aubin JE (2001) Regulation of osteoblast formation and function. Rev Endocr Metab Disord 2(1):81–94

Bancroft GN, Sikavitsas VI, van den Dolder J, Sheffield TL, Ambrose CG, Jansen JA, Mikos AG (2002) Fluid flow increases mineralized matrix deposition in 3D perfusion culture of marrow stromal osteoblasts in a dose-dependent manner. Proc Natl Acad Sci USA 99(20):12600–12605

Barou O, Palle S, Vico L, Alexandre C, Lafage-Proust MH (1998) Hindlimb unloading in rat decreases preosteoblast proliferation assessed in vivo with BrdU incorporation. Am J Physiol 274(1 Pt 1):E108–E114

Basso N, Heersche JN (2002) Characteristics of in vitro osteoblastic cell loading models. Bone 30(2):347–351

Bateman TA, Dunstan CR, Ferguson VL, Lacey DL, Ayers RA, Simske SJ (2000) Osteoprotegerin mitigates tail suspension-induced osteopenia. Bone 26(5):443–449

Bikle DD (2008) Integrins, insulin like growth factors, and the skeletal response to load. Osteoporos Int 19(9):1237–1246

Bikle DD, Halloran BP (1999) The response of bone to unloading. J Bone Miner Metab 17(4):233–244

Bikle DD, Harris J, Halloran BP, Morey-Holton E (1994) Altered skeletal pattern of gene expression in response to spaceflight and hindlimb elevation. Am J Physiol 267(6 Pt 1):E822–E827

Bonewald LF, Johnson ML (2008) Osteocytes, mechanosensing and Wnt signaling. Bone 42(4):606–615

Boppart MD, Kimmel DB, Yee JA, Cullen DM (1998) Time course of osteoblast appearance after in vivo mechanical loading. Bone 5:409–415

Boutahar N, Guignandon A, Vico L, Lafage-Proust MH (2004) Mechanical strain on osteoblasts activates autophosphorylation of focal adhesion kinase and proline-rich tyrosine kinase 2 tyrosine sites involved in ERK activation. J Biol Chem 279(29):30588–30599

Burger EH, Klein-Nulend J (1998) Microgravity and bone cell mechanosensitivity. Bone 22 (5 Suppl):127S–130S

Burr DB, Robling AG, Turner CH (2002) Effects of biomechanical stress on bones in animals. Bone 30(5):781–786

Capulli M, Rufo A, Teti A, Rucci N (2009) Global transcriptome analysis in mouse calvarial osteoblasts highlights sets of genes regulated by modeled microgravity and identifies a "mechanoresponsive osteoblast gene signature". J Cell Biochem 107(2):240–252

Carvalho RS, Scott JE, Suga DM, Yen EH (1994) Stimulation of signal transduction pathways in osteoblasts by mechanical strain potentiated by parathyroid hormone. J Bone Miner Res 9(7):999–1011

Carvalho RS, Schaffer JL, Gerstenfeld LC (1998) Osteoblasts induce osteopontin expression in response to attachment on fibronectin: demonstration of a common role for integrin receptors in the signal transduction processes of cell attachment and mechanical stimulation. J Cell Biochem 70(3):376–390

Carvalho RS, Bumann A, Schaffer JL, Gerstenfeld LC (2002) Predominant integrin ligands expressed by osteoblasts show preferential regulation in response to both cell adhesion and mechanical perturbation. J Cell Biochem 84(3):497–508

Case N, Ma M, Sen B, Xie Z, Gross TS, Rubin J (2008) Beta-catenin levels influence rapid mechanical responses in osteoblasts. J Biol Chem 283(43):29196–29205

Chen KD, Li YS, Kim M, Li S, Yuan S, Chien S, Shyy JY (1999) Mechanotransduction in response to shear stress. Roles of receptor tyrosine kinases, integrins, and Shc. J Biol Chem 274(26):18393–18400

Chen NX, Geist DJ, Genetos DC, Pavalko FM, Duncan RL (2003) Fluid shear-induced NFkappaB translocation in osteoblasts is mediated by intracellular calcium release. Bone 33(3):399–410

Cheng MZ, Rawlinson SC, Pitsillides AA, Zaman G, Mohan S, Baylink DJ, Lanyon LE (2002) Human osteoblasts' proliferative responses to strain and 17beta-estradiol are mediated by the estrogen receptor and the receptor for insulin-like growth factor I. J Bone Miner Res 17(4):593–602

Cherian PP, Siller-Jackson AJ, Gu S, Wang X, Bonewald LF, Sprague E, Jiang JX (2005) Mechanical strain opens connexin 43 hemichannels in osteocytes: a novel mechanism for the release of prostaglandin. Mol Biol Cell 16(7):3100–3106

Cowin SC (1998) On mechanosensation in bone under microgravity. Bone 22:119S–125S

David V, Lafage-Proust MH, Laroche N, Christian A, Ruegsegger P, Vico L (2006) Two-week longitudinal survey of bone architecture alteration in the hindlimb-unloaded rat model of bone loss: sex differences. Am J Physiol Endocrinol Metab 290(3):E440–E447

David V, Martin A, Lafage-Proust MH, Malaval L, Peyroche S, Jones DB, Vico L, Guignandon A (2007) Mechanical loading down-regulates peroxisome proliferator-activated receptor gamma in bone marrow stromal cells and favors osteoblastogenesis at the expense of adipogenesis. Endocrinology 148(5):2553–2562

Davidson RM, Tatakis DW, Auerbach AL (1990) Multiple forms of mechanosensitive ion channels in osteoblast-like cells. Pflugers Arch 416(6):646–651

Donahue HJ (2000) Gap junctions and biophysical regulation of bone cell differentiation. Bone 26(5):417–422

Drissi H, Lomri A, Lasmoles F, Holy X, Zerath E, Marie PJ (1999) Skeletal unloading induces biphasic changes in insulin like growth factor-I mRNA levels and osteoblast activity. Exp Cell Res 251:275–284

Dufour C, Holy X, Marie PJ (2007) Skeletal unloading induces osteoblast apoptosis and targets alpha5beta1-PI3K-Bcl-2 signaling in rat bone. Exp Cell Res 313(2):394–403

Dufour C, Holy X, Marie PJ (2008) Transforming growth factor-beta prevents osteoblast apoptosis induced by skeletal unloading via PI3K/Akt, Bcl-2, and phospho-Bad signaling. Am J Physiol Endocrinol Metab 294(4):E794–E801

Duncan RL, Turner CH (1995) Mechanotransduction and the functional response of bone to mechanical strain. Calcif Tissue Int 57:344–358

El Haj AJ, Minter SL, Rawlinson SCF, Suswillo R, Lanyon LE (1990) Cellular responses to mechanical loading in vitro. J Bone Miner Res 5:923–932

Fan X, Roy E, Zhu L, Murphy TC, Ackert-Bicknell C, Hart CM, Rosen C, Nanes MS, Rubin J (2004) Nitric oxide regulates receptor activator of nuclear factor-kappaB ligand and osteoprotegerin expression in bone marrow stromal cells. Endocrinology 145(2):751–759

Forwood MR, Turner CH (1995) Skeletal adaptations to mechanical usage: results from tibial loading studies in rats. Bone 17(4 Suppl):197S–205S

Faure C, Linossier MT, Malaval L, Lafage-Proust MH, Peyroche S, Vico L, Guignandon A (2008) Mechanical signals modulated vascular endothelial growth factor-A (VEGF-A) alternative splicing in osteoblastic cells through actin polymerisation. Bone 42(6):1092–1101

Frost HM (2003) Bone's mechanostat: a 2003 update. Anat Rec A Discov Mol Cell Evol Biol 275(2):1081–1101

Genetos DC, Geist DJ, Liu D, Donahue HJ, Duncan RL (2005) Fluid shear-induced ATP secretion mediates prostaglandin release in MC3T3-E1 osteoblasts. J Bone Miner Res 20(1):41–49

Geng WD, Boskovic G, Fultz ME, Li C, Niles RM, Ohno S, Wright GL (2001) Regulation of expression and activity of four PKC isozymes in confluent and mechanically stimulated UMR-108 osteoblastic cells. J Cell Physiol 189(2):216–228

Globus RK, Bikle DD, Morey-Holton E (1986) The temporal response of bone to unloading. Endocrinology 118(2):733–742

Granet C, Boutahar N, Vico L, Alexandre C, Lafage-Proust MH (2001) MAPK and SRC-kinases control EGR-1 and NF-kappa B inductions by changes in mechanical environment in osteoblasts. Biochem Biophys Res Commun 284(3):622–631

Granet C, Vico AG, Alexandre C, Lafage-Proust MH (2002) MAP and src kinases control the induction of AP-1 members in response to changes in mechanical environment in osteoblastic cells. Cell Signal 14(8):679–688

Grano M, Mori G, Minielli V, Barou O, Colucci S, Giannelli G, Alexandre C, Zallone AZ, Vico L (2002) Rat hindlimb unloading by tail suspension reduces osteoblast differentiation, induces IL-6 secretion, and increases bone resorption in ex vivo cultures. Calcif Tissue Int 70(3):176–185

Grimston SK, Screen J, Haskell JH, Chung DJ, Brodt MD, Silva MJ, Civitelli R (2006) Role of connexin43 in osteoblast response to physical load. Ann N Y Acad Sci 1068:214–224

Grimston SK, Brodt MD, Silva MJ, Civitelli R (2008) Attenuated response to in vivo mechanical loading in mice with conditional osteoblast ablation of the connexin43 gene (Gja1). J Bone Miner Res 23(6):879–886

Guignandon A, Usson Y, Laroche N, Lafage-Proust MH, Sabido O, Alexandre C, Vico L (1997) Effects of intermittent or continuous gravitational stresses on cell- matrix adhesion: quantitative analysis of focal contacts in osteoblastic ROS 17/2.8 cells. Exp Cell Res 236:66–75

Harter LV, Hruska KA, Duncan RL (1995) Human osteoblast-like cells respond to mechanical strain with increased bone matrix protein production independent of hormonal regulation. Endocrinology 136(2):528–535

Hens JR, Wilson KM, Dann P, Chen X, Horowitz MC, Wysolmerski JJ (2005) TOPGAL mice show that the canonical Wnt signaling pathway is active during bone development and growth and is activated by mechanical loading in vitro. J Bone Miner Res 20(7):1103–1113

Hillam RA, Skerry TM (1995) Inhibition of bone resorption and stimulation of formation by mechanical loading of the modeling rat ulna in vivo. J Bone Miner Res 10(5):683–689

Hino K, Nakamoto T, Nifuji A, Morinobu M, Yamamoto H, Ezura Y, Noda M (2007) Deficiency of CIZ, a nucleocytoplasmic shuttling protein, prevents unloading-induced bone loss through the enhancement of osteoblastic bone formation in vivo. Bone 40(4):852–860

Huiskes R, Ruimerman R, van Lenthe GH, Janssen JD (2000) Effects of mechanical forces on maintenance and adaptation of form in trabecular bone. Nature 405(6787):704–706

Inoue D, Kido S, Matsumoto T (2004) Transcriptional induction of FosB/DeltaFosB gene by mechanical stress in osteoblasts. J Biol Chem 279(48):49795–49803

Ishijima M, Rittling SR, Yamashita T, Tsuji K, Kurosawa H, Nifuji A, Denhardt DT, Noda M (2001) Enhancement of osteoclastic bone resorption and suppression of osteoblastic bone formation in response to reduced mechanical stress do not occur in the absence of osteopontin. J Exp Med 193(3):399–404

Jackson WM, Jaasma MJ, Tang RY, Keaveny TM (2008) Mechanical loading by fluid shear is sufficient to alter the cytoskeletal composition of osteoblastic cells. Am J Physiol Cell Physiol 295(4):C1007–C1015

Jagger CJ, Chow JW, Chambers TJ (1996) Estrogen suppresses activation but enhances formation phase of osteogenic response to mechanical stimulation in rat bone. J Clin Invest 98(10):2351–2357

Jalali S, del Pozo MA, Chen K, Miao H, Li Y, Schwartz MA, Shyy JY, Chien S (2001) Integrin-mediated mechanotransduction requires its dynamic interaction with specific extracellular matrix (ECM) ligands. Proc Natl Acad Sci USA 98(3):1042–1046

Jessop HL, Sjoberg M, Cheng MZ, Zaman G, Wheeler-Jones CP, Lanyon LE (2001) Mechanical strain and estrogen activate estrogen receptor alpha in bone cells. J Bone Miner Res 16(6):1045–1055

Jones D, Leivseth G, Tenbosch J (1995) Mechano-reception in osteoblast-like cells. Biochem Cell Biol 73:525–534

Kanematsu M, Yoshimura K, Takaoki M, Sato A (2002) Vector-averaged gravity regulates gene expression of receptor activator of NF-kappaB (RANK) ligand and osteoprotegerin in bone marrow stromal cells via cyclic AMP/protein kinase A pathway. Bone 30(4):553–558

Kapur S, Baylink DJ, Lau KH (2003) Fluid flow shear stress stimulates human osteoblast proliferation and differentiation through multiple interacting and competing signal transduction pathways. Bone 3:241–251

Kaspar D, Seidl W, Neidlinger-Wilke C, Beck A, Claes L, Ignatius A (2002) Proliferation of human-derived osteoblast-like cells depends on the cycle number and frequency of uniaxial strain. J Biomech 35(7):873–880

Katsumi A, Orr AW, Tzima E, Schwartz MA (2004) Integrins in mechanotransduction. J Biol Chem 279(13):12001–12004

Kawamura S, Miyamoto S, Brown JH (2003) Initiation and transduction of stretch-induced RhoA and Rac1 activation through caveolae: cytoskeletal regulation of ERK translocation. J Biol Chem 278(33):31111–31117

Keila S, Pitaru S, Grosskopf A, Weinreb M (1994) Bone marrow from mechanically unloaded rat bones expresses reduced osteogenic capacity in vitro. J Bone Miner Res 9(3):321–327

Khatiwala CB, Kim PD, Peyton SR, Putnam AJ (2009) ECM compliance regulates osteogenesis by influencing MAPK signaling downstream of RhoA and ROCK. J Bone Miner Res 24(5):886–898

Klein-Nulend J, Roelofsen J, Sterck JG, Semeins CM, Burger EH (1995) Mechanical loading stimulates the release of transforming growth factor- beta activity by cultured mouse calvariae and periosteal cells. J Cell Physiol 163:115–119

Klein-Nulend J, Burger EH, Semeins CM, Raisz LG, Pilbeam CC (1997) Pulsating fluid flow stimulates prostaglandin release and inducible prostaglandin G/H synthase mRNA expression in primary mouse bone cells. J Bone Miner Res 12:45–51

Kletsas D, Basdra EK, Papavassiliou AG (2002) Effect of protein kinase inhibitors on the stretch-elicited c-Fos and c-Jun up-regulation in human PDL osteoblast-like cells. J Cell Physiol 190(3):313–321

Kostenuik PJ, Halloran BP, Morey-Holton ER, Bikle DD (1997) Skeletal unloading inhibits the in vitro proliferation and differentiation of rat osteoprogenitor cells. Am J Physiol 273(6 Pt 1): E1133–E1139

Lau KH, Kapur S, Kesavan C, Baylink DJ (2006) Up-regulation of the Wnt, estrogen receptor, insulin-like growth factor-I, and bone morphogenetic protein pathways in C57BL/6J osteo-blasts as opposed to C3H/HeJ osteoblasts in part contributes to the differential anabolic response to fluid shear. J Biol Chem 281(14):9576–9588

Li Z, Zhou Z, Yellowley CE, Donahue HJ (1999) Inhibiting gap junctional intercellular communication alters expression of differentiation markers in osteoblastic cells. Bone 25(6):661–666

Li J, Chen G, Zheng L, Luo S, Zhao Z (2007) Osteoblast cytoskeletal modulation in response to compressive stress at physiological levels. Mol Cell Biochem 304(1–2):45–52

Machwate M, Zerath E, Holy X, Hott H, Modrowski D, Malouvier A, Marie PJ (1993) Skeletal unloading in rat decreases proliferation of rat bone and marrow-derived osteoblastic cells. Am J Physiol Endocrinol Metab 27:E790–E799

Machwate M, Zerath E, Holy X, Hott M, Pastoureau P, Marie PJ (1994) Insulin-like growth factor-I increases trabecular bone formation and osteoblastic cell proliferation in unloaded rats. Endocrinology 134(3):1031–1038

Machwate M, Zerath E, Holy X, Hott M, GodetD LA, Marie PJ (1995) Systemic administration of transforming growth factor-β 2 prevents the impaired bone formation and osteopenia induced by unloading in rats. J Clin Invest 96:1245–1253

Malone AM, Batra NN, Shivaram G, Kwon RY, You L, Kim CH, Rodriguez J, Jair K, Jacobs CR (2007a) The role of actin cytoskeleton in oscillatory fluid flow-induced signaling in MC3T3-E1 osteoblasts. Am J Physiol Cell Physiol 292(5):C1830–1836

Malone AM, Anderson CT, Tummala P, Kwon RY, Johnston TR, Stearns T, Jacobs CR (2007b) Primary cilia mediate mechanosensing in bone cells by a calcium-independent mechanism. Proc Natl Acad Sci USA 104(33):13325–13330

Marie PJ (2008) Transcription factors controlling osteoblastogenesis. Arch Biochem Biophys 473(2):98–105

Marie PJ, Kaabeche K (2006) PPAR gamma activity and control of bone mass in skeletal unloading. PPAR Res 2006:64807

Marie PJ, Zerath E (2000) Role of growth factors in osteoblast alterations induced by skeletal unloading in rats. Growth Factors 18(1):1–10

Marie PJ, Jones D, Vico L, Zallone A, Hinsenkamp M, Cancedda R (2000) Osteobiology, strain and microgravity. Part I: Studies at the cellular level. Calcif Tissue Int 67(1):2–9

Mason DJ (2004) Glutamate signalling and its potential application to tissue engineering of bone. Eur Cell Mater 7:12–25

McAllister TN, Frangos JA (1999) Steady and transient fluid shear stress stimulate NO release in osteoblasts through distinct biochemical pathways. J Bone Miner Res 14(6):930–936

McBeath R, Pirone DM, Nelson CM, Bhadriraju K, Chen CS (2004) Cell shape, cytoskeletal tension, and RhoA regulate stem cell lineage commitment. Dev Cell 6(4):483–495

Meyers VE, Zayzafoon M, Gonda SR, Gathings WE, McDonald JM (2004) Modeled microgravity disrupts collagen I/integrin signaling during osteoblastic differentiation of human mesenchymal stem cells. J Cell Biochem 93(4):697–707

Meyers VE, Zayzafoon M, Douglas JT, McDonald JM (2005) RhoA and cytoskeletal disruption mediate reduced osteoblastogenesis and enhanced adipogenesis of human mesenchymal stem cells in modeled microgravity. J Bone Miner Res 20(10):1858–1866

Miles RR, Turner CH, Santerre R, Tu Y, McClelland P, Argot J, DeHoff BS, Mundy CW, Rosteck PR Jr, Bidwell J, Sluka JP, Hock J, Onyia JE (1998) Analysis of differential gene expression in rat tibia after an osteogenic stimulus in vivo: mechanical loading regulates osteopontin and myeloperoxidase. J Cell Biochem 68(3):355–365

Morey-Holton ER, Globus RK (1998) Hindlimb unloading of growing rats: a model for predicting skeletal changes during space flight. Bone 22(5 Suppl): 83S–88S

Motokawa M, Kaku M, Tohma Y, Kawata T, Fujita T, Kohno S, Tsutsui K, Ohtani J, Tenjo K, Shigekawa M, Kamada H, Tanne K (2005) Effects of cyclic tensile forces on the expression of vascular endothelial growth factor (VEGF) and macrophage-colony-stimulating factor (M-CSF) in murine osteoblastic MC3T3-E1 cells. J Dent Res 84(5): 422–427

Mullender MG, Dijcks SJ, Bacabac RG, Semeins CM, Van Loon JJ, Klein-Nulend J (2006) Release of nitric oxide, but not prostaglandin E2, by bone cells depends on fluid flow frequency. J Orthop Res 24(6): 1170–1177

Nomura S, Takano-Yamamoto T (2000) Molecular events caused by mechanical stress in bone. Matrix Biol 19(2): 91–96

Norvell SM, Ponik SM, Bowen DK, Gerard R, Pavalko FM (2004) Fluid shear stress induction of COX-2 protein and prostaglandin release in cultured MC3T3-E1 osteoblasts does not require intact microfilaments or microtubules. J Appl Physiol 96(3): 957–966

Owan I, Burr DB, Turner CH, Qiu J, Tu Y, Onyia JE, Duncan RL (1997) Mechanotransduction in bone: osteoblasts are more responsive to fluid forces than mechanical strain. Am J Physiol 273(3 Pt 1): C810–C815

Ozawa H, Imamura K, Abe E, Takahashi N, Hiraide T, Shibasaki Y, Fukuhara T, Suda T (1990) Effect of a continuously applied compressive pressure on mouse osteoblast-like cells (MC3T3-E1) in vitro. J Cell Physiol 142(1): 177–185

Pavalko FM, Chen NX, Turner CH, Burr DB, Atkinson S, Hsieh YF, Qiu J, Duncan RL (1998) Fluid shear-induced mechanical signaling in MC3T3-E1 osteoblasts requires cytoskeleton-integrin interactions. Am J Physiol 275(6 Pt 1): C1591–C1601

Pavalko FM, Norvell SM, Burr DB, Turner CH, Duncan RL, Bidwell JP (2003a) A model for mechanotransduction in bone cells: the load-bearing mechanosomes. J Cell Biochem 88(1): 104–112

Pavalko FM, Gerard RL, Ponik SM, Gallagher PJ, Jin Y, Norvell SM (2003b) Fluid shear stress inhibits TNF-alpha-induced apoptosis in osteoblasts: a role for fluid shear stress-induced activation of PI3-kinase and inhibition of caspase-3. J Cell Physiol 194(2): 194–205

Pavlin D, Zadro R, Gluhak-Heinrich J (2001) Temporal pattern of stimulation of osteoblast-associated genes during mechanically-induced osteogenesis in vivo: early responses of osteocalcin and type I collagen. Connect Tissue Res 42(2): 135–148

Peake MA, Cooling LM, Magnay JL, Thomas PB, El Haj AJ (2000) Selected contribution: regulatory pathways involved in mechanical induction of c-fos gene expression in bone cells. J Appl Physiol 89(6): 2498–2507

Pommerenke H, Schmidt C, Durr F, Nebe B, Luthen F, Muller P, Rychly J (2002) The mode of mechanical integrin stressing controls intracellular signaling in osteoblasts. J Bone Miner Res 17(4): 603–611

Rath B, Nam J, Knobloch TJ, Lannutti JJ, Agarwal S (2008) Compressive forces induce osteogenic gene expression in calvarial osteoblasts. J Biomech 41(5): 1095–1103

Rawlinson SC, el-Haj AJ, Minter SL, Tavares IA, Bennett A, Lanyon LE (1991) Loading-related increases in prostaglandin production in cores of adult canine cancellous bone in vitro: a role for prostacyclin in adaptive bone remodeling? J Bone Miner Res 6(12): 1345–1351

Rawlinson SC, Mohan S, Baylink DJ, Lanyon LE (1993) Exogenous prostacyclin, but not prostaglandin E2, produces similar responses in both G6PD activity and RNA production as mechanical loading, and increases IGF-II release, in adult cancellous bone in culture. Calcif Tissue Int 53(5): 324–329

Rawlinson SC, Mosley JR, Suswillo RF, Pitsillides AA, Lanyon LE (1995) Calvarial and limb bone cells in organ and monolayer culture do not show the same early responses to dynamic mechanical strain. J Bone Miner Res 10(8): 1225–1232

Rawlinson SC, Pitsillides AA, Lanyon LE (1996) Involvement of different ion channels in osteoblasts' and osteocytes' early responses to mechanical strain. Bone 19(6): 609–614

Reich KM, Frangos JA (1991) Effect of flow on prostaglandin E2 and inositol trisphosphate levels in osteoblasts. Am J Physiol 261(3 Pt 1): C428–C432

Robinson JA, Chatterjee-Kishore M, Yaworsky PJ, Cullen DM, Zhao W, Li C, Kharode Y, Sauter L, Babij P, Brown EL, Hill AA, Akhter MP, Johnson ML, Recker RR, Komm BS, Bex FJ (2006) Wnt/beta-catenin signaling is a normal physiological response to mechanical loading in bone. J Biol Chem 281(42):31720–31728

Robling AG, Castillo AB, Turner CH (2006) Biomechanical and molecular regulation of bone remodeling. Annu Rev Biomed Eng 8:455–498

Robling AG, Niziolek PJ, Baldridge LA, Condon KW, Allen MR, Alam I, Mantila SM, Gluhak-Heinrich J, Bellido TM, Harris SE, Turner CH (2008) Mechanical stimulation of bone in vivo reduces osteocyte expression of Sost/sclerostin. J Biol Chem 283(9):5866–5875

Rubin J, Murphy T, Nanes MS, Fan X (2000) Mechanical strain inhibits expression of osteoclast differentiation factor by murine stromal cells. Am J Physiol Cell Physiol 278(6):C1126–C1132

Rubin C, Turner AS, Bain S, Mallinckrodt C, Anabolism McLeod K (2001) Low mechanical signals strengthen long bones. Nature 412(6847):603–604

Rubin J, Murphy TC, Fan X, Goldschmidt M, Taylor WR (2002) Activation of extracellular signal-regulated kinase is involved in mechanical strain inhibition of RANKL expression in bone stromal cells. J Bone Miner Res 17(8):1452–1460

Rubin J, Rubin C, Jacobs CR (2006) Molecular pathways mediating mechanical signaling in bone. Gene 367:1–16

Rucci N, Rufo A, Alamanou M, Teti A (2007) Modeled microgravity stimulates osteoclastogenesis and bone resorption by increasing osteoblast RANKL/OPG ratio. J Cell Biochem 100(2):464–473

Ryder KD, Duncan RL (2000) Parathyroid hormone modulates the response of osteoblast-like cells to mechanical stimulation. Calcif Tissue Int 67(3):241–246

Sakai A, Sakata T, Tanaka S, Okazaki R, Kunugita N, Norimura T, Nakamura T (2002) Disruption of the p53 gene results in preserved trabecular bone mass and bone formation after mechanical unloading. J Bone Miner Res 17(1):119–127

Salingcarnboriboon R, Tsuji K, Komori T, Nakashima K, Ezura Y, Noda M (2006) Runx2 is a target of mechanical unloading to alter osteoblastic activity and bone formation in vivo. Endocrinology 147(5):2296–2305

Salter DM, Robb JE, Wright MO (1997) Electrophysiological responses of human bone cells to mechanical stimulation: evidence for specific integrin function in mechanotransduction. Bone Miner Res 12(7):1133–1141

Saunders MM, Taylor AF, Du C, Zhou Z, Pellegrini VD Jr, Donahue HJ (2006) Mechanical stimulation effects on functional end effectors in osteoblastic MG-63 cells. J Biomech 39(8):1419–1427

Sawakami K, Robling AG, Ai M, Pitner ND, Liu D, Warden SJ, Li J, Maye P, Rowe DW, Duncan RL, Warman ML, Turner CH (2006) The Wnt co-receptor LRP5 is essential for skeletal mechanotransduction but not for the anabolic bone response to parathyroid hormone treatment. J Biol Chem 281(33):23698–23711

Sen B, Xie Z, Case N, Ma M, Rubin C, Rubin J (2008) Mechanical strain inhibits adipogenesis in mesenchymal stem cells by stimulating a durable beta-catenin signal. Endocrinology 149(12):6065–6075

Shyy JY, Chien S (1997) Role of integrins in cellular responses to mechanical stress and adhesion. Curr Opin Cell Biol 9(5):707–713

Sikavitsas VI, Bancroft GN, Holtorf HL, Jansen JA, Mikos AG (2003) Mineralized matrix deposition by marrow stromal osteoblasts in 3D perfusion culture increases with increasing fluid shear forces. Proc Natl Acad Sci USA 100(25):14683–14688

Skerry TM (2008) The response of bone to mechanical loading and disuse: fundamental principles and influences on osteoblast/osteocyte homeostasis. Arch Biochem Biophys 473(2):117–123

Skerry TM, Genever PG (2001) Glutamate signalling in non-neuronal tissues. Trends Pharmacol Sci 22(4):174–181

Smalt R, Mitchell FT, Howard RL, Chambers TJ (1997) Induction of NO and prostaglandin E2 in osteoblasts by wall-shear stress but not mechanical strain. Am J Physiol 273(4 Pt 1): E751–E758

Szczesniak AM, Gilbert RW, Mukhida M, Anderson GI (2005) Mechanical loading modulates glutamate receptor subunit expression in bone. Bone 37(1):63–73

Tanabe Y, Koga M, Saito M, Matsunaga Y, Nakayama K (2004) Inhibition of adipocyte differentiation by mechanical stretching through ERK-mediated downregulation of PPARgamma2. J Cell Sci 117(Pt 16):3605–3614

Terai K, Takano-Yamamoto T, Ohba Y, Hiura K, Sugimoto M, Sato M, Kawahata H, Inaguma N, Kitamura Y, Nomura S (1999) Role of osteopontin in bone remodeling caused by mechanical stress. J Bone Miner Res 14(6):839–849

Toma CD, Ashkar S, Gray ML, Schaffer JL, Gerstenfeld LC (1997) Signal transduction of mechanical stimuli is dependent on microfilament integrity: identification of osteopontin as a mechanically induced gene in osteoblasts. J Bone Miner Res 12:1626–1636

Turner CH (1992) Functional determinants of bone structure: beyond Wolff's law of bone transformation. Bone 13(6):403–409

Turner CH (1998) Three rules for bone adaptation to mechanical stimuli. Bone 23(5):399–407

Turner CH, Owan I, Alvey T, Hulman J, Hock JM (1998) Recruitment and proliferative responses of osteoblasts after mechanical loading in vivo determined using sustained-release bromodeoxyuridine. Bone 22(5):463–469

Turner CH, Warden SJ, Bellido T, Plotkin LI, Kumar N, Jasiuk I, Danzig J, Robling AG (2009) Mechanobiology of the skeleton. Sci Signal 2(68):pt3

Vico L, Hinsenkamp M, Jones D, Marie PJ, Zallone A, Cancedda R (2001) Osteobiology, strain and microgravity. Part II: Studies at the tissue level. Calcif Tissue Int 68(1):1–10

Wadhwa S, Godwin SL, Peterson DR, Epstein MA, Raisz LG, Pilbeam CC (2002) Fluid flow induction of cyclo-oxygenase 2 gene expression in osteoblasts is dependent on an extracellular signal-regulated kinase signaling pathway. J Bone Miner Res 17(2):266–274

Wakley GK, Portwood JS, Turner RT (1992) Disuse osteopenia is accompanied by downregulation of gene expression for bone proteins in growing rats. Am J Physiol 263(6 Pt 1):E1029–E1034

Weyts FA, Li YS, van Leeuwen J, Weinans H, Chien S (2002) ERK activation and alpha v beta 3 integrin signaling through Shc recruitment in response to mechanical stimulation in human osteoblasts. J Cell Biochem 87(1):85–92

Xiao Z, Zhang S, Mahlios J, Zhou G, Magenheimer BS, Guo D, Dallas SL, Maser R, Calvet JP, Bonewald L, Quarles LD (2006) Cilia-like structures and polycystin-1 in osteoblasts/osteocytes and associated abnormalities in skeletogenesis and Runx2 expression. J Biol Chem 281(41):30884–30895

Yamaguchi M, Kishi S (1994) Differential effects of insulin and insulin-like growth factor-I in the femoral tissues of rats with skeletal unloading. Calcif Tissue Int 55(5):363–367

Yao Z, Lafage-Proust MH, Plouët J, Bloomfield S, Alexandre C, Vico L (2004) Increase of both angiogenesis and bone mass in response to exercise depends on VEGF. J Bone Miner Res 19(9):1471–1480

Yeh CK, Rodan GA (1984) Tensile forces enhance prostaglandin E synthesis in osteoblastic cells grown on collagen ribbons. Calcif Tissue Int 36(Suppl 1):S67–S71

Yellowley CE, Li Z, Zhou Z, Jacobs CR, Donahue HJ (2000) Functional gap junctions between osteocytic and osteoblastic cells. J Bone Miner Res 15(2):209–217

You J, Reilly GC, Zhen X, Yellowley CE, Chen Q, Donahue HJ, Jacobs CR (2001) Osteopontin gene regulation by oscillatory fluid flow via intracellular calcium mobilization and activation of mitogen-activated protein kinase in MC3T3-E1 osteoblasts. J Biol Chem 276(16):13365–13371

Zaman G, Suswillo RF, Cheng MZ, Tavares IA, Lanyon LE (1997) Early responses to dynamic strain change and prostaglandins in bone-derived cells in culture. J Bone Miner Res 12(5):769–777

Zaman G, Pitsillides AA, Rawlinson SC, Suswillo RF, Mosley JR, Cheng MZ, Platts LA, Hukkanen M, Polak JM, Lanyon LE (1999) Mechanical strain stimulates nitric oxide production by rapid activation of endothelial nitric oxide synthase in osteocytes. J Bone Miner Res 14(7):1123–1131

Zaman G, Cheng MZ, Jessop HL, White R, Lanyon LE (2000) Mechanical strain activates estrogen response elements in bone cells. Bone 27(2):233–239

Zayzafoon M, Gathings WE, McDonald JM (2004) Modeled microgravity inhibits osteogenic differentiation of human mesenchymal stem cells and increases adipogenesis. Endocrinology 145(5):2421–2432

Zerath E, Holy X, Noël B, Malouvier A, Hott M, Marie PJ (1998) Effects of BMP-2 on osteoblastic cells and on skeletal growth and bone formation in unloaded rats. Growth Horm IGF Res 8(2):141–149

Zhang R, Supowit SC, Klein GL, Lu Z, Christiensen MD, Lozano R, Simmons DJ (1995) Rat tail suspension reduces messenger RNA level for growth factors and osteopontin and decreases the osteoblastic differentiation of bone marrow stromal cells. J Bone Miner Res 10:415–423

Ziambaras K, Lecanda F, Steinberg TH, Civitelli R (1998) Cyclic stretch enhances gap junctional communication between osteoblastic cells. J Bone Miner Res 13(2):218–228

Ziros PG, Gil AP, Georgakopoulos T, Habeos I, Kletsas D, Basdra EK, Papavassiliou AG (2002) The bone-specific transcriptional regulator Cbfa1 is a target of mechanical signals in osteoblastic cells. J Biol Chem 277(26):23934–23941

Chapter 9
Osteocytes in Mechanosensing: Insights from Mouse Models and Human Patients

Ken Watanabe and Kyoji Ikeda

9.1 Introduction

The development and homeostatic maintenance of skeletal tissue is regulated through multiple systems involving hormonal, neuronal, immune and mechanical signals (Zaidi 2007). Many of these signals impinge on the activities of two major cell types, osteoblasts and osteoclasts, that are effector cells performing bone formation and resorption, respectively (Karsenty and Wagner 2002).

Osteocytes are terminally differentiated from osteoblasts and are encased individually in lacunae surrounded by mineralized matrix (Bonewald and Johnson 2008). Although osteocytes exist in a quiescent state, they are thought to be actively engaged in signal transduction through numerous long cell processes in contact with neighboring osteocytes, and also with osteoblasts/lining cells on bone surface.

Emerging evidence suggests that osteocytes play a critical role in skeletal homeostasis through mechanosensing and mechanotransductory functions. In this article, we review the pathophysiological roles osteocytes play in bone remodeling, focusing mainly on in vivo findings in model mice as well as in human patients.

9.2 Osteocytes in Aging and Disease

Although the precise pathophysiology of osteocytes in aging and disease remains to be determined, decreased numbers of living osteocytes have been documented in various clinical situations with reduced bone volume and/or strength. Sex steroids, estrogens and androgens, are essential hormones for skeletal growth and maintenance, and their decline with aging ultimately leads to bone loss and fracture.

K. Watanabe (✉) and K. Ikeda
Department of Bone and Joint Disease, National Center for Geriatrics and Gerontology (NCGG), 35 Gengo, Morioka, Obu, Aichi 474-8511, Japan
e-mail: kwatanab@ncgg.go.jp; kikeda@ncgg.go.jp

M. Noda (ed.), *Mechanosensing Biology*,
DOI 10.1007/978-4-431-89757-6_9, © Springer 2011

It has been reported that, in cases of estrogen depletion following the treatment of women with a gonadotropin-releasing hormone analog, osteocyte viability was reduced due to increased apoptosis (Tomkinson et al. 1997). This increased apoptosis of osteocytes was recapitulated in the ovariectomized (OVX) rat model of osteoporosis, and attenuated by treatment with 17β-estradiol or a selective estrogen receptor modulator (SERM) in OVX rats (Tomkinson et al. 1998). Orchidectomy also increased the apoptosis of osteocytes as well as osteoblasts in mice (Kousteni et al. 2001).

Whether sex steroids modulate the mechanical response of bone metabolism remains an issue of considerable controversy. Mice deficient in *Esr1*, a gene encoding estrogen receptor α, exhibited a decrease in the anabolic response to mechanical loading on their ulna shafts, compared with wild-type controls, suggesting that estrogen signaling is required for mechanotransduction (Lee et al. 2003). Other studies in women with estrogen deficiency or OVX animals, however, show that estrogen is required for maximal osteogenesis, but not for mechanical responsiveness, suggesting that the mechanical response of bone is not entirely estrogen dependent (Dalsky et al. 1988; Jagger et al. 1996).

Proper mechanical stress seems to be required for the maintenance of live osteocytes. In the cortical bone of rats which underwent space flight for 12.5 days, degeneration of osteocytes adjacent to the vasculature was observed (Doty et al. 1990). Increased incidence of osteocyte apoptosis was also reported during unloading by hindlimb suspension of mice (Aguirre et al. 2006). On the other hand, overloading damages the bone architecture and induces apoptosis in osteocytes (Verborgt et al. 2000). Thus, in addition to the roles of osteocytes in mechanotransduction, mechanical stress itself is a regulator of osteocyte fate.

Glucocorticoid-induced osteoporosis is a major cause of fragility fracture. The glucocorticoid receptor (GR) is expressed in osteoblasts and osteocytes of human bone (Abu et al. 2000), and it has been demonstrated that glucocorticoid induces apoptosis of osteoblasts and osteocytes, resulting in reduced bone formation in mice (Weinstein et al. 2000; O'Brien et al. 2004). Representative antiresorptive drugs, estrogen and bisphosphonates, have been shown to ameliorate glucocorticoid-induced apoptosis of osteocytes in vitro (Plotkin et al. 2005). Thus, glucocorticoid signaling in osteocytes as well as in osteoblasts appears to be important for cell survival and skeletal health.

Parathyroid hormone (PTH) has been suggested as acting on osteocytes and influence the mechanical response of bone. Lindgren reported that immobilization-induced bone loss was attenuated by thyroparathyroidectomy (TPTX) in rats (Lindgren 1976). Bone formation activated by mechanical stimulation was abrogated in TPTX rats, which was restored following a single injection of PTH, suggesting that PTH is required for the mechanical response of bone (Chow et al. 1998). Interestingly, serum PTH levels were reduced during space flight, but not during bed rest or hindlimb suspension in rodents (Bikle and Halloran 1999). It has been demonstrated that PTH decreases the apoptosis of osteoblasts and osteocytes in mice, while in transgenic mice expressing constitutively active PTH receptor specifically in osteocytes, bone turnover is heightened and bone volume is increased (O'Brien et al. 2008).

Collectively, these observations suggest that the mechanical response of bone is, at least in part, mediated through PTH receptor signaling in osteocytes.

A decline in the number of osteocytes has been observed with aging (Frost 1960a, b; Vashishth et al. 2000). Together with the observation that the osteogenic response to mechanical stress is blunted with aging, it is suggested that decreased osteocyte number may underlie declines in mechanical stimulation and bone-forming activity with aging. Klotho, a mouse model of accelerated aging, displays multiple aging phenotypes, including osteopenia. In this model, osteocyte distribution is observably altered, with an increased number of empty lacunae (Suzuki et al. 2005).

Not only the number but also the morphology of osteocytes seems to be altered in certain pathologies. According to a recent report (van Hove et al. 2009), the osteocytes in the proximal tibia of a patient with osteopenia were observed to be round-shaped, osteocytes in osteopetrotic bone were small and disk-shaped, and osteocytes in osteoarthritic bone were large and elongated. Thus, there may exist a close relationship between the morphology of osteocytes and the surrounding bone matrix architecture.

9.3 Osteocytes in Genetically Modified Mice

Defects in osteocyte function have been described in genetically modified mouse models. DMP1 is a member of the small integrin-binding ligand, N-linked glycoprotein (SIBLING) family that is expressed in osteocytes and odontoblasts of adult mice. Mutations in the *DMP1* gene have been found in patients with autosomal recessive hypophosphatemia, and deletion of the gene in mice causes hypophosphatemia with rickets and osteomalacia (Lorenz-Depiereux et al. 2006). In the *Dmp1* knockout (KO) mice, the morphology of osteocytes as well as the lacunocanalicular structure have been reported to be disorganized (Feng et al. 2006). This is an important finding providing compelling evidence that osteocytes play a critical role in the local control of mineralization defining the lacunocanalicular border.

Runx2 is a master regulator of osteoblast differentiation (Ducy et al. 1997), and loss of function mutations cause cleidocranial dysplasia in humans (Mundlos et al. 1997) and a lethal, "bone-less" phenotype in homozygous KO mice (Komori et al. 1997; Otto et al. 1997). It has been reported that overexpression of Runx2 in mice results in an osteopenic phenotype with multiple fractures (Liu et al. 2001). In the Runx2 transgenic mice, terminal differentiation of osteoblasts was impaired, with a marked reduction in the number of osteocytes, which indicates that, in the absence of mature osteoblasts, osteocytes cannot be generated, which is compatible with the notion that osteocytes are derived from mature osteoblasts. These results also imply that, although Runx2 is essential for osteoblastogenesis, it is not sufficient for the terminal differentiation process from osteoblasts to osteocytes, and this suggests that a distinct transcriptional program may drive "osteocytogenesis."

TGFβ-inducible early gene 1 (TIEG) is an Sp/KLF family transcription factor that is expressed in osteoblasts. Mice deficient in the *TIEG* gene exhibit osteopenia with reduced bone strength (Subramaniam et al. 2005; Bensamoun et al. 2006; Hawse et al. 2008). Although the number of osteoblasts was markedly increased in these mice, the number of osteocytes was decreased. As in the Runx2 transgenic mice, the terminal differentiation from osteoblasts to osteocytes may be impaired in the TIEG-deficient mouse, lending further support to the concept that the cell fate decision from osteoblasts to osteocytes is regulated by a distinct set of transcription factors. Also, the findings above can be taken as evidence suggesting that, even in the case of increased osteoblasts, a reduction in osteocytes is associated with decreased bone formation, which implies that osteocytes are actively engaged in bone formation.

An autosomal recessive form of multicentric osteolysis with severe osteoporosis is caused by loss of function mutations in the *MMP2* gene, which encodes an extracellular matrix-degrading enzyme (Martignetti et al. 2001). Although a reduced bone density was observed in the cancellous and cortical bones of the *Mmp2* KO mice, the calvarial bone density displayed an age-dependent increase. In the mutant calvaria, the proportion of empty lacunae was significantly elevated, while the number of osteocytic processes and their connections were significantly decreased, indicating that the osteocytic network is impaired in *Mmp2* KO mice (Inoue et al. 2006). The impairment of the osteocytic network preceded osteocytic death, suggesting that such network formation is required for osteocyte survival.

Although osteocytic differentiation and/or function are abnormal in the mouse models describe above, the mechanical response in the skeleton is not known. Furthermore, since the models exhibited osteoblastic dysfunction and/or systemic metabolic alterations, even if some defect in the skeletal response to mechanical stimulation is observed, it is difficult to conclude that the osteocytes per se are responsible.

9.4   Insights from a Mouse Model Lacking Osteocytes

To address specifically whether osteocytes are involved in mechanotransduction in vivo, Tatsumi et al. generated a transgenic mouse model expressing diphtheria toxin (DT) receptor in osteocytes using the mouse *Dmp1* promoter (Tatsumi et al. 2007). The receptor is the human HB-EGF protein; DT does not bind the mouse protein, so that mice are naturally resistant to DT (Saito et al. 2001). Using the human-specific toxin-receptor system, the *Dmp1* promoter-driven DT receptor was expected to induce osteocyte death in a DT-dependent manner.

As expected, the majority of osteocytes (~70%) died within a couple of days following a single injection of DT in the DT-receptor transgenic mice. In a week or so, enlarged empty lacunae and intracortical cavities with fissures reminiscent of "microcracks" were observed in the cortical bone. Osteocyte ablation also caused a mineralization defect, but the trabecular bone mass did not change significantly in the short term. Forty days after the acute ablation of osteocytes, bone mass markedly decreased and fat cells increased in the marrow space; these features are reminiscent of those often seen in the skeleton of the elderly.

Most importantly, the osteocyte-ablated mice exhibited almost complete resistance to the bone loss induced by mechanical unloading (Tatsumi et al. 2007). Upon unloading of hindlimbs by tail suspension, RANKL expression usually increases within several days; however, the increase in RANKL expression in response to unloading was attenuated in the osteocyte-ablated mice, and the increase in the number of osteoclasts during tail suspension was not observed either, suggesting that osteocytes are somehow involved in the elevation of bone resorption upon unloading.

Interestingly, the bone mass recovery during a 2-week period of reambulation was preserved when osteocyte ablation was targeted specifically during the reloading phase (Tatsumi et al. 2007), implying that the robust anabolic and anticatabolic response during the reloading period takes place even when the osteocytic network is severely disrupted, and by inference, that cells other than osteocytes may compensate for the response.

9.5   Osteocyte-Derived Factors that Regulate Bone Metabolism

Emerging evidence supports the concept that osteocytes control bone formation through factors that are secreted from osteocytes and regulate osteoblastic function (Fig. 9.1). *Sost*, encoding sclerostin, was identified as the causative gene of the autosomal recessive disease sclerostenosis, and has been shown to be a negative regulator of bone formation through its anti-Wnt function (Balemans et al. 2001; Brunkow et al. 2001; Winkler et al. 2003; van Bezooijen et al. 2004). *Sost* is expressed specifically in osteocytes, and *Sost* KO mice exhibit high bone mass and increased bone strength without apparent skeletal malformation (Winkler et al. 2003; Li et al. 2008). PTH has been demonstrated to suppress *Sost* expression via the cAMP/PKA pathway (Bellido et al. 2005; Keller and Kneissel 2005; Leupin et al. 2007), suggesting that sclerostin plays an important role in the anabolic action of PTH. Thus, there exists a negative feedback regulation mechanism in that terminally differentiated cells (osteocytes) control the activity of their progenitors (osteoblasts) through sclerostin (Fig. 9.1).

The mechanotransduction in bone seems to utilize this feedback link between osteocytes and osteoblasts. Mechanical loading reduces the expression of *Sost* in osteocytes, while unloading upregulates the negative regulator, suggesting that sclerostin is located somewhere along the mechanotransduction pathway in osteocytes and is involved in the mechanical response of bone (Robling et al. 2008; Lin et al. 2009) (Fig. 9.1).

Sclerostin is not the sole osteoblast regulator derived from osteocytes. Dkk1 is another antagonist of Wnt signaling that is predominantly (although not exclusively) expressed in osteocytes (Li et al. 2005). Inhibition of Dkk1 results in increased bone mass in mice (Morvan et al. 2006; MacDonald et al. 2007). Dkk1 as well as Wif1, yet another Wnt antagonist, has been shown to be upregulated by glucocorticoid and downregulated by treatment with PTH (Yao et al. 2008). In addition to these negative regulators, osteocytes are assumed to send positive

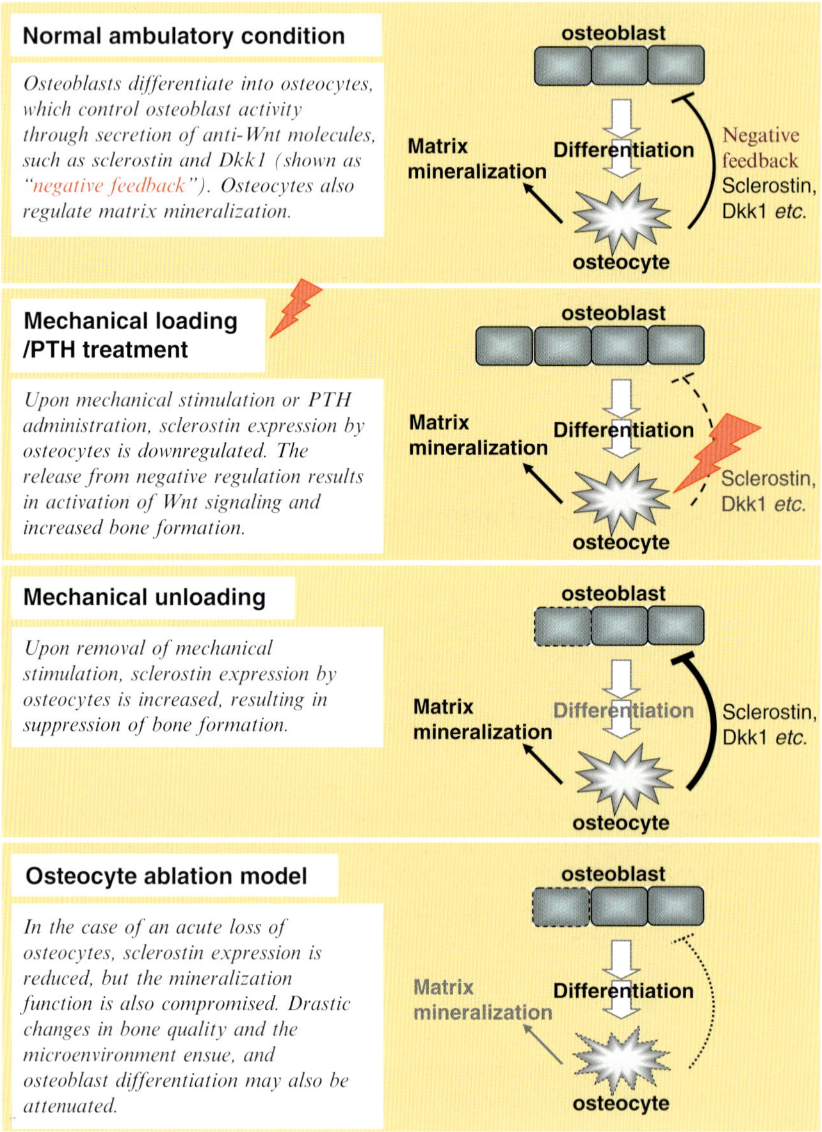

Fig. 9.1 Communication between osteoblasts and osteocytes under various mechanical conditions. Osteocytes differentiate from osteoblasts and control osteoblastic function by a variety of mechanisms

signals to osteoblasts as well. Osteocytes express IGF-1, a well-known stimulator of bone formation, and pleiotrophin, which has been shown to stimulate osteoblastic function (Lean et al. 1995, 1996; Zhao et al. 2000; Imai et al. 2009). These growth factors and cytokines are factors that are apparently secreted from osteocytes and that support the proliferation, activation and/or survival of osteoblasts; elimination of these positive regulators of bone formation as a result of osteocyte ablation may account for the inhibition of bone formation and osteopenic phenotype observed in the model lacking osteocytes (Tatsumi et al. 2007).

Nonprotein factors, such as prostaglandin E_2 (PGE_2) and nitric oxide (NO), may also play roles in osteocyte-mediated mechanotransduction (Chambers et al. 1999; Zaman et al. 1999; Watanuki et al. 2002). Although the in vivo functions of these molecules in mechanotransduction have been studied using pharmacological and genetic approaches, it remains elusive whether the PGE_2 and NO produced in osteocytes are directly involved in the control of osteoblastic function.

Osteocytes seem to regulate bone resorption under normal ambulatory conditions as well as in mechanical unloading (Fig. 9.2). In the areas surrounding microcracks, apoptotic osteocytes were observed concomitantly with osteoclastic resorption (Verborgt et al. 2000), and osteocytic death preceded intracortical bone remodeling (Noble et al. 2003). The absence of osteocytes results in increased osteoclast recruitment, which implies that, under normal ambulatory conditions, the presence of osteocytes prevents osteoclasts from aberrant activation. The suppression of bone resorption by osteocytes may hypothetically involve one or more factors with repulsive activity against osteoclast guidance (Fig. 9.2), in view of the fact that osteoclast recruitment resulting from osteocytic death is a regional event. Thus far, only a few factors have been shown to downregulate RANKL expression in osteoblasts, such as estrogen and TGF-β (Heino et al. 2002); it remains to be determined whether these factors are involved in the suppression of osteoclast recruitment by osteocytes.

Alternatively, dying osteocytes may attract osteoclast migration. HMGB1 is a nonhistone nuclear protein that is released from dying cells and acts as a cytokine in association with autoimmune diseases (Wang et al. 1999). *Hmgb1* KO mice exhibited impaired endochondral ossification with defects in the invasion of osteoclasts and endothelial cells (Taniguchi et al. 2007). HMGB1 is secreted from developing cartilage, and also expressed in osteoblastic and osteocytic cell lines in vitro (Yang et al. 2008). Thus, the recruitment of osteoclasts may be regulated by a molecule, such as HMGB, with repulsive activity and the conveyance of a death signal.

The notion that, where osteocytes are lacking, osteoclastic bone resorption is activated is consistent with the observations that excessive bone resorption ensues following osteocyte ablation (Tatsumi et al. 2007). However, the opposite was found in the same osteocyte-less mice under the condition of unloading (Fig. 9.2). Upon the shifting of osteocyte-less mice to an unloading state by tail suspension, increases in osteoclastic bone resorption and RANKL expression were attenuated. The simplest interpretation of these findings is that osteocytes are required for osteoclast activation via the induction of RANKL upon removal of mechanical load on bone. It is tempting to speculate, therefore, that osteocytes express a mechano-sensitive accelerator of RANKL expression. Although PGE_2 is known as an inducer of RANKL expression (Tsukii et al. 1998) and has been implicated as a mediator of mechanotransduction, the expression in bone is upregulated upon loading, not during unloading; thus it does not fit into the scheme.

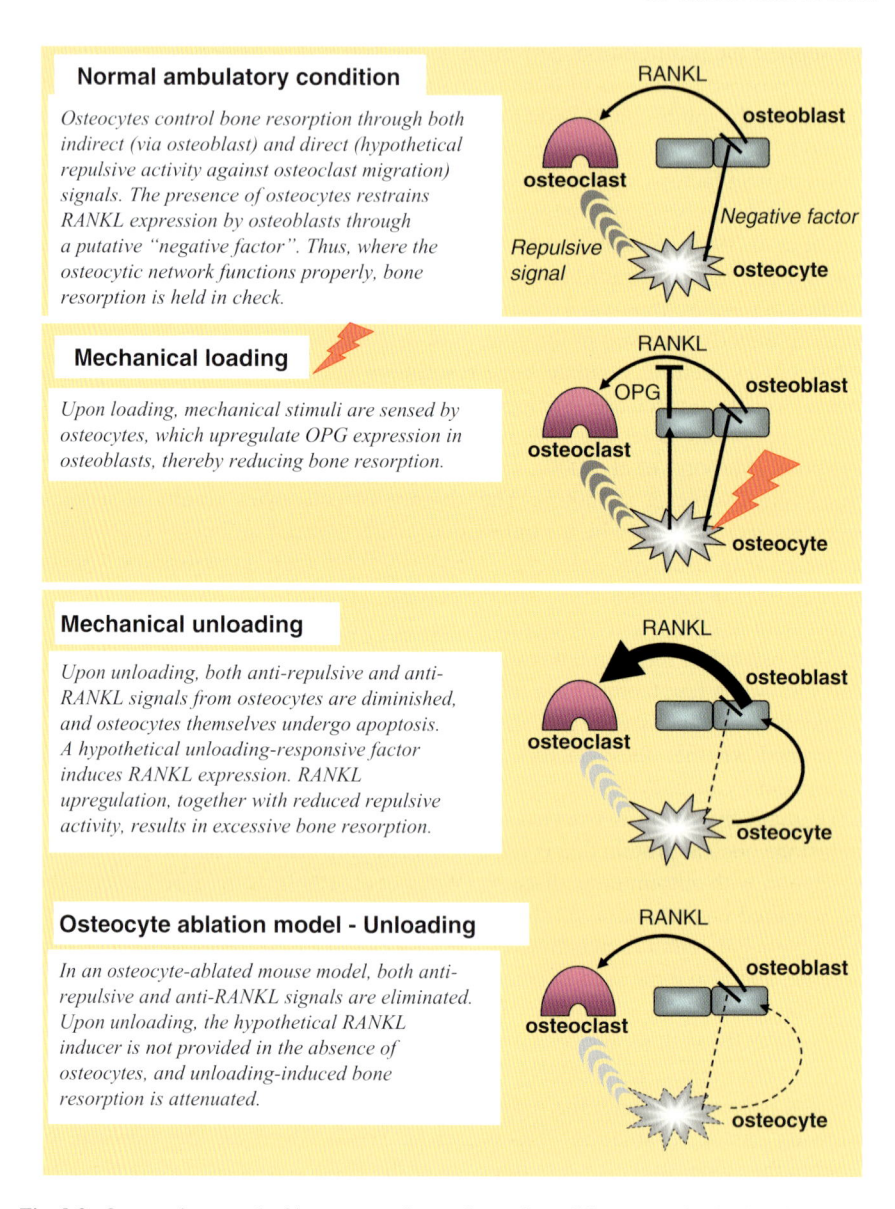

Normal ambulatory condition

Osteocytes control bone resorption through both indirect (via osteoblast) and direct (hypothetical repulsive activity against osteoclast migration) signals. The presence of osteocytes restrains RANKL expression by osteoblasts through a putative "negative factor". Thus, where the osteocytic network functions properly, bone resorption is held in check.

Mechanical loading

Upon loading, mechanical stimuli are sensed by osteocytes, which upregulate OPG expression in osteoblasts, thereby reducing bone resorption.

Mechanical unloading

Upon unloading, both anti-repulsive and anti-RANKL signals from osteocytes are diminished, and osteocytes themselves undergo apoptosis. A hypothetical unloading-responsive factor induces RANKL expression. RANKL upregulation, together with reduced repulsive activity, results in excessive bone resorption.

Osteocyte ablation model - Unloading

In an osteocyte-ablated mouse model, both anti-repulsive and anti-RANKL signals are eliminated. Upon unloading, the hypothetical RANKL inducer is not provided in the absence of osteocytes, and unloading-induced bone resorption is attenuated.

Fig. 9.2 Osteocytic control of bone resorption under various different mechanical environments

Although osteoblasts are the main RANKL-expressing cells, they alone cannot upregulate RANKL expression in response to unloading, in view of the findings in the osteocyte ablation model that bone with intact osteoblasts, but no osteocytes, does not respond to unloading with increased RANKL expression. Thus, the flow of mechanosignals through an osteocyte–osteoblast–osteoclast relay orchestrates the skeletal response of bone in various mechanical environments.

9.6 Conclusion

Emerging experimental evidence is consistent with the concept that bone remodeling is explicitly regulated through osteocyte–osteoblast–osteoclast triangular interaction, and that the osteocytic impact on bone turnover mostly takes place via osteoblasts. Whether osteocytes have any direct control over osteoclasts remains to be determined. Mechanical forces are an important regulator of the triangular interaction, and most likely sensed by osteocytes. Among the outstanding central questions of bone biology, it remains to be elucidated how osteocytes sense the mechanical environment so as to shape bone, the identity of the mechanoreceptor in osteocytes, and how the mechanical signals are transduced.

Acknowledgments We thank Sunao Takeshita (NCGG) for comments on the manuscript. This work was supported by a grant from the program Promotion of Fundamental Studies in Health Sciences of the National Institute of Biomedical Innovation (NIBIO) of Japan (#06-31 to K.I.) and a grant from the Mitsubishi Foundation (to K.I.). Pacific Edit reviewed the manuscript prior to submission.

References

Abu EO, Horner A, Kusec V, Triffitt JT, Compston JE (2000) The localization of the functional glucocorticoid receptor alpha in human bone. J Clin Endocrinol Metab 85(2):883–889

Aguirre JI, Plotkin LI, Stewart SA, Weinstein RS, Parfitt AM, Manolagas SC, Bellido T (2006) Osteocyte apoptosis is induced by weightlessness in mice and precedes osteoclast recruitment and bone loss. J Bone Miner Res 21(4):605–615

Balemans W, Ebeling M, Patel N, Van Hul E, Olson P, Dioszegi M, Lacza C, Wuyts W, Van Den Ende J, Willems P, Paes-Alves AF, Hill S, Bueno M, Ramos FJ, Tacconi P, Dikkers FG, Stratakis C, Lindpaintner K, Vickery B, Foernzler D, Van Hul W (2001) Increased bone density in sclerosteosis is due to the deficiency of a novel secreted protein (SOST). Hum Mol Genet 10(5):537–543

Bellido T, Ali AA, Gubrij I, Plotkin LI, Fu Q, O'Brien CA, Manolagas SC, Jilka RL (2005) Chronic elevation of parathyroid hormone in mice reduces expression of sclerostin by osteocytes: a novel mechanism for hormonal control of osteoblastogenesis. Endocrinology 146(11):4577–4583

Bensamoun SF, Hawse JR, Subramaniam M, Ilharreborde B, Bassillais A, Benhamou CL, Fraser DG, Oursler MJ, Amadio PC, An KN, Spelsberg TC (2006) TGFbeta inducible early gene-1 knockout mice display defects in bone strength and microarchitecture. Bone 39(6):1244–1251

Bikle DD, Halloran BP (1999) The response of bone to unloading. J Bone Miner Metab 17(4):233–244

Bonewald LF, Johnson ML (2008) Osteocytes, mechanosensing and Wnt signaling. Bone 42(4):606–615

Brunkow ME, Gardner JC, Van Ness J, Paeper BW, Kovacevich BR, Proll S, Skonier JE, Zhao L, Sabo PJ, Fu Y, Alisch RS, Gillett L, Colbert T, Tacconi P, Galas D, Hamersma H, Beighton P, Mulligan J (2001) Bone dysplasia sclerosteosis results from loss of the SOST gene product, a novel cystine knot-containing protein. Am J Hum Genet 68(3):577–589

Chambers TJ, Fox S, Jagger CJ, Lean JM, Chow JW (1999) The role of prostaglandins and nitric oxide in the response of bone to mechanical forces. Osteoarthr Cartil 7(4):422–423

Chow JW, Fox S, Jagger CJ, Chambers TJ (1998) Role for parathyroid hormone in mechanical responsiveness of rat bone. Am J Physiol 274(1 Pt 1):E146–E154

Dalsky GP, Stocke KS, Ehsani AA, Slatopolsky E, Lee WC, Birge SJ Jr (1988) Weight-bearing exercise training and lumbar bone mineral content in postmenopausal women. Ann Intern Med 108(6):824–828

Doty SB, Morey-Holton ER, Durnova GN, Kaplansky AS (1990) Cosmos 1887: morphology, histochemistry, and vasculature of the growing rat tibia. FASEB J 4(1):16–23

Ducy P, Zhang R, Geoffroy V, Ridall AL, Karsenty G (1997) Osf2/Cbfa1: a transcriptional activator of osteoblast differentiation. Cell 89(5):747–754

Feng JQ, Ward LM, Liu S, Lu Y, Xie Y, Yuan B, Yu X, Rauch F, Davis SI, Zhang S, Rios H, Drezner MK, Quarles LD, Bonewald LF, White KE (2006) Loss of DMP1 causes rickets and osteomalacia and identifies a role for osteocytes in mineral metabolism. Nat Genet 38(11):1310–1315

Frost HM (1960a) In vivo osteocyte death. J Bone Joint Surg Am 42-A:138–143

Frost HM (1960b) Micropetrosis. J Bone Joint Surg Am 42-A:144–150

Hawse JR, Iwaniec UT, Bensamoun SF, Monroe DG, Peters KD, Ilharreborde B, Rajamannan NM, Oursler MJ, Turner RT, Spelsberg TC, Subramaniam M (2008) TIEG-null mice display an osteopenic gender-specific phenotype. Bone 42(6):1025–1031

Heino TJ, Hentunen TA, Vaananen HK (2002) Osteocytes inhibit osteoclastic bone resorption through transforming growth factor-beta: enhancement by estrogen. J Cell Biochem 85(1):185–197

Imai S, Heino TJ, Hienola A, Kurata K, Buki K, Matsusue Y, Vaananen HK, Rauvala H (2009) Osteocyte-derived HB-GAM (pleiotrophin) is associated with bone formation and mechanical loading. Bone 44(5):785–794

Inoue K, Mikuni-Takagaki Y, Oikawa K, Itoh T, Inada M, Noguchi T, Park JS, Onodera T, Krane SM, Noda M, Itohara S (2006) A crucial role for matrix metalloproteinase 2 in osteocytic canalicular formation and bone metabolism. J Biol Chem 281(44):33814–33824

Jagger CJ, Chow JW, Chambers TJ (1996) Estrogen suppresses activation but enhances formation phase of osteogenic response to mechanical stimulation in rat bone. J Clin Invest 98(10):2351–2357

Karsenty G, Wagner EF (2002) Reaching a genetic and molecular understanding of skeletal development. Dev Cell 2(4):389–406

Keller H, Kneissel M (2005) SOST is a target gene for PTH in bone. Bone 37(2):148–158

Komori T, Yagi H, Nomura S, Yamaguchi A, Sasaki K, Deguchi K, Shimizu Y, Bronson RT, Gao YH, Inada M, Sato M, Okamoto R, Kitamura Y, Yoshiki S, Kishimoto T (1997) Targeted disruption of Cbfa1 results in a complete lack of bone formation owing to maturational arrest of osteoblasts. Cell 89(5):755–764

Kousteni S, Bellido T, Plotkin LI, O'Brien CA, Bodenner DL, Han L, Han K, DiGregorio GB, Katzenellenbogen JA, Katzenellenbogen BS, Roberson PK, Weinstein RS, Jilka RL, Manolagas SC (2001) Nongenotropic, sex-nonspecific signaling through the estrogen or androgen receptors: dissociation from transcriptional activity. Cell 104(5):719–730

Lean JM, Jagger CJ, Chambers TJ, Chow JW (1995) Increased insulin-like growth factor I mRNA expression in rat osteocytes in response to mechanical stimulation. Am J Physiol 268(2 Pt 1):E318–E327

Lean JM, Mackay AG, Chow JW, Chambers TJ (1996) Osteocytic expression of mRNA for c-fos and IGF-I: an immediate early gene response to an osteogenic stimulus. Am J Physiol 270(6 Pt 1):E937–E945

Lee K, Jessop H, Suswillo R, Zaman G, Lanyon L (2003) Endocrinology: bone adaptation requires oestrogen receptor-alpha. Nature 424(6947):389

Leupin O, Kramer I, Collette NM, Loots GG, Natt F, Kneissel M, Keller H (2007) Control of the SOST bone enhancer by PTH using MEF2 transcription factors. J Bone Miner Res 22(12):1957–1967

Li X, Liu P, Liu W, Maye P, Zhang J, Zhang Y, Hurley M, Guo C, Boskey A, Sun L, Harris SE, Rowe DW, Ke HZ, Wu D (2005) Dkk2 has a role in terminal osteoblast differentiation and mineralized matrix formation. Nat Genet 37(9):945–952

Li X, Ominsky MS, Niu QT, Sun N, Daugherty B, D'Agostin D, Kurahara C, Gao Y, Cao J, Gong J, Asuncion F, Barrero M, Warmington K, Dwyer D, Stolina M, Morony S, Sarosi I, Kostenuik PJ, Lacey DL, Simonet WS, Ke HZ, Paszty C (2008) Targeted deletion of the sclerostin gene in mice results in increased bone formation and bone strength. J Bone Miner Res 23(6):860–869

Lin C, Jiang X, Dai Z, Guo X, Weng T, Wang J, Li Y, Feng G, Gao X, He L (2009) Sclerostin mediates bone response to mechanical unloading via antagonizing Wnt/beta-catenin signaling. J Bone Miner Res 24:1651–1661

Lindgren JU (1976) Studies of the calcium accretion rate of bone during immobilization in intact and thyroparathyroidectomized adult rats. Calcif Tissue Res 22(1):41–47

Liu W, Toyosawa S, Furuichi T, Kanatani N, Yoshida C, Liu Y, Himeno M, Narai S, Yamaguchi A, Komori T (2001) Overexpression of Cbfa1 in osteoblasts inhibits osteoblast maturation and causes osteopenia with multiple fractures. J Cell Biol 155(1):157–166

Lorenz-Depiereux B, Bastepe M, Benet-Pages A, Amyere M, Wagenstaller J, Muller-Barth U, Badenhoop K, Kaiser SM, Rittmaster RS, Shlossberg AH, Olivares JL, Loris C, Ramos FJ, Glorieux F, Vikkula M, Juppner H, Strom TM (2006) DMP1 mutations in autosomal recessive hypophosphatemia implicate a bone matrix protein in the regulation of phosphate homeostasis. Nat Genet 38(11):1248–1250

MacDonald BT, Joiner DM, Oyserman SM, Sharma P, Goldstein SA, He X, Hauschka PV (2007) Bone mass is inversely proportional to Dkk1 levels in mice. Bone 41(3):331–339

Martignetti JA, Aqeel AA, Sewairi WA, Boumah CE, Kambouris M, Mayouf SA, Sheth KV, Eid WA, Dowling O, Harris J, Glucksman MJ, Bahabri S, Meyer BF, Desnick RJ (2001) Mutation of the matrix metalloproteinase 2 gene (MMP2) causes a multicentric osteolysis and arthritis syndrome. Nat Genet 28(3):261–265

Morvan F, Boulukos K, Clement-Lacroix P, Roman Roman S, Suc-Royer I, Vayssiere B, Ammann P, Martin P, Pinho S, Pognonec P, Mollat P, Niehrs C, Baron R, Rawadi G (2006) Deletion of a single allele of the Dkk1 gene leads to an increase in bone formation and bone mass. J Bone Miner Res 21(6):934–945

Mundlos S, Otto F, Mundlos C, Mulliken JB, Aylsworth AS, Albright S, Lindhout D, Cole WG, Henn W, Knoll JH, Owen MJ, Mertelsmann R, Zabel BU, Olsen BR (1997) Mutations involving the transcription factor CBFA1 cause cleidocranial dysplasia. Cell 89(5):773–779

Noble BS, Peet N, Stevens HY, Brabbs A, Mosley JR, Reilly GC, Reeve J, Skerry TM, Lanyon LE (2003) Mechanical loading: biphasic osteocyte survival and targeting of osteoclasts for bone destruction in rat cortical bone. Am J Physiol Cell Physiol 284(4):C934–C943

O'Brien CA, Jia D, Plotkin LI, Bellido T, Powers CC, Stewart SA, Manolagas SC, Weinstein RS (2004) Glucocorticoids act directly on osteoblasts and osteocytes to induce their apoptosis and reduce bone formation and strength. Endocrinology 145(4):1835–1841

O'Brien CA, Plotkin LI, Galli C, Goellner JJ, Gortazar AR, Allen MR, Robling AG, Bouxsein M, Schipani E, Turner CH, Jilka RL, Weinstein RS, Manolagas SC, Bellido T (2008) Control of bone mass and remodeling by PTH receptor signaling in osteocytes. PLoS One 3(8):e2942

Otto F, Thornell AP, Crompton T, Denzel A, Gilmour KC, Rosewell IR, Stamp GW, Beddington RS, Mundlos S, Olsen BR, Selby PB, Owen MJ (1997) Cbfa1, a candidate gene for cleidocranial dysplasia syndrome, is essential for osteoblast differentiation and bone development. Cell 89(5):765–771

Plotkin LI, Aguirre JI, Kousteni S, Manolagas SC, Bellido T (2005) Bisphosphonates and estrogens inhibit osteocyte apoptosis via distinct molecular mechanisms downstream of extracellular signal-regulated kinase activation. J Biol Chem 280(8):7317–7325

Robling AG, Niziolek PJ, Baldridge LA, Condon KW, Allen MR, Alam I, Mantila SM, Gluhak-Heinrich J, Bellido TM, Harris SE, Turner CH (2008) Mechanical stimulation of bone in vivo reduces osteocyte expression of Sost/sclerostin. J Biol Chem 283(9):5866–5875

Saito M, Iwawaki T, Taya C, Yonekawa H, Noda M, Inui Y, Mekada E, Kimata Y, Tsuru A, Kohno K (2001) Diphtheria toxin receptor-mediated conditional and targeted cell ablation in transgenic mice. Nat Biotechnol 19(8):746–750

Subramaniam M, Gorny G, Johnsen SA, Monroe DG, Evans GL, Fraser DG, Rickard DJ, Rasmussen K, van Deursen JM, Turner RT, Oursler MJ, Spelsberg TC (2005) TIEG1 null mouse-derived osteoblasts are defective in mineralization and in support of osteoclast differentiation in vitro. Mol Cell Biol 25(3):1191–1199

Suzuki H, Amizuka N, Oda K, Li M, Yoshie H, Ohshima H, Noda M, Maeda T (2005) Histological evidence of the altered distribution of osteocytes and bone matrix synthesis in klotho-deficient mice. Arch Histol Cytol 68(5):371–381

Taniguchi N, Yoshida K, Ito T, Tsuda M, Mishima Y, Furumatsu T, Ronfani L, Abeyama K, Kawahara K, Komiya S, Maruyama I, Lotz M, Bianchi ME, Asahara H (2007) Stage-specific secretion of HMGB1 in cartilage regulates endochondral ossification. Mol Cell Biol 27(16):5650–5663

Tatsumi S, Ishii K, Amizuka N, Li M, Kobayashi T, Kohno K, Ito M, Takeshita S, Ikeda K (2007) Targeted ablation of osteocytes induces osteoporosis with defective mechanotransduction. Cell Metab 5(6):464–475

Tomkinson A, Reeve J, Shaw RW, Noble BS (1997) The death of osteocytes via apoptosis accompanies estrogen withdrawal in human bone. J Clin Endocrinol Metab 82(9):3128–3135

Tomkinson A, Gevers EF, Wit JM, Reeve J, Noble BS (1998) The role of estrogen in the control of rat osteocyte apoptosis. J Bone Miner Res 13(8):1243–1250

Tsukii K, Shima N, Mochizuki S, Yamaguchi K, Kinosaki M, Yano K, Shibata O, Udagawa N, Yasuda H, Suda T, Higashio K (1998) Osteoclast differentiation factor mediates an essential signal for bone resorption induced by 1 alpha, 25-dihydroxyvitamin D3, prostaglandin E2, or parathyroid hormone in the microenvironment of bone. Biochem Biophys Res Commun 246(2):337–341

van Bezooijen RL, Roelen BA, Visser A, van der Wee-Pals L, de Wilt E, Karperien M, Hamersma H, Papapoulos SE, ten Dijke P, Lowik CW (2004) Sclerostin is an osteocyte-expressed negative regulator of bone formation, but not a classical BMP antagonist. J Exp Med 199(6):805–814

van Hove RP, Nolte PA, Vatsa A, Semeins CM, Salmon PL, Smit TH, Klein-Nulend J (2009) Osteocyte morphology in human tibiae of different bone pathologies with different bone mineral density–is there a role for mechanosensing? Bone 45:321–329

Vashishth D, Verborgt O, Divine G, Schaffler MB, Fyhrie DP (2000) Decline in osteocyte lacunar density in human cortical bone is associated with accumulation of microcracks with age. Bone 26(4):375–380

Verborgt O, Gibson GJ, Schaffler MB (2000) Loss of osteocyte integrity in association with microdamage and bone remodeling after fatigue in vivo. J Bone Miner Res 15(1):60–67

Wang H, Bloom O, Zhang M, Vishnubhakat JM, Ombrellino M, Che J, Frazier A, Yang H, Ivanova S, Borovikova L, Manogue KR, Faist E, Abraham E, Andersson J, Andersson U, Molina PE, Abumrad NN, Sama A, Tracey KJ (1999) HMG-1 as a late mediator of endotoxin lethality in mice. Science 285(5425):248–251

Watanuki M, Sakai A, Sakata T, Tsurukami H, Miwa M, Uchida Y, Watanabe K, Ikeda K, Nakamura T (2002) Role of inducible nitric oxide synthase in skeletal adaptation to acute increases in mechanical loading. J Bone Miner Res 17(6):1015–1025

Weinstein RS, Nicholas RW, Manolagas SC (2000) Apoptosis of osteocytes in glucocorticoid-induced osteonecrosis of the hip. J Clin Endocrinol Metab 85(8):2907–2912

Winkler DG, Sutherland MK, Geoghegan JC, Yu C, Hayes T, Skonier JE, Shpektor D, Jonas M, Kovacevich BR, Staehling-Hampton K, Appleby M, Brunkow ME, Latham JA (2003) Osteocyte control of bone formation via sclerostin, a novel BMP antagonist. EMBO J 22(23):6267–6276

Yang J, Shah R, Robling AG, Templeton E, Yang H, Tracey KJ, Bidwell JP (2008) HMGB1 is a bone-active cytokine. J Cell Physiol 214(3):730–739

Yao W, Cheng Z, Pham A, Busse C, Zimmermann EA, Ritchie RO, Lane NE (2008) Glucocorticoid-induced bone loss in mice can be reversed by the actions of parathyroid hormone and risedronate on different pathways for bone formation and mineralization. Arthritis Rheum 58(11):3485–3497

Zaidi M (2007) Skeletal remodeling in health and disease. Nat Med 13(7):791–801

Zaman G, Pitsillides AA, Rawlinson SC, Suswillo RF, Mosley JR, Cheng MZ, Platts LA, Hukkanen M, Polak JM, Lanyon LE (1999) Mechanical strain stimulates nitric oxide production by rapid activation of endothelial nitric oxide synthase in osteocytes. J Bone Miner Res 14(7):1123–1131

Zhao G, Monier-Faugere MC, Langub MC, Geng Z, Nakayama T, Pike JW, Chernausek SD, Rosen CJ, Donahue LR, Malluche HH, Fagin JA, Clemens TL (2000) Targeted overexpression of insulin-like growth factor I to osteoblasts of transgenic mice: increased trabecular bone volume without increased osteoblast proliferation. Endocrinology 141(7):2674–2682

Chapter 10
Osteocyte Mechanosensation and Transduction

Lynda Faye Bonewald

10.1 Introduction

Osteocytes are unique cells in the body. Analogies have been made to other cells in the body such as neuronal cells, but few consistent similarities with other cells in the body have been identified. Osteocytes share their closest properties with osteo-blasts from which they are descended, yet still retain their uniqueness. This unique-ness stems from their intricate, convoluted network that penetrates the hard mineralized bone matrix. These cells are connected to each other, to cells on the bone surface such as lining cells, osteoblasts, osteoclasts, and their precursors, to cells of the vasculature such as vascular epithelial cells, and these cells even have the capacity to send their dendritic processes into marrow spaces. Clearly in young bone, these cells have the capacity to extend and retract their processes, connecting and disconnecting with other cells, but it is not known if the cells maintain this capacity with aging. Osteocytes appear to be orchestrators of bone formation and resorption and play a role in the regulation of both calcium and phosphate homeo-stasis. However, one of their earliest functions was postulated to be as the mecha-nosensory cells of bone. (For review of osteocyte characteristics and function see Bonewald (2006, 2007), Klein-Nulend and Bonewald (2008)).

It has long been known that both growing and adult bone has the capacity to accommodate loading or lack of loading by modifying bone mass (Wolff 1982). This capacity appears to decrease with age. Osteocytes due to their location in bone and their complex dendritic network were hypothesized to be the cells most likely to sense mechanical loading or lack of loading. These cells compose 90–95% of all bone cells in the adult skeleton. In the growing postnatal skeleton, the bone surface is covered by osteoblasts and osteoclasts modeling bone, however in the adult skeleton, the majority of cells covering the bone surface are lining cells, with 2–5% osteoblasts and

L.F. Bonewald (✉)
Department of Oral Biology, University of Missouri at Kansas City School of Dentistry,
650 East 25th Street, Kansas City, MO 64108-2784, USA
e-mail: bonewaldl@umkc.edu

M. Noda (ed.), *Mechanosensing Biology*,
DOI 10.1007/978-4-431-89757-6_10, © Springer 2011

even fewer osteoclasts. The lining cell is an ignored cell that most likely has a role as a mechanical sensor on the bone surface, but as very little is known about this bone cell type, nothing is known regarding its role in mechanosensation. In this chapter, the focus will be on the role of the osteocyte in mechanotransduction in the adult skeleton as the majority of studies are performed in adult animals.

10.2   The Osteocyte as a Mechanosensory Cell

In vivo loading has been well characterized. Frequency, intensity, and timing of mechanical loading affect the response of bone. Bone mass is regulated by peak applied strain (Rubin 1984), bone formation rate is related to loading rate (Turner et al. 1994), and bone subjected to loading with frequent rest periods undergoes increased bone formation compared to bone subjected to a single bout of mechanical loading. Bone structure and strength is greater if loading is applied in shorter versus longer increments (Robling et al. 2002). It has been proposed that the response of the osteocyte to load is responsible for these changes and adaptation to load. The challenge has been to identify suitable in vitro models that replicate in vivo results.

The osteocyte cell body is located within a lacunae and its dendritic processes are localized within small tunnels called canaliculi. A 'bone fluid' baths the osteocyte and its processes. Little is known about this bone fluid, except that it appears that molecules larger than 70 kDa cannot travel through the osteocyte lacunocanalicular system (Wang et al. 2004). Therefore a sieve or molecular 'cut-off' must exist between blood and the bone fluid. Clearly blood flow is connected to and regulates the flow of the bone fluid as injection of dye into the tail vein of a mouse results in penetration of the osteocyte lacunocanalicular system with dye within minutes. Bone loss due to hindlimb unloading is restored with restored blood flow (Bergula et al. 1999). It has been proposed that bone fluid flow is driven by extravascular pressure as well as applied cyclic mechanical loading of osteocytes (Weinbaum et al. 1994).

A number of genes in osteocytes have been shown to be regulated by loading or by unloading. Early studies using turkey ulna showed that immediately following a 6-min period of intermittent loading, the number of osteocytes expressing glucose-6-phosphate dehydrogenase activity was increased in relation to local strain magnitude (Skerry et al. 1989). Osteocytes in intact bone change their enzyme activity and RNA synthesis rapidly after mechanical loading (Klein-Nulend and Bonewald 2008). E11/gp38, a marker for the early osteocyte, was shown to be upregulated in response to mechanical loading (Zhang et al. 2006), as is Phex, MEPE, DMP1, all regulators of phosphate homeostasis (Gluhak-Heinrich et al. 2003, 2007; Yang et al. 2005). Efforts are being made to correlate magnitude of strain and compression or tension with osteocyte gene expression. Sost/sclerostin, a marker for the late osteocyte and an inhibitor of osteoblast function, is down-regulated by anabolic mechanical loading and increased in response to hindlimb unloading (Lin et al. 2009;

Robling et al. 2006). These observations suggest that mechanical loading can regulate other functions of osteocytes other than bone remodeling, such as mineral homeostasis and dendrite formation.

10.3 In Vitro Cell Culture Models

Clearly, any cell will respond to some form of mechanical loading. A question in the bone field has been – what type of mechanical loading is the osteocyte subjected to? Whereas early in vitro experiments were conducted using hydrostatic pressure and substrate stretching, it is fairly well accepted at the present time that the osteocyte is most likely subjected to fluid flow shear stress. Osteocytes are more sensitive than osteoblasts to fluid flow shear stress and more sensitive to fluid flow shear stress than to substrate stretching (Klein-Nulend et al. 1995a, b). MLO-Y4 osteocyte-like cells are several orders of magnitude more sensitive to fluid flow shear stress than 2T3 osteoblast-like cells (Kamel et al. 2010). Fluid flow shear stress has numerous biological effects as demonstrated on primary osteocytes and MLO-Y4 cells. Fluid flow shear stress induces the release of nitric oxide, ATP and prostaglandins, opens hemichannels and gap junctions, promotes dendrite elongation, bends cilia, prevents apoptosis, initiates signaling pathways such as the Wnt/b-catenin pathway, PKA, and others, induces b-catenin translocation to the nucleus, activates gene transcription and translation, etc. Many of these effects will be expanded upon in this chapter.

Early cell cultures utilized osteoblast cell lines as models for osteocytes until the osteocyte-like cell line MLO-Y4 became available (For review see Bonewald (2007), Bonewald and Johnson (2008)). Cell lines were preferable for in vitro cultures as primary osteocytes were difficult to isolate in any purity or numbers. Primary cultures of young osteocytes can be prepared by sequential alternating collagenase digestions with EDTA of fetal rat or chick calvaria (Mikuni-Takagaki et al. 1996; van der Plas and Nijweide 1992). These primary osteocyte culture systems have been and continue to be useful in the study of osteocyte markers and function. Mice in which the 8 kb Dentin Matrix Protein 1, DMP1 promoter, (Dmp1 is a gene highly expressed in early osteocytes that regulates mineral homeostasis and matrix formation), driving GFP expression in osteocytes have been generated and fluorescence activated cell sorting can be performed to obtain a highly purified GFP positive population (Kalajzic et al. 2004). However, the yields and viability continue to be low making cell lines preferable for large scale experiments.

Numerous laboratories have used the murine MLO-Y4 osteocyte-like cell line as a model for early osteocytes (Search PubMed for MLO-Y4 or MLOY-4). MLO-Y4 cells exhibit properties of osteocytes including a dendritic morphology, high expression of osteocalcin, low expression of alkaline phosphatase, high expression of connexin 43 and the antigen E11/gp38, a known marker of early osteocytes. Like primary osteocytes these cells are more sensitive to fluid flow shear stress than osteoblast cell lines especially with regards to the release of prostaglandin (Kamel et al. 2010).

Like primary osteocytes, MLO-Y4 cells release prostaglandin, ATP, and nitric oxide in response to low magnitude fluid flow shear stress.

It has been straightforward to study osteoblast and osteoclast function, but not osteocyte function. Clearly one can examine resorption and pit formation by osteoclasts and matrix formation by osteoblasts, but what is a specific osteocyte function? The ideal function proposed for osteocytes is sensitivity to fluid flow shear stress with regards to ATP, nitric oxide, and prostaglandin production. Therefore, many in vitro experiments have utilized the generation of these small molecules to study osteocyte mechanosensation and signaling. In vivo prostaglandin enhances new bone formation in response to mechanical load and indomethacin has been shown to block the effects of anabolic loading in vivo (Forwood 1996). Prostaglandin appears to be released through hemichannels in response to shear stress (Cherian et al. 2005).

Nitric oxide (NO) was suggested as another mechanical mediator correlating with PGE_2 release from osteocytes (Klein-Nulend et al. 1995a), which was supported by the detection of endothelial nitric oxide synthase (eNOS) in osteocytes (Zaman et al. 1999). In bone, NO inhibits resorption and promotes bone formation. Both osteoblasts and osteocytes release NO in response to mechanical strain or fluid flow shear stress (Bakker et al. 2001).

ATP and intracellular calcium can also be released from osteocytes in response to extracellular calcium or mechanical stimulation (Genetos et al. 2007; Kamioka et al. 1995). Not only osteocytes but also neural progenitors and neurons, astrocytes, heart, and osteoblasts have functional hemichannels formed by Cx43 where the opening of hemichannels appears to provide a mechanism for ATP and NAD^+ release, which raises intracellular Ca^{2+} levels (Cherian et al. 2003, 2005). The P2X7 nucleotide receptor, an ATP-gated ion channel expressed in many cell types, may play a role in mechanosensation as deletion resulted in a 70% reduction in bone anabolic response in mice (Li et al. 2005). Fluid flow shear stress did not induce prostaglandin release in cells isolated from these mice. Blockers of P2X7 receptors suppressed prostaglandin release, whereas agonists enhance release in bone cells suggesting that the P2X7 receptor is necessary for release of prostaglandin in response to mechanical load.

Voltage-operated calcium channels (VOCC) that can be regulated by hormones, were shown to be expressed in osteoblasts and osteocytes (Gu et al. 2001b; Shao et al. 2005). Expression of potassium (K^+) channels during differentiation from osteoblasts to osteocytes leads to different K^+ currents between osteocytes and osteoblasts (Gu et al. 2001a). Because of their contribution to the maintenance of the cell membrane potential, and their fast response in 20 ms, K^+ channels and other ion channels may be involved in the earliest initiation of mechanical response in osteocytes. Other ion channels include L-type voltage gated and gadolinium-sensitive cation channels (Ajubi et al. 1999; Mikuni-Takagaki 1999).

Questions now being pursued are: what is the type of fluid flow, steady, pulsatile, oscillatory or combination? What is the magnitude, frequency, duration? How does one replicate in vivo loading in vitro? What are the ideal cells to use for in vitro studies? Theoretical models have been used to predict osteocyte wall shear stresses resulting from peak physiologic loads in-vivo to be in the range of 8–30 dynes/cm^2

(Weinbaum et al. 1994). It will be a significant advance in the field to be able to actually visualize, characterize, and quantitate bone fluid flow within the osteocyte lacunocanalicular system in vivo.

10.4 Cell Body, Cell Process, Cilia

The osteocyte may have several means to sense and transduce loading (Bonewald 2006). It has been proposed that the osteocyte only senses load through its processes (Han et al. 2004; Wang et al. 2007), or through both the cell body and the processes (Nicolella et al. 2008), and/or through cilia, a flagellar-like structure found on every cell (Malone et al. 2007; Xiao et al. 2006). See Fig. 10.1.

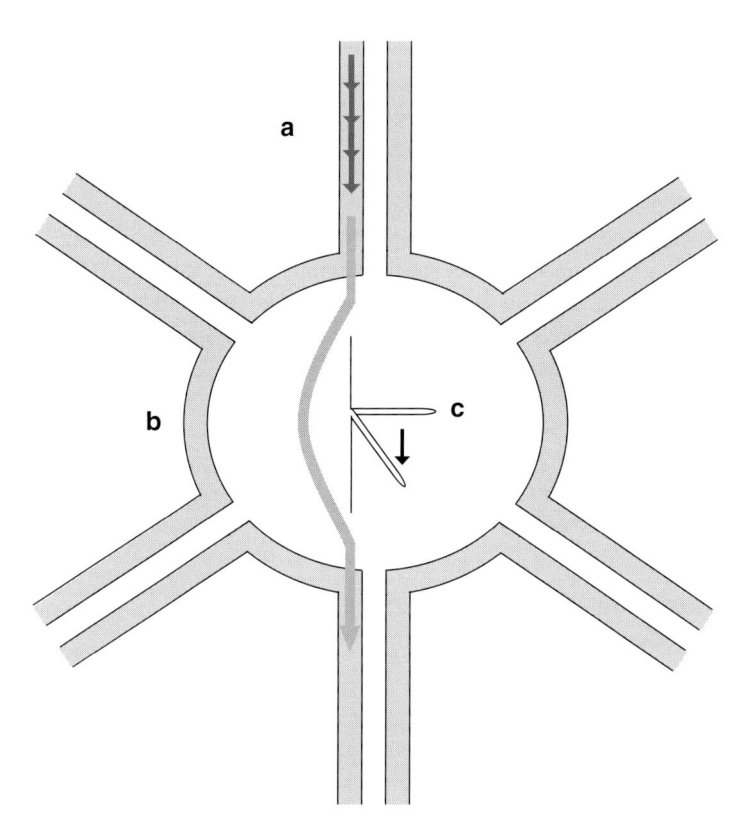

Fig. 10.1 Cartoon showing potential means by which an osteocyte may sense fluid flow shear stress. (**a**) Fluid flow shear stress (*red arrow*) could perturb tethering elements between the canalicular wall and the cell membrane. (**b**) Fluid flow shear stress (*blue arrow*) may also affect the cell body, causing cell deformation. (**c**) Fluid flow may perturb primary cilia leading to mechanosensation. Both matrix and cell deformation are also proposed to play a role in mechanosensation. (Modification of a figure with permission to use from Bonewald (2006) http://www.bonekey-ibms.org/cgi/content/full/ibmske;3/10/7 DOI: 10.1138/20060233)

Wang and colleagues proposed that osteocytes experience amplified local strains based on conical structures protruding from canalicular walls to make contact with the dendrite through integrins. Therefore they proposed that it is the osteocyte dendrite and not the cell body that senses mechanical load. In contrast, Nicolella and colleagues propose that it is the strain concentrations that occur at the lacunae and consequently the cell body that influences mechanosensation by the osteocyte.

In renal epithelial cells, polycystin 1 (PC1) encoded by the gene Pkd1 is part of a mechanosensing complex involving polycystin 2 and primary cilia, therefore it was hypothesized that this complex may play a similar role in osteocytes. PC1, PC2 and primary cilia are expressed in osteocytes and mice with impaired PC1 function develop osteopenia associated with impaired osteoblast-mediated bone formation (Xiao et al. 2006). To determine if PC1 plays a role in osteocyte mechanosensation, conditional deletion of *Pkd1* was performed using Dentin Matrix Protein 1 (*Dmp1*)-Cre mice. Only minor differences were observed in 16 week old *Pkd1*[Dmp1-cKO] mice by DEXA and μCT analysis but a dramatic decrease in response to anabolic loading was observed showing that PC1 in osteocytes is essential for the bone anabolic response to load (Xiao et al. 2009).

Cilia in bone cells do not mediate calcium flux in response to fluid flow, therefore the mechanism used by cilia in bone cells is distinct from that of kidney cells (Malone et al. 2007). Reducing the number of cilia in both MC3T3 and MLO-Y4 osteocyte cells reduces the induction of prostaglandin and in MLO-Y4 cells the increase of COX2 and OPG/RANKL ratio in response to fluid flow shear stress. The regulation of the OPG/RANKL ratio in osteocytes in response to mechanical loading may be a means by which bone formation and resorption can be maintained in equilibrium.

It is not clear if the cell body, cell processes and cilia work separately or in conjunction to sense and transmit mechanical stimuli. Just deleting only one of the three early small molecules released in the response to shear stress, NO, ATP, or prostaglandin, will inhibit bone's anabolic response to loading. Elimination of any one of the three cellular components may also abrogate the mechanical response. For review see Bonewald (2006), Chen et al. (2010). See Fig. 10.1.

10.5 Signaling Pathways Used by Osteocytes in Response to Loading

The Wnt/b-catenin pathway is important for osteoblast differentiation, proliferation and matrix production, whereas in osteocytes this pathway plays a role in transmitting signals of mechanical loading to cells on the bone surface (Bonewald and Johnson 2008). It was hypothesized by Johnson and co-workers that Lrp5/Wnt/b-catenin pathway was involved in osteocyte response to mechanical stimuli in (Johnson et al. 2002) and later shown that mice with deletion of Lrp5 did not fully respond to anabolic loading (Sawakami et al. 2004).

Negative regulators of the Wnt/b-catenin pathway such as Dkk1 and sclerostin/Sost are highly expressed in osteocytes; only sclerostin exclusively in osteocytes in the adult skeleton (Poole et al. 2005). Sclerostin is expressed in mature, not early, osteocytes and inhibits Wnt/b-catenin signaling by binding to Lrp5/6 and preventing the binding of Wnt. Mutations or deletions in *Sost* and *Dkk-1* in humans and/or mice have been shown to result in increased bone mass. Clinical trial studies using antibodies to sclerostin have also been shown to result in increased bone mass, suggesting that targeting of these negative regulators of Wnt/b-catenin signaling pathway might be anabolic treatments for diseases such as osteoporosis. Mechanical loading has been shown to reduce sclerostin levels in bone whereas hindlimb unloading has been shown to increase sclerostin expression (For review see Bonewald and Johnson (2008), ten Dijke et al. (2008)). Downregulation of Dkk1 and Sost may create a permissive environment in which Wnt proteins already present can activate this pathway.

Mechanical loading of the ulna of the TOPGAL mouse, which carries a *LacZ* reporter gene driven by Tcf consensus promoter sequences, results in activation of β-catenin signaling within 1 h after a single loading session only in osteocytes. By 24 h post loading, Wnt/b-catenin signaling activation was observed in cells on the bone surfaces, suggesting that a signal originating in the osteocyte was propagated to surrounding osteocytes and eventually to the bone surface. These data support the concept that the osteocyte is the primary mechanosensory cell in bone (Kim-Weroha et al. 2008).

How could the Wnt/b-catenin pathway be activated within an hour of loading? This is most likely too rapid for the down-regulation of a negative regulator of this pathway to have an effect. Clues came from a publication by Castellone and coworkers showing that PGE_2 could crosstalk with the Wnt/b-catenin pathway in colon cancer cells (Castellone et al. 2005). EP2 receptor activated G proteins activated PI3-kinase, which in turn activated Akt, phosphorylating GSK-3β, thereby inhibiting phosphorylation of β-catenin. At the same time the G_α subunit bound to Axin to induce dissociation of the β-catenin degradation complex. β-catenin signaling was occurring independent of Wnt binding. Secondly, the release of PGE_2 by bone in response to mechanical loading has long been known to be one of the earliest responses to loading. PGE_2 and fluid flow shear stress treatment of MLO-Y4 osteocytes results in increased phosphorylation of GSK-3β, β-catenin nuclear translocation and changes in the expression of β-catenin target genes. Therefore, it was proposed that prostaglandin rapidly activates the Wnt/b-catenin pathway in osteocytes (Bonewald and Johnson 2008). This hypothesis is currently being tested.

Mechanical loading of the MLO-Y4 cells by fluid flow shear stress can protect against dexamethasone induced apoptosis (Kitase et al. 2006). The mechanism of this protective effect of mechanical loading appears to be partially mediated through prostaglandin E2 (PGE_2) crosstalk with β-catenin signaling (Kitase et al. 2010). Therefore the Wnt/b-catenin signaling pathway plays a role not only in bone response to loading but also plays an important in osteocyte apoptosis.

Other signaling pathways are activated in response to mechanical loading and may also crosstalk with the Wnt/b-catenin signaling pathway. Lanyon and co-workers suggests that the estrogen receptor alpha isoform (ER-α) may play a role in shuttling β-catenin into the nucleus in response to mechanical strain in osteoblasts

(Zaman et al. 2006). This may in part explain how estrogen regulates bone mass through a functional intersection through ER-α with Wnt/b-catenin signaling.

Fluid flow-induced shear stress stimulates gap junction function and increases Cx43 expression by the release of prostaglandin. In an autocrine/paracrine fashion, prostaglandin works through the EP_2 receptor, increasing intracellular cAMP and activating PKA. Connexins not only form gap junctions but can function as un-apposed halves of gap junction channels called hemichannels, not requiring physical contact with adjacent cells. Hemichannels expressed in MLO-Y4 cells function as transducers of the anti-apoptotic effects of bisphosphonates and directly serve as a portal or pathway for the exit of intracellular PGE_2 in osteocytes induced by fluid flow shear stress. Therefore, the release of prostaglandin in response to shear stress through hemichannels has an autocrine and paracrine effect on osteocytes. For review see Bonewald (2007), Jiang et al. (2007).

10.6   Role of Integrins in Osteocyte Mechanotransduction

One of the first adhesion receptor to be described on osteocytes was CD44, highly expressed on the osteocyte surface (Hughes et al. 1994). One of the first acting associating protein described fairly specific for osteocytes compared to osteoblasts was the actin-bundling protein fimbrin, expressed where the dendritic processes branch (Tanaka-Kamioka et al. 1998). Both are known to connect to and influence the cell cytoskeleton. However, it is becoming more evident that integrins, found on many cell types, may play a major role in osteocyte mechanosensation. It was shown in cultured endothelial cells, that mechanical perturbation of the cell membrane using integrin-bound beads induces cytoskeletal rearrangements (Wang and Ingber 1994) suggesting that the integrin-cytoskeleton complex may play a role as an intracellular signal transducer for stress signals. Deletion of b1 integrin in osteo-blasts and consequently osteocytes resulted in a skeleton that did not lose bone in response to disuse. This suggests that integrin is important in both osteoblast and osteocyte mechanosensation (Phillips et al. 2008).

Integrins, comprised of heterodimers of α and β subunits, serve as the major receptors/transducers that connect the cytoskeleton to the extracellular matrix and have been proposed to be the candidate mechanosensors in bone cells (Salter et al. 1997; Plotkin et al. 2005). Integrins appear to bridge osteocyte processes to their canalicular wall (You et al. 2004). McNamara and colleagues reported that the osteocyte canalicular wall appears to have a corrugated appearance where projec-tions appear to make contact with the cell membrane of dendritic processes (McNamara et al. 2009). They also described integrin avb3 at these attachment sites. This observation led to the hypothetical model that axial strains in the dendrite are several orders of magnitude greater than whole tissue strains (Wang et al. 2008). These authors therefore propose that the osteocyte dendritic process is the major cellular component that senses loading.

Recently it was found that Integrin α5β1 interacts with Cx43 to mediate the opening of hemichannels in response to mechanical stimulation. Surprisingly, this interaction was independent of the integrin's association with fibronectin and interaction with the extracellular matrix (Burra in review). The interaction between integrin a5 and Cx43 is important in the regulation and extracellular release of prostaglandin, an essential transducer of the effects of anabolic loading. Therefore, fluid flow shear stress may have two major effects on the osteocyte mediated through integrins. The first is the well known kinetics of the integrin acting as a linker between the extracellular matrix and the intracellular cytoskeleton. The second, novel effect is through the opening of hemichannels releasing small molecules with autocrine and paracrine receptor mediated effects. The combination of these two effects and perhaps others, may synergize to produce the dramatic effects of mechanical loading on the cell. See Fig. 10.2.

Two Roles for Integrin in Osteocyte Mechanotransduction

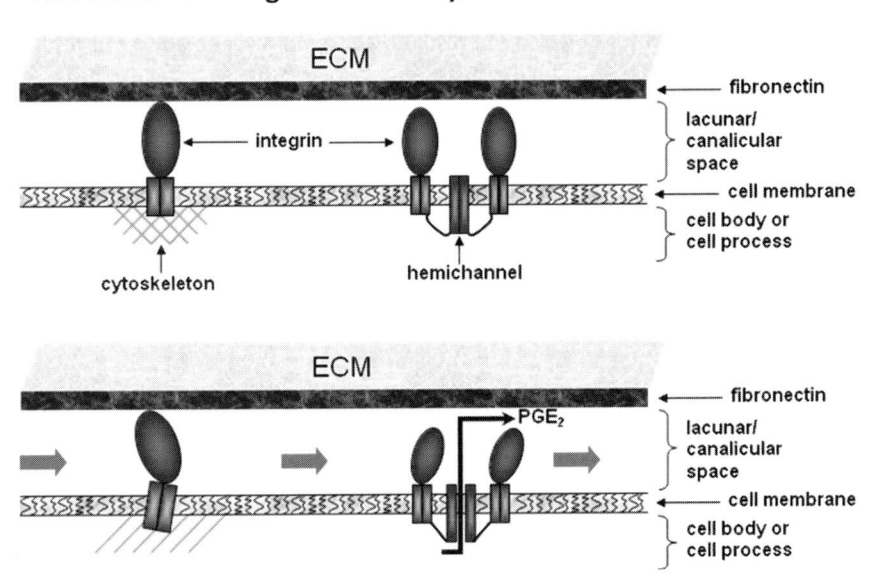

Fig. 10.2 Two potential roles for integrin in osteocyte mechanotransduction. First role: Integrin may form a bridge from the cell membrane to the canalicular wall. With the application of mechanical strain in the form of fluid flow shear stress (*red arrows*) the integrin connection with the cytoskeleton is perturbed causing a rearrangement and signaling. Second role: Integrin can be unattached to the canalicular wall, but still perturbed by shear stress in a manner causing interaction with connexin 43 hemichannels in a manner to open the hemichannel. This allows the release of prostaglandin and potentially other small signaling molecules which can signal in an autocrine and/or paracrine fashion. Therefore, fluid flow shear stress can induce mechanotransduction in the osteocyte by two means, one through rearrangement of the cytoskeleton and the second through receptor mediated signaling pathways

10.7   Influence of the Osteocyte Perilacunar Matrix on Mechanosensation

The osteocyte appears to have the capacity to modify its microenvironment. The removal of the its perilacunar matrix has been referred to as "osteolytic osteolysis" and was initially used to describe the enlarged lacunae in patients with hyperparathyroidism (Belanger 1969) and later in immobilized rats (Kremlien et al. 1976). The term "osteocyte halos" was used by Heuck (Heuck 1970) to describe pericanicular demineralization in rickets and later by others to describe periosteocytic lesions in X-linked hypophosphatemic rickets (Marie and Glorieux 1983), a condition due to an inactivating mutation in Phex. Such periosteocytic lesions are not present in other chronic hypophosphatemic states. Mice injected with pellets releasing prednisolone showed an enlargement of osteocyte lacunae in trabecular bone and the generation of a surrounding sphere of hypomineralized bone (Lane et al. 2006). As lacunae act as stress concentrators in bone, it was proposed that these changes in properties of perilacunar bone matrix may influence fracture risk in glucocorticoid-treated patients.

Qing and coworkers found that the osteocyte can not only remove, but replace its perilacunar and pericanalicular matrix in response to lactation followed by forced weaning (Qing et al. 2009). This capacity has also been described in egg-laying hens (Zambonin Zallone et al. 1983). These observations suggest that the healthy osteocyte can both add and remove mineral from its lacunae and canaliculi. In contrast to the young, healthy osteocyte, the aging osteocyte may allow hypermineralization of its perilacunar matrix (Potter et al. 2009) and micropetrosis has been described in aging bone where the lacunae fill in with mineral. The capacity to deposit or remove mineral from lacunae and canaliculi, modification of perilacunar matrix properties, and infilling of lacunae and/or canaliculi has important implications with regards to magnitude of fluid shear stress applied to the osteocyte in addition to the mechanical properties of bone.

Nicolella and colleagues reported that the osteocyte lacuna acts as a strain concentrator that amplifies the macroscopic strain applied to the whole bone and this amplification factor is a function of the local peri-lacuna bone tissue material properties (Nicolella et al. 2006, 2008). Using a microstructural finite element analysis model, they found that changes in the osteocyte cell body and cell process modulus had little effect on the maximum strain in the osteocyte, the average strain in the cell process, or on the maximum strain in the lacuna. However, changing the material properties of the perilacunar matrix had the greatest impact on the strain transmitted to the osteocyte, with the maximum osteocyte strain relating inversely to the perilacunar tissue modulus. Therefore, any mechanism that changes the material properties of the perilacunar matrix will have consequences on mechanosensation by osteocytes. See Fig. 10.3.

If mechanical loading is a major regulator of osteocyte function, what does this mean for the diseased osteocyte in hypophosphatemic rickets or exposed to glucocorticoid? What happens to the osteocyte that is hormonally regulated to remove

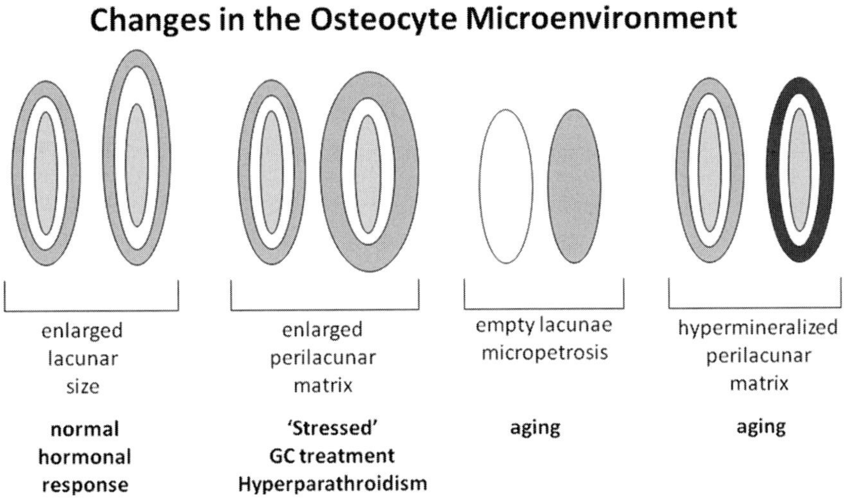

Changes in the Osteocyte Microenvironment

enlarged lacunar size	enlarged perilacunar matrix	empty lacunae micropetrosis	hypermineralized perilacunar matrix
normal hormonal response	**'Stressed' GC treatment Hyperparathroidism**	**aging**	**aging**

Fig. 10.3 Changes in the osteocyte microenvironment. It appears that the osteocyte has the capacity to remodel its extracellular matrix. With lactation in mammals or with egg-laying in hens, the osteocyte appears to have the capacity to remove and replace mineralized matrix. This appears to be a normal, healthy hormonal response. However, under conditions of stress such as glucocorticoid treatment or with hyperparathyroidism, the osteocyte continues to remove its perilacunar matrix and the surrounding area of hypomineralized matrix is increased. It is well known that with aging, some osteocyte lacunae either become empty or are filled with mineral by a process called micropetrosis. Recently it has been proposed that the osteocyte may also actually increase the mineral content of its perilacunar matrix. All of these states have implications with regards to how the osteocyte will sense mechanical loading

and replace both its perilacunar and pericanalicular matrix and in lactation and weaning? Are the effects of mechanical loading ignored or overridden in the case of enlarged lacunae with lactation? Are the effects of mechanical loading diminished with enlargement of lacunae and hypomineralization of the perilacunar matrix? How does hypermineralization of the perilacunar matrix and infilling of lacunae that occurs with age affect response to mechanical load? These are questions that remain to be answered.

10.8   Looking to the Future

Mechanobiology of the osteocyte remains a field with many unanswered questions. To answer these questions new in vitro and in vivo experimental approaches are required. Better cell lines that more closely mimic the properties and functions of primary osteocytes are required as are three dimensional matrices that more closely resemble the bone matrix. Better technology is required to visualize the osteocyte and its processes within living bone and to monitor bone fluid flow. Reporter mice with promoters from E11/gp38, Mepe, Phex, Dmp1, Sost and other osteocyte selective/specific markers need to be generated to determine the effects of different forms of loading and unloading on gene expression. Each of these genes may be differentially regulated by mechanical strain and these genes may have different loading thresholds for transcription. Targeted deletion of genes in osteocytes will need to be performed. Using the10 kb Dmp1-Cre, genes can be deleted in early osteocytes and using the Sost promoter, genes can be deleted in late osteocytes. Genes may have very different functions in osteocytes compared to other bone cells. Inducible deletion or expression of genes in osteocytes using the Dmp1 and Sost promoters will allow removal of genes at specific ages in osteocytes. Early and late osteocytes may have different functions. In addition to predicted or hypothesized results, it will be the totally unpredicted results that will ignite the imagination.

Acknowledgment   The author's work in osteocyte biology is supported by the National Institutes of Health AR-46798.

References

Ajubi NE, Klein-Nulend J, Alblas MJ, Burger EH, Nijweide PJ (1999) Signal transduction pathways involved in fluid flow-induced PGE2 production by cultured osteocytes. Am J Physiol 276:E171–E178

Bakker AD, Soejima K, Klein-Nulend J, Burger EH (2001) The production of nitric oxide and prostaglandin E(2) by primary bone cells is shear stress dependent. J Biomech 34:671–677

Belanger LF (1969) Osteocytic osteolysis. Calcif Tissue Res 4:1–12

Bergula AP, Huang W, Frangos JA (1999) Femoral vein ligation increases bone mass in the hindlimb suspended rat. Bone 24:171–177

Bonewald LF (2006) Mechanosensation and transduction in osteocytes. Bonekey Osteovision 3:7–15

Bonewald L (2007) Osteocytes. In: Marcus DFR, Nelson D, Rosen C (eds) Osteoporosis. Elsevier, pp 169–190

Bonewald LF, Johnson ML (2008) Osteocytes, mechanosensing, and Wnt signaling. Bone 42:606–615

Burra S, Siller-Jackson A, Harris SE, Weber G, DeSimone D, Dallas S, Bonewald LF, Sprague E, Schwartz MA, Jiang J Role of integrin a5 in the regulation of osteocyte mechanotransduction via opening of Cx43 hemichannels (in review)

Castellone MD, Teramoto H, Williams BO, Druey KM, Gutkind JS (2005) Prostaglandin E2 promotes colon cancer cell growth through a Gs-axin-beta-catenin signaling axis. Science 310:1504–1510

Chen JH, Liu C, You L, Simmons CA (2010) Boning up on Wolff's law: mechanical regulation of the cells that make and maintain bone. J Biomech 43:108–118

Cherian PP, Cheng B, Gu S, Sprague E, Bonewald LF, Jiang JX (2003) Effects of mechanical strain on the function of gap junctions in osteocytes are mediated through the prostaglandin EP2 receptor. J Biol Chem 278:43146–43156

Cherian PP, Siller-Jackson AJ, Gu S, Wang X, Bonewald LF, Sprague E, Jiang JX (2005) Mechanical strain opens connexin 43 hemichannels in osteocytes: a novel mechanism for the release of prostaglandin. Mol Biol Cell 16:3100–3106

Forwood MR (1996) Inducible cyclo-oxygenase (COX-2) mediates the induction of bone formation by mechanical loading in vivo. J Bone Miner Res 11:1688–1693

Genetos DC, Kephart CJ, Zhang Y, Yellowley CE, Donahue HJ (2007) Oscillating fluid flow activation of gap junction hemichannels induces ATP release from MLO-Y4 osteocytes. J Cell Physiol 212:207–214

Gluhak-Heinrich J, Ye L, Bonewald LF, Feng JQ, MacDougall M, Harris SE, Pavlin D (2003) Mechanical loading stimulates dentin matrix protein 1 (DMP1) expression in osteocytes in vivo. J Bone Miner Res 18:807–817

Gluhak-Heinrich J, Pavlin D, Yang W, Macdougall M, Harris SE (2007) MEPE expression in osteocytes during orthodontic tooth movement. Arch Oral Biol 52:684–690

Gu Y, Preston MR, El Haj AJ, Howl JD, Publicover SJ (2001a) Three types of K(+) currents in murine osteocyte-like cells (MLO-Y4). Bone 28:29–37

Gu Y, Preston MR, Magnay J, El Haj AJ, Publicover SJ (2001b) Hormonally-regulated expression of voltage-operated Ca(2+) channels in osteocytic (MLO-Y4) cells. Biochem Biophys Res Commun 282:536–542

Han Y, Cowin SC, Schaffler MB, Weinbaum S (2004) Mechanotransduction and strain amplification in osteocyte cell processes. Proc Natl Acad Sci USA 101:16689–16694

Heuck F (1970) Comparative investigations of the function of osteocytes in bone resorption. Calcif Tissue Res Suppl:148–149

Hughes DE, Salter DM, Simpson R (1994) CD44 expression in human bone: a novel marker of osteocytic differentiation. J Bone Miner Res 9:39–44

Jiang JX, Siller-Jackson AJ, Burra S (2007) Roles of gap junctions and hemichannels in bone cell functions and in signal transmission of mechanical stress. Front Biosci 12:1450–1462

Johnson ML, Picconi JL, Recker RR (2002) The gene for high bone mass. Endocrinologist 12:445–453

Kalajzic I, Braut A, Guo D, Jiang X, Kronenberg MS, Mina M, Harris MA, Harris SE, Rowe DW (2004) Dentin matrix protein 1 expression during osteoblastic differentiation, generation of an osteocyte GFP-transgene. Bone 35:74–82

Kamel MA, Picconi JL, Lara-Castillo N, Johnson ML (2010) Activation of β-catenin signaling in MLO-Y4 osteocytic cells versus 2T3 osteoblastic cells by fluid flow shear stress and PGE(2): Implications for the study of mechanosensation in bone. Bone 47:872–81. Epub 2010 Aug 14

Kamioka H, Miki Y, Sumitani K, Tagami K, Terai K, Hosoi K, Kawata T (1995) Extracellular calcium causes the release of calcium from intracellular stores in chick osteocytes. Biochem Biophys Res Commun 212:692–696

Kim-Weroha NA, Ferris A, Holladay BA, Kotha SP, Kamel MA, Johnson ML (2008) In vivo load activated propagation of b-catenin signaling in osteocytes through coordinated downregulation of inhibitors of Lrp5. J Bone Miner Res 23(Suppl 1):S13 (abstract 1041)

Kitase Y, Jiang JX, Bonewald L (2006) The anti-apoptotic effects of mechanical strain on osteocytes are mediated by PGE2 and monocyte chemotactic protein, (MCP-3); selective protection my MCP3 against glucocorticoid (GC) and not TNF-a induced apoptosis. J Bone Miner Res 21:S48

Kitase Y, Barragan L, Jiang JX, Johnson ML, Bonewald LF (2010) Mechanical induction of PGE(2) in osteocytes blocks glucocorticoid induced apoptosis through both the β-catenin and PKA pathways. J Bone Miner Res PMID:20578217

Klein-Nulend J, Bonewald LF (2008) The osteocyte. In: Bilezikian JP, Raisz LG (eds) Principles of bone biology. Academic, San Diego

Klein-Nulend J, Semeins CM, Ajubi NE, Nijweide PJ, Burger EH (1995a) Pulsating fluid flow increases nitric oxide (NO) synthesis by osteocytes but not periosteal fibroblasts–correlation with prostaglandin upregulation. Biochem Biophys Res Commun 217:640–648

Klein-Nulend J, van der Plas A, Semeins CM, Ajubi NE, Frangos JA, Nijweide PJ, Burger EH (1995b) Sensitivity of osteocytes to biomechanical stress in vitro. FASEB J 9:441–445

Kremlien B, Manegold C, Ritz E, Bommer J (1976) The influence of immobilization on osteocyte morphology: osteocyte differential count and electron microscopic studies. Virchows Arch A Pathol Anat Histol 370:55–68

Lane NE, Yao W, Balooch M, Nalla RK, Balooch G, Habelitz S, Kinney JH, Bonewald LF (2006) Glucocorticoid-treated mice have localized changes in trabecular bone material properties and

osteocyte lacunar size that are not observed in placebo-treated or estrogen-deficient mice. J Bone
Miner Res 21:466–476

Li J, Liu D, Ke HZ, Duncan RL, Turner CH (2005) The P2X7 nucleotide receptor mediates skeletal
mechanotransduction. J Biol Chem 280:42952–42959

Lin C, Jiang X, Dai Z, Guo X, Weng T, Wang J, Li Y, Feng G, Gao X, He L (2009) Sclerostin
mediates bone response to mechanical unloading through antagonizing Wnt/beta-catenin signaling. J Bone Miner Res 24:1651–1661

Malone AM, Anderson CT, Tummala P, Kwon RY, Johnston TR, Stearns T, Jacobs CR (2007)
Primary cilia mediate mechanosensing in bone cells by a calcium-independent mechanism.
Proc Natl Acad Sci USA 104:13325–13330

Marie PJ, Glorieux FH (1983) Relation between hypomineralized periosteocytic lesions and bone
mineralization in vitamin D-resistant rickets. Calcif Tissue Int 35:443–448

McNamara LM, Majeska RJ, Weinbaum S, Friedrich V, Schaffler MB (2009) Attachment of
osteocyte cell processes to the bone matrix. Anat Rec (Hoboken) 292:355–363

Mikuni-Takagaki Y (1999) Mechanical responses and signal transduction pathways in stretched
osteocytes. J Bone Miner Metab 17:57–60

Mikuni-Takagaki Y, Suzuki Y, Kawase T, Saito S (1996) Distinct responses of different populations of bone cells to mechanical stress. Endocrinology 137:2028–2035

Nicolella DP, Moravits DE, Gale AM, Bonewald LF, Lankford J (2006) Osteocyte lacunae tissue
strain in cortical bone. J Biomech 39:1735–1743

Nicolella DP, Feng JQ, Moravits DE, Bonivitch AR, Wang Y, Dusecich V, Yao W, Lane N, Bonewald
LF (2008) Effects of nanomechanical bone tissue properties on bone tissue strain: implications
for osteocyte mechanotransduction. J Musculoskelet Neuronal Interact 8:330–331

Phillips JA, Almeida EA, Hill EL, Aguirre JI, Rivera MF, Nachbandi I, Wronski TJ, van der
Meulen MC, Globus RK (2008) Role for beta1 integrins in cortical osteocytes during acute
musculoskeletal disuse. Matrix Biol 27:609–618

Plotkin LI, Mathov I, Aguirre JI, Parfitt AM, Manolagas SC, Bellido T (2005) Mechanical stimulation prevents osteocyte apoptosis: requirement of integrins, Src kinases, and ERKs. Am J
Physiol Cell Physiol. 289:C633–643

Poole KE, van Bezooijen RL, Loveridge N, Hamersma H, Papapoulos SE, Lowik CW, Reeve J
(2005) Sclerostin is a delayed secreted product of osteocytes that inhibits bone formation.
FASEB J 19:1842–1844

Potter R, Miller M, Moravits D, Havill L, Bonewald LF, Nyman J, Nicolella D (2009) Raman
spectroscopic characterization of bone tissue material properties around the osteocyte lacuna:
effect of aging. J Bone Miner Res Suppl 1:Su0266

Qing H, Ardeshirpour L, Dusevich V, Wysolmerski J, Bonewald LF (2009) Osteocyte perilacunar
remodeling is regulated hormonally, but not by mechanical unloading. J Bone Miner Res Suppl
1:Mo0255

Robling AG, Hinant FM, Burr DB, Turner CH (2002) Shorter, more frequent mechanical loading
sessions enhance bone mass. Med Sci Sports Exerc 34:196–202

Robling AG, Bellido T, Turner CH (2006) Mechanical stimulation in vivo reduces osteocyte
expression of sclerostin. J Musculoskelet Neuronal Interact 6:354

Rubin C (1984) Skeletal strain and the functional significance of bone architecture. Calcif Tissue
Int 36:S11–S18

Salter DM, Robb JE, Wright MO (1997) Electrophysiological responses of human bone cells to
mechanical stimulation: evidence for specific integrin function in mechanotransduction.
J Bone Min Res 12:1133–1141

Sawakami K, Robling AG, Pitner ND, Warden SJ, Li J, Warman ML, Turner CH (2004) Site-
specific osteopenia and decreased mechanoreactivity in Lrp5 mutant mice. J Bone Miner Res
19(Suppl 1):S38

Shao Y, Alicknavitch M, Farach-Carson MC (2005) Expression of voltage sensitive calcium channel (VSCC) L-type Cav1.2 (alpha1C) and T-type Cav3.2 (alpha1H) subunits during mouse
bone development. Dev Dyn 234:54–62

Skerry TM, Bitensky L, Chayen J, Lanyon LE (1989) Early strain-related changes in enzyme activity in osteocytes following bone loading in vivo. J Bone Miner Res 4:783–788

Tanaka-Kamioka K, Kamioka H, Ris H, Lim SS (1998) Osteocyte shape is dependent on actin filaments and osteocyte processes are unique actin-rich projections. J Bone Miner Res 13:1555–1568

ten Dijke P, Krause C, de Gorter DJ, Lowik CW, van Bezooijen RL (2008) Osteocyte-derived sclerostin inhibits bone formation: its role in bone morphogenetic protein and Wnt signaling. J Bone Joint Surg Am 90(Suppl 1):31–35

Turner CH, Forwood MR, Otter MW (1994) Mechanotransduction in bone: do bone cells act as sensors of fluid flow? FASEB J 8:875–878

van der Plas A, Nijweide PJ (1992) Isolation and purification of osteocytes. J Bone Miner Res 7:389–396

Wang N, Ingber DE (1994) Control of cytoskeletal mechanics by extracellular matrix, cell shape, and mechanical tension. Biophys J 66:2181–2189

Wang L, Ciani C, Doty SB, Fritton SP (2004) Delineating bone's interstitial fluid pathway in vivo. Bone 34:499–509

Wang Y, McNamara LM, Schaffler MB, Weinbaum S (2007) A model for the role of integrins in flow induced mechanotransduction in osteocytes. Proc Natl Acad Sci USA 104:15941–15946

Wang Y, McNamara LM, Schaffler MB, Weinbaum S (2008) Strain amplification and integrin based signaling in osteocytes. J Musculoskelet Neuronal Interact 8:332–334

Weinbaum S, Cowin SC, Zeng Y (1994) A model for the excitation of osteocytes by mechanical loading-induced bone fluid shear stresses. J Biomech 27:339–360

Wolff J (1982) Das Gesetz der Transformation der Knochen. A Hirschwald, Berlin

Xiao Z, Zhang S, Mahlios J, Zhou G, Magenheimer BS, Guo D, Dallas SL, Maser R, Calvet JP, Bonewald L, Quarles LD (2006) Cilia-like structures and polycystin-1 in osteoblasts/osteocytes and associated abnormalities in skeletogenesis and Runx2 expression. J Biol Chem 281:30884–30895

Xiao Z, Dallas M, Zhang S, Nicolella D, He D, Qiu N, Cao L, Johnson M, Bonewald LF, Quarles D (2009) Conditional deletion and/or disruption of Pkd1 in osteocytes results in a significant reduction in anabolic response to mechanical loading. J Bone Miner Res Suppl 1:1042

Yang W, Lu Y, Kalajzic I, Guo D, Harris MA, Gluhak-Heinrich J, Kotha S, Bonewald LF, Feng JQ, Rowe DW, Turner CH, Robling AG, Harris SE (2005) Dentin matrix protein 1 gene cis-regulation: use in osteocytes to characterize local responses to mechanical loading in vitro and in vivo. J Biol Chem 27:20680–20690

You LD, Weinbaum S, Cowin SC, Schaffler MB (2004) Ultrastructure of the osteocyte process and its pericellular matrix. Anat Rec A Discov Mol Cell Evol Biol. 278:505–513

Zaman G, Pitsillides AA, Rawlinson SC, Suswillo RF, Mosley JR, Cheng MZ, Platts LA, Hukkanen M, Polak JM, Lanyon LE (1999) Mechanical strain stimulates nitric oxide production by rapid activation of endothelial nitric oxide synthase in osteocytes. J Bone Miner Res 14:1123–1131

Zaman G, Jessop HL, Muzylak M, De Souza RL, Pitsillides AA, Price JS, Lanyon LL (2006) Osteocytes use estrogen receptor alpha to respond to strain but their ERalpha content is regulated by estrogen. J Bone Miner Res 21:1297–1306

Zambonin Zallone A, Teti A, Primavera MV, Pace G (1983) Mature osteocytes behaviour in a repletion period: the occurrence of osteoplastic activity. Basic Appl Histochem 27:191–204

Zhang K, Barragan-Adjemian C, Ye L, Kotha S, Dallas M, Lu Y, Zhao S, Harris M, Harris SE, Feng JQ, Bonewald LF (2006) E11/gp38 selective expression in osteocytes: regulation by mechanical strain and role in dendrite elongation. Mol Cell Biol 26:4539–4552

Chapter 11
Mechanosensing and Signaling Crosstalks

Toshio Matsumoto, Rika Kuriwaka-Kido, and Shinsuke Kido

Abbreviations

AP-1	Activator protein-1
ATP	Adenosine triphosphate
BMP	Bone morphogenetic protein
BR-Smads	BMP-specific receptor-regulated Smads
cNOS	Constitutive nitric oxide synthase
CREB	Cyclic AMP response element-binding protein
Dkk	Dickkopf
ERK	Extracellular signal-regulated kinase
FSS	Fluid shear stress
IL-11	Interleukin-11
IP_3	Inositol triphosphate
LRP	Low-density lipoprotein receptor-related protein
OPG	Osteoprotegerin
PGE2	Prostaglandin E2
PKCδ	Protein kinase Cδ
PLC	Phospholipase C
RANKL	Receptor activator of nuclear factor-κB ligand
SA-Cat	Stress-activated cation channel
SBE	Smad-binding element
TGF-β	Transforming growth factor-β

T. Matsumoto (✉), R. Kuriwaka-Kido, and S. Kido
Department of Medicine and Bioregulatory Sciences,
The University of Tokushima and Graduate School of Medical Sciences,
3-18-15 Kuramoto-cho, Tokushima 770-8503, Japan
e-mail: toshimat@clin.med.tokushima-u.ac.jp

M. Noda (ed.), *Mechanosensing Biology*,
DOI 10.1007/978-4-431-89757-6_11, © Springer 2011

11.1   Introduction

Mechanical stress to bone plays an important role in the maintenance of bone homeostasis. Mechanical unloading by prolonged bed rest, immobilization, or microgravity in space causes a marked loss of bone due to an imbalance between bone formation and resorption. While enhanced bone resorption in the endosteal surface is a major feature of unloading-induced bone loss in mature animals and humans, the impairment of bone formation in the periosteal surface constitutes an important mechanism for unloading-induced bone loss especially in the growing stage (Jaworski et al. 1980; Kodama et al. 1997; Morey and Baylink 1978). Therefore, it is important to understand the mechanism whereby mechanical loading enhances and unloading reduces bone formation.

11.2   Mechanosensors and Signaling Systems

In order to respond to mechanical stress, cells need to be equipped with mechanosensors. There are several systems in osteoblastic cells that are reported to perceive mechanical stress signals. These include mechanosensitive channel and adenosine triphosphate (ATP) release to respond to fluid shear stress (FSS), and integrins and cytoskeletal proteins to respond to tensile stress (Pavalko et al. 1998). Although fluid flow along cell surfaces produces not only FSS but also electric potential as stress-generated potential, cells appear to be more sensitive to fluid shear than to electric potential (Hung et al. 1996), and it is unknown whether electric potential activates different signaling systems from those caused by FSS. After osteoblastic cells sense mechanical stress, they respond to the stress by activating the downstream signaling systems. These include Ca^{2+} influx via mechanosensitive stress-activated cation channel (SA-Cat) which activates extracellular signal-regulated kinase (ERK), ATP release from cells which activates G protein-coupled ATP receptors, ionotropic P2X7 and metabotropic P2Y receptors. P2X7 stimulates prostaglandin E2 (PGE2) synthesis, while P2Y activates phospholipase C (PLC) which induces inositol triphosphate (IP_3) release to stimulate Ca^{2+} efflux from intracellular stores (Genetos et al. 2005; Nishii et al. 2009). Early changes in osteoblastic cells in response to mechanical stress to bone are summarized in Table 11.1.

   Among those mechanosensing and signaling systems, the increase in the intracellular Ca^{2+} can activate constitutive nitric oxide synthase (cNOS) to increase NO production. NO is a strong inhibitor of bone resorption, and is reported to suppress the expression of receptor activator of nuclear factor-κB ligand (RANKL) and increases osteoprotegerin (OPG) expression in bone marrow stromal cells (Fan et al. 2004). Therefore, the increase in NO production may be important for the suppression of bone resorption by mechanical stress. In contrast, overexpression of some members of Fos family proteins including ΔFosB, Fra-1 and Fra-2 is shown to enhance bone formation (Jochum et al. 2000; Sabatakos et al. 2000), and the expression of not only c-Fos but also FosB/ΔFosB is rapidly enhanced after mechanical stress (Inoue et al. 2004). In addition, PGE2 signaling via EP4 receptor is also shown to enhance bone formation (Yoshida et al. 2002). Thus, these signaling systems appear to play important roles in mechanical stress-induced stimulation of bone formation. Because EP4-mediated signals also activate cyclic AMP response element-binding protein (CREB) signaling (Fujino et al. 2005), this signal may merge with the Ca^{2+}–ERK–CREB–FosB/ΔFosB signal cascade. Therefore, we will mainly discuss the Fos family-mediated signaling and its crosstalks with other signaling systems in response to mechanical stress.

11.3 Fos Family Gene Expression in Response to Mechanical Stress

Mechanical stress to bone causes a rapid fluid flow surrounding osteoblasts and osteocytes, and elicits FSS to these cells. FSS is shown to be one of the most important signal transduction mechanisms to enhance osteoblast differentiation and bone formation in response to mechanical loading to bone (Burr et al. 2002; Knothe Tate 2003). FSS rapidly stimulates intracellular signaling cascade in cells of the osteoblast lineage: stimulation of gadolinium-sensitive SA-Cat with an increase in intracellular calcium, activation of ERK, and phosphorylation of CREB by ERK (Kido et al. 2009; Liu et al. 2008; You et al. 2001).

Table 11.1 Early response to mechanical stress in osteoblastic cells

Mechanical stress	Mechanosensor	Receptor/ mediator	Signaling system	Downstream signals	References
Fluid shear	Mechanosensitive channel	Ca^{2+} influx	ERK–CREB– Fos family	IL-11- canonical Wnt signal	Inoue et al. (2004), Kido et al. (2009), Liu et al. (2008), You et al. (2001)
–	ATP release	P2X7 receptor	PGE2	EP4	Genetos et al. (2005), Yoshida et al. (2002), Fujino et al. (2005)
		P2Y receptor	Gq-PLC-IP_3-Ca^{2+} release	NOS	Fan et al. (2004), Nishii et al. (2009)
Stretch	Integrins Cytoskeletal proteins	Amplifier of mechanical signals?	–	–	Pavalko et al. (1998)
Electric potential	?	–	–	–	Hung et al. (1996)

One of the earliest responding factors induced by mechanical loading in bone cells is an activator protein (AP)-1 family transcription factor, c-Fos (Lean et al. 1996). However, because ubiquitous overexpression of c-Fos in transgenic mice develops osteosarcoma without evidence for increased bone formation (Grigoriadis et al. 1993), c-Fos is unlikely to be a factor mediating mechanical stress to bone formation. Mechanical loading to bone in vivo or FSS to osteoblasts in vitro also induces *fosB* gene transcription with a similar time course to that in *c-fos* gene transcription. The increase in *fosB* gene transcription is mediated via the Ca^{2+}–ERK–CREB signaling pathway, and activated CREB stimulates *fosB* gene transcription through binding to a putative CRE site in *fosB* gene promoter (Fig. 11.1) (Inoue et al. 2004). Enhanced *fosB* gene transcription by mechanical stress causes an accumulation of mainly ΔFosB protein, a short splice variant of FosB lacking C-terminal transactivation domain (Nakabeppu and Nathans 1991). Transgenic overexpression of ΔFosB has been shown to stimulate bone formation and cause osteosclerosis in mice (Sabatakos et al. 2000). Because ΔFosB protein has a long half-life which enables this protein to accumulate after transient stimulations (Chen et al. 1997), ΔFosB can be a suitable mediator of intermittent mechanical loading signal to sustained bone formation signal.

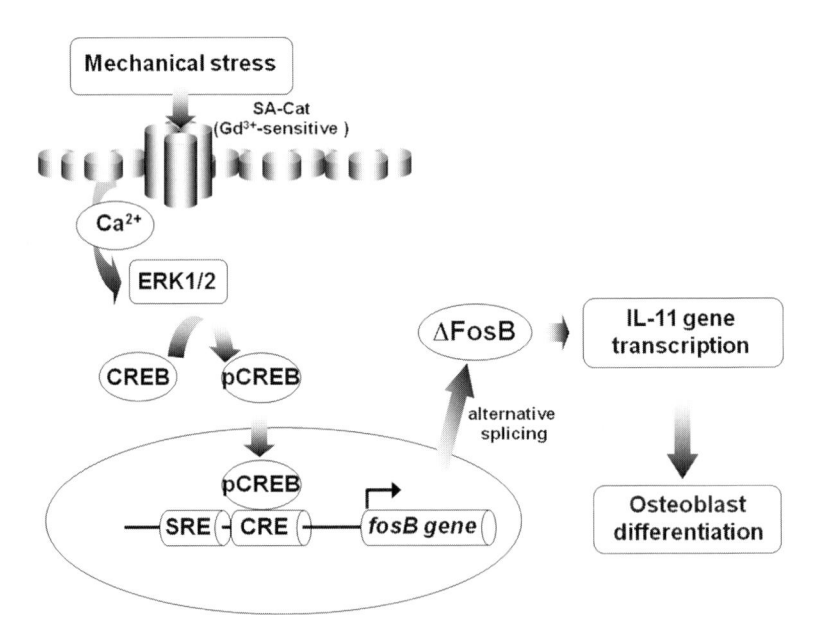

Fig. 11.1 *FosB* gene transcription mediated via Ca^{2+}–ERK–CREB signaling pathway. Fluid shear stress to osteoblasts causes Ca^{2+} influx by activating gadolinium-sensitive stress-activated cation channel (SA-Cat), which activates ERK to phosphorylate CREB. Phosphorylated CREB binds to the *fosB* gene promoter to enhance its transcriptional activity, and upregulates the expression of mainly ΔFosB, a C-terminal truncated splice variant of FosB

11.4   Enhanced IL-11 Expression by Mechanical Stress

Interleukin (IL)-11 is expressed in bone marrow stromal cells and is involved in the regulation of multiple biological processes such as enhancement of myeloid cell growth (Du and Williams 1997; Keller et al. 1993), inhibition of adipogenesis (Kawashima et al. 1991; Keller et al. 1993), and stimulation of osteoblastogenesis (Kodama et al. 1998). Mechanical unloading causes a marked reduction in the expression of IL-11 mRNA in the unloaded bone, and mechanical loading causes a rapid and robust increase in IL-11 mRNA expression in the loaded bone (Fig. 11.2) (Kido et al. 2009). The expression of IL-11 is reduced with decreased JunD binding to the *IL-11* gene promoter by aging (Tohjima et al. 2003). In contrast, transgenic mice overexpressing IL-11 exhibit increased bone mass with increased bone formation without change in bone resorption, and are protected from aging-associated bone loss (Takeuchi et al. 2002). These observations are consistent with the notion that the enhanced IL-11 expression may mediate the stimulation of bone formation in response to mechanical stress, whereas a reduction in IL-11 expression by aging may play a role in the age-related reduction in bone formation.

11.5   Mechanical Stress Upregulates IL-11 Via ΔFosB/JunD Binding to the AP-1 Site

IL-11 gene promoter contains two tandem AP-1 sites located upstream from the TATA box which confer transcriptional activation by transforming growth factor (TGF)-β and other stimuli (Tang et al. 1998). Of the two AP-1 sites, site-directed

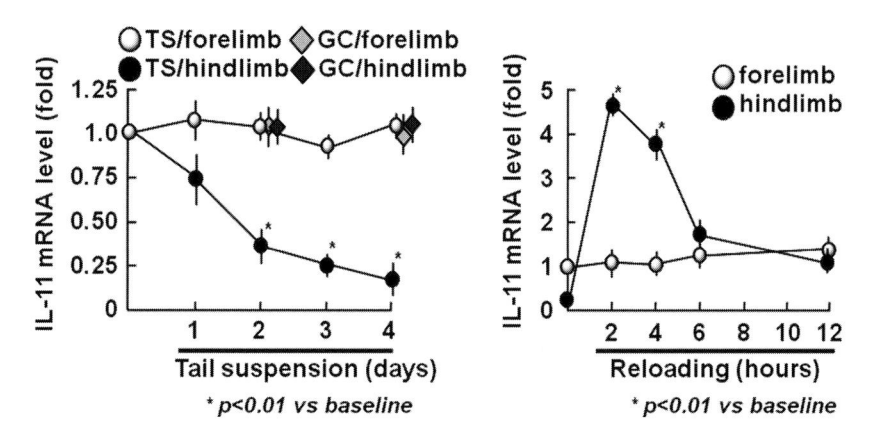

Fig. 11.2 IL-11 mRNA expression in hindlimb by mechanical unloading and loading. Mechanical unloading causes a marked reduction in the expression of IL-11 mRNA in the unloaded bone, and mechanical loading causes a rapid and robust increase in IL-11 mRNA expression in the loaded bone (Kido et al. 2009). *TS* Tail suspension, *GC* ground control

mutagenesis revealed that 5' upstream AP-1 site confers the transcriptional activation of *IL-11* gene by mechanical stress. Mechanical loading enhances the expression of ΔFosB, and the upregulated ΔFosB forms heterodimers on the 5'AP-1 site of the *IL-11* gene promoter with JunD, which is bound to the 5'AP-1 site regardless of mechanical stimuli. Binding of the ΔFosB/JunD heterodimer to the 5'AP-1 site of *IL-11* gene promoter causes an enhanced transcription of *IL-11* gene (Kido et al. 2009). Thus, downregulation of ΔFosB/JunD expression by siRNA reduces, and overexpression of ΔFosB/JunD enhances, *IL-11* gene promoter activity in osteoblasts.

11.6   Smad Signaling in Response to Mechanical Stress

Bone morphogenetic proteins (BMPs) play pivotal roles in the regulation of osteoblast differentiation and bone formation (Urist 1965; Wozney et al. 1988). However, the role of BMPs in mediating mechanical stress signal to osteogenic signal is controversial. Although compressive forces to osteoblastic cell cultures (Mitsui et al. 2006; Rath et al. 2008) or tensile forces to cranial suture of neonatal calvaria in culture (Ikegame et al. 2001) caused an elevation of BMP-4 expression in osteoblasts, FSS to SaOS-2 cells in culture was reported to reduce BMP-4 expression (McCormick et al. 2006). In addition, FSS to osteoblasts in culture or mechanical loading to hindlimbs in vivo did not increase BMP-2 expression (Kido et al. 2010). Nevertheless, not only mechanical stress but also BMP-2 stimulates IL-11 expression in osteoblastic cells (Kido et al. 2010) (Fig. 11.3).

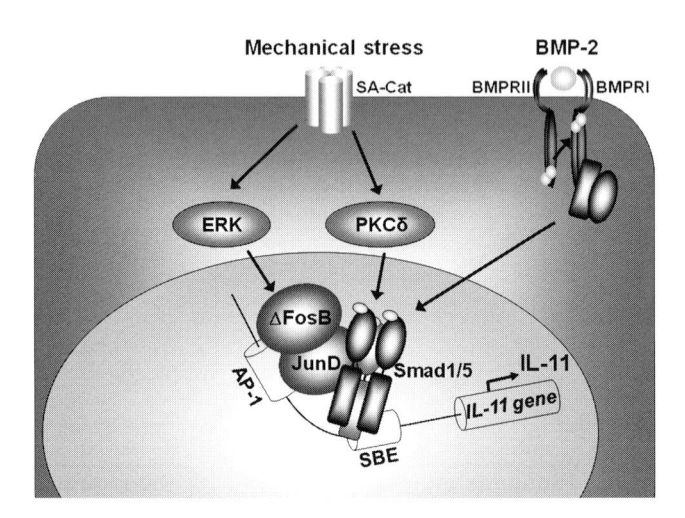

Fig. 11.3 AP-1 and Smad1/5 signaling cooperatively stimulate *IL-11* gene expression in response to mechanical stress. FSS induces phosphorylation of Smad1/5, but not of Smad8 or Smad2/3, in osteoblasts. The stimulation of Smad1/5 phosphorylation by FSS was mediated via tyrosine phosphorylation and activation of protein kinase Cδ (PKCδ) in osteoblasts. BMP-2 also activates Smad1/5 and stimulates IL-11 expression in osteoblastic cells. FSS-activated Smad1 is bound to Smad-binding element (SBE) and forms a complex with ΔFosB/JunD heterodimer which is bound to the 5'AP-1 site on the *IL-11* gene promoter

BMP signals are transmitted via phosphorylation by type I BMP receptor of BMP-specific receptor-regulated Smads (BR-Smads), Smad1, 5 and 8. Phosphorylated BR-Smads then form a heteromeric complex with Smad4, a common Smad, and translocate into the nucleus, where they regulate transcription of various target genes (Miyazono et al. 2005). Thus, the possibility remains that BR-Smad signaling is activated by mechanical stress to bone and is also involved in the enhancement of osteoblast differentiation in response to mechanical stress. In order to test this possibility, we investigated the effect of mechanical stress on BR-Smad phosphorylation in osteoblastic cells. We found that FSS induced phosphorylation of Smad1/5, but not of Smad8 or Smad2/3, in osteoblasts. The stimulation of BR-Smad phosphorylation by FSS was mediated via tyrosine phosphorylation and activation of protein kinase Cδ (PKCδ) in osteoblasts (Kido et al. 2010).

11.7 AP-1 and Smad Signaling Pathways Merge on IL-11 Gene Promoter

Mouse *IL-11* gene promoter contains a putative Smad-binding element (SBE) along with the AP-1 sites. Studies with site-directed mutagenesis at the putative SBE and 5′AP-1 sites revealed that both SBE and AP-1 sites were required for full activation of *IL-11* gene promoter activity by FSS. FSS-activated Smad1 is bound to SBE and forms a complex with ΔFosB/JunD heterodimer which is bound to the 5′AP-1 site on the *IL-11* gene promoter (Fig. 11.3) (Kido et al. 2010). These observations demonstrate that the Ca^{2+}–ERK–CREB–ΔFosB signaling and PKCδ-Smad1/5 signaling pathways merge together on the *IL-11* gene promoter, and that AP-1 and BR-Smad signaling cooperatively stimulate *IL-11* gene expression in response to mechanical stress.

11.8 Stimulation of Canonical Wnt Signaling Downstream IL-11

Wnt/β-catenin signaling pathway plays an important role in the regulation of bone formation. Low-density lipoprotein receptor-related protein (LRP) 5 and LRP6 are coreceptors for Wnt with the frizzled family of receptors, and are involved in signaling through the canonical Wnt/β-catenin pathway (Moon et al. 2002; Wehrli et al. 2000). Inactivating mutations in LRP5 results in osteoporosis pseudoglioma syndrome (Gong et al. 2001), and gain of function mutations in *LRP5* gene gives rise to a high bone mass phenotype in humans (Boyden et al. 2002; Little et al. 2002; Van Wesenbeeck et al. 2003) as well as in mice (Babij et al. 2003; Kato et al. 2002). Wnt/β-catenin signaling is also a physiological response to mechanical loading, and mechanical unloading enhances, and mechanical stress suppresses, the expression of sclerostin, an inhibitor of Wnt/β-catenin signaling expressed in osteocytes (Robling et al. 2008). These results suggest that modulation of sclerostin expression offers a finely tuned regulatory system in which osteocytes coordinate regional osteogenesis in areas with mechanical stress. However, the upstream signal that reduced sclerostin expression in response to mechanical stress remained unknown.

Wnt/β-catenin signaling can be inhibited by not only sclerostin but also dickkopf (Dkk) 1 and 2. Using murine primary osteoblasts that do not express sclerostin but express Dkk1 and 2, we demonstrated that FSS enhances Wnt/β-catenin signaling by suppressing Dkk1 and 2 expression (Kido et al. 2009). The reduction of Dkk1 and 2 is mediated via the increased expression of IL-11 by mechanical stress, and knockdown of IL-11 expression by siRNA enhances, and overexpression of IL-11 suppresses, Dkk1 and 2. These observations demonstrate that stimulation of Wnt/β-catenin signaling is a major pathway mediating mechanical stress signal to osteogenic signal, and that enhanced IL-11 expression by mechanical stress is at least one of the upstream signal to enhance Wnt/β-catenin signaling.

References

Babij P, Zhao W, Small C, Kharode Y, Yaworsky PJ et al (2003) High bone mass in mice expressing a mutant LRP5 gene. J Bone Miner Res 18:960–974

Boyden LM, Mao J, Belsky J, Mitzner L, Farhi A et al (2002) High bone density due to a mutation in LDL-receptor-related protein 5. N Engl J Med 346:1513–1521

Burr DB, Robling AG, Turner CH (2002) Effects of biomechanical stress on bones in animals. Bone 30:781–786

Chen J, Kelz MB, Hope BT, Nakabeppu Y, Nestler EJ (1997) Chronic Fos-related antigens: stable variants of deltaFosB induced in brain by chronic treatments. J Neurosci 17:4933–4941

Du X, Williams DA (1997) Interleukin-11: review of molecular, cell biology, and clinical use. Blood 89:3897–3908

Fan X, Roy E, Zhu L, Murphy TC, Ackert-Bicknell C et al (2004) Nitric oxide regulates receptor activator of nuclear factor-kappaB ligand and osteoprotegerin expression in bone marrow stromal cells. Endocrinology 145:751–759

Fujino H, Salvi S, Regan JW (2005) Differential regulation of phosphorylation of the cAMP response element-binding protein after activation of EP2 and EP4 prostanoid receptors by prostaglandin E2. Mol Pharmacol 68:251–259

Genetos DC, Geist DJ, Liu D, Donahue HJ, Duncan RL (2005) Fluid shear-induced ATP secretion mediates prostaglandin release in MC3T3-E1 osteoblasts. J Bone Miner Res 20:41–49

Gong Y, Slee RB, Fukai N, Rawadi G, Roman-Roman S et al (2001) LDL receptor-related protein 5 (LRP5) affects bone accrual and eye development. Cell 107:513–523

Grigoriadis AE, Schellander K, Wang ZQ, Wagner EF (1993) Osteoblasts are target cells for transformation in c-fos transgenic mice. J Cell Biol 122:685–701

Hung CT, Allen FD, Pollack SR, Brighton CT (1996) What is the role of the convective current density in the real-time calcium response of cultured bone cells to fluid flow? J Biomech 29:1403–1409

Ikegame M, Ishibashi O, Yoshizawa T, Shimomura J, Komori T et al (2001) Tensile stress induces bone morphogenetic protein 4 in preosteoblastic and fibroblastic cells, which later differentiate into osteoblasts leading to osteogenesis in the mouse calvariae in organ culture. J Bone Miner Res 16:24–32

Inoue D, Kido S, Matsumoto T (2004) Transcriptional induction of FosB/DeltaFosB gene by mechanical stress in osteoblasts. J Biol Chem 279:49795–49803

Jaworski ZF, Liskova-Kiar M, Uhthoff HK (1980) Effect of long-term immobilisation on the pattern of bone loss in older dogs. J Bone Joint Surg Br 62-B:104–110

Jochum W, David JP, Elliott C, Wutz A, Plenk H Jr et al (2000) Increased bone formation and osteosclerosis in mice overexpressing the transcription factor Fra-1. Nat Med 6:980–984

Kato M, Patel MS, Levasseur R, Lobov I, Chang BH et al (2002) Cbfa1-independent decrease in osteoblast proliferation, osteopenia, and persistent embryonic eye vascularization in mice deficient in Lrp5, a Wnt coreceptor. J Cell Biol 157:303–314

Kawashima I, Ohsumi J, Mita-Honjo K, Shimoda-Takano K, Ishikawa H et al (1991) Molecular cloning of cDNA encoding adipogenesis inhibitory factor and identity with interleukin-11. FEBS Lett 283:199–202

Keller DC, Du XX, Srour EF, Hoffman R, Williams DA (1993) Interleukin-11 inhibits adipogenesis and stimulates myelopoiesis in human long-term marrow cultures. Blood 82:1428–1435

Kido S, Kuriwaka-Kido R, Imamura T, Ito Y, Inoue D et al (2009) Mechanical stress induces interleukin-11 expression to stimulate osteoblast differentiation. Bone 45:1125–1132

Kido S, Kuriwaka-Kido R, Umino-Miyatani Y, Endo I, Inoue D et al (2010) Mechanical stress activates Smad pathway through PKCδ to enhance interleukin-11 gene transcription in osteoblasts. PLoS ONE 5:e13090

Knothe Tate ML (2003) "Whither flows the fluid in bone?" An osteocyte's perspective. J Biomech 36:1409–1424

Kodama Y, Nakayama K, Fuse H, Fukumoto S, Kawahara H et al (1997) Inhibition of bone resorption by pamidronate cannot restore normal gain in cortical bone mass and strength in tail-suspended rapidly growing rats. J Bone Miner Res 12:1058–1067

Kodama Y, Takeuchi Y, Suzawa M, Fukumoto S, Murayama H et al (1998) Reduced expression of interleukin-11 in bone marrow stromal cells of senescence-accelerated mice (SAMP6): relationship to osteopenia with enhanced adipogenesis. J Bone Miner Res 13:1370–1377

Lean JM, Mackay AG, Chow JW, Chambers TJ (1996) Osteocytic expression of mRNA for c-fos and IGF-I: an immediate early gene response to an osteogenic stimulus. Am J Physiol 270:E937–E945

Little RD, Carulli JP, Del Mastro RG, Dupuis J, Osborne M et al (2002) A mutation in the LDL receptor-related protein 5 gene results in the autosomal dominant high-bone-mass trait. Am J Hum Genet 70:11–19

Liu D, Genetos DC, Shao Y, Geist DJ, Li J et al (2008) Activation of extracellular-signal regulated kinase (ERK1/2) by fluid shear is Ca(2+)- and ATP-dependent in MC3T3-E1 osteoblasts. Bone 42:644–652

McCormick SM, Saini V, Yazicioglu Y, Demou ZN, Royston TJ (2006) Interdependence of pulsed ultrasound and shear stress effects on cell morphology and gene expression. Ann Biomed Eng 34:436–445

Mitsui N, Suzuki N, Maeno M, Yanagisawa M, Koyama Y et al (2006) Optimal compressive force induces bone formation via increasing bone morphogenetic proteins production and decreasing their antagonists production by Saos-2 cells. Life Sci 78:2697–2706

Miyazono K, Maeda S, Imamura T (2005) BMP receptor signaling: transcriptional targets, regulation of signals, and signaling cross-talk. Cytokine Growth Factor Rev 16:251–263

Moon RT, Bowerman B, Boutros M, Perrimon N (2002) The promise and perils of Wnt signaling through beta-catenin. Science 296:1644–1646

Morey ER, Baylink DJ (1978) Inhibition of bone formation during space flight. Science 201:1138–1141

Nakabeppu Y, Nathans D (1991) A naturally occurring truncated form of FosB that inhibits Fos/Jun transcriptional activity. Cell 64:751–759

Nishii N, Nejime N, Yamauchi C, Yanai N, Shinozuka K et al (2009) Effects of ATP on the intracellular calcium level in the osteoblastic TBR31-2 cell line. Biol Pharm Bull 32:18–23

Pavalko FM, Chen NX, Turner CH, Burr DB, Atkinson S et al (1998) Fluid shear-induced mechanical signaling in MC3T3-E1 osteoblasts requires cytoskeleton-integrin interactions. Am J Physiol 275:C1591–1601

Rath B, Nam J, Knobloch TJ, Lannutti JJ, Agarwal S (2008) Compressive forces induce osteogenic gene expression in calvarial osteoblasts. J Biomech 41:1095–1103

Robling AG, Niziolek PJ, Baldridge LA, Condon KW, Allen MR et al (2008) Mechanical stimulation of bone in vivo reduces osteocyte expression of Sost/sclerostin. J Biol Chem 283:5866–5875

Sabatakos G, Sims NA, Chen J, Aoki K, Kelz MB et al (2000) Overexpression of DeltaFosB transcription factor(s) increases bone formation and inhibits adipogenesis. Nat Med 6:985–990

Takeuchi Y, Watanabe S, Ishii G, Takeda S, Nakayama K et al (2002) Interleukin-11 as a stimulatory factor for bone formation prevents bone loss with advancing age in mice. J Biol Chem 277:49011–49018

Tang W, Yang L, Yang YC, Leng SX, Elias JA (1998) Transforming growth factor-beta stimulates interleukin-11 transcription via complex activating protein-1-dependent pathways. J Biol Chem 273:5506–5513

Tohjima E, Inoue D, Yamamoto N, Kido S, Ito Y et al (2003) Decreased AP-1 activity and interleukin-11 expression by bone marrow stromal cells may be associated with impaired bone formation in aged mice. J Bone Miner Res 18:1461–1470

Urist MR (1965) Bone: formation by autoinduction. Science 150:893–899

Van Wesenbeeck L, Cleiren E, Gram J, Beals RK, Benichou O et al (2003) Six novel missense mutations in the LDL receptor-related protein 5 (LRP5) gene in different conditions with an increased bone density. Am J Hum Genet 72:763–771

Wehrli M, Dougan ST, Caldwell K, O'Keefe L, Schwartz S et al (2000) Arrow encodes an LDL-receptor-related protein essential for wingless signalling. Nature 407:527–530

Wozney JM, Rosen V, Celeste AJ, Mitsock LM, Whitters MJ et al (1988) Novel regulators of bone formation: molecular clones and activities. Science 242:1528–1534

Yoshida K, Oida H, Kobayashi T, Maruyama T, Tanaka M et al (2002) Stimulation of bone formation and prevention of bone loss by prostaglandin E EP4 receptor activation. Proc Natl Acad Sci USA 99:4580–4585

You J, Reilly GC, Zhen X, Yellowley CE, Chen Q et al (2001) Osteopontin gene regulation by oscillatory fluid flow via intracellular calcium mobilization and activation of mitogen-activated protein kinase in MC3T3-E1 osteoblasts. J Biol Chem 276:13365–13371

Chapter 12
Osteoblast Development in Bone Loss Due to Skeletal Unloading

Akinori Sakai and Toshitaka Nakamura

12.1 Introduction

Physiological skeletal loading is essential for the maintenance of bone mass in loaded bones. Skeletal unloading, such as space flight and long-term bed rest, induces bone loss in loaded bone in humans (Morey and Baylink 1978). Tail-suspended rats and mice are frequently used for the investigation of cellular and molecular mechanisms concerning bone loss after unloading. Using this animal model, the tibia of the elevated hind limb shows that trabecular bone volume is rapidly reduced within 7 days (Sakata et al. 1999; Vico et al. 1991). Unloading inhibits proliferation and differentiation of osteoprogenitor cells in vitro (Kostenuik et al. 1997). The basic mechanism underlying inhibition of proliferation and differentiation of osteoprogenitor cells after skeletal unloading has not been elucidated.

Apoptosis as well as p53 gene expression is observed in bone marrow cells (Wlodarski et al. 1998). Apoptosis has also been seen in osteoblasts and osteocytes (Weinstein et al. 1998). We proposed that this rapid bone loss due to unloading relates to facilitation of p53 apoptotic signal in bone marrow cells. We investigated the effects of the p53 gene status on trabecular bone turnover and osteoblast development (Sakai et al. 2002).

Vascular factors, such as vascular endothelial growth factor (VEGF) and its receptors, Flt-1 (VEGF receptor-1) and Flk-1 (VEGF receptor-2), play a role in bone development and regeneration (Uchida et al. 2003). Recently, it has been reported that hematopoietic stem cell interacts with osteoblastic niche (Yoshihara et al. 2007). Platelet endothelial cell adhesion molecule-1 (PECAM-1, CD31) is a cell adhesion molecule localized to the inter-endothelial cell adhesion site. Fujiwara and colleagues (Fujiwara et al. 2001; Osawa et al. 1997, 2002) reported that PECAM-1 is tyrosine-phosphorylated when endothelial cells are exposed to physiological levels of fluid shear stress. These findings suggest the possible role of PECAM-1 in

A. Sakai (✉) and T. Nakamura
Department of Orthopaedic Surgery, School of Medicine, University of Occupational and Environmental Health, 1-1 Iseigaoka, Yahatanishi-ku, Kitakyushu, 807-8555, Japan
e-mail: a-sakai@med.uoeh-u.ac.jp

M. Noda (ed.), *Mechanosensing Biology*,
DOI 10.1007/978-4-431-89757-6_12, © Springer 2011

mechanosensing by endothelial cells. We proposed that bone adaptation to unloading through PECAM-1 signaling has important implications for local regulation of trabecular bone turnover (Sakuma-Zenke et al. 2005).

Thus, we hypothesized that reduction of bone formation due to unloading is closely related to the facilitation of p53 signaling, and that unloading inhibits local factors related to angiogenesis and cell adhesion in bone marrow cells, thereby reducing osteoblast development.

12.2   Materials and Methods

12.2.1   Experimental Design

To determine the effect of p53 gene status on bone mass and bone formation after 7-day unloading and normal loading, we used p53 gene knockout (p53$^{-/-}$) mice and p53 wild-type (p53$^{+/+}$) mice (Sakai et al. 2002). Eight-week-old male p53$^{+/+}$ and p53$^{-/-}$ mice were respectively assigned to two groups: normally loaded mice and hind limb-unloaded mice by tail suspension described previously by our laboratory (Sakata et al. 1999).

To determine the effect of unloading on changes in molecules related to angiogenesis in unloaded limbs, C57BL/6NCrj 8-week-old male mice were divided randomly into two groups: normally loaded mice and hind limb-unloaded mice by tail suspension (Sakuma-Zenke et al. 2005).

All experimental protocols were approved in advance by the Ethics Review Committee for Animal Experimentation of the University of Occupational and Environmental Health.

12.2.2   Histomorphometry

Histomorphometric analyses were done at the secondary spongiosa of the proximal tibia as described previously (Sakai et al. 1996). Bone labeling with an intraperitoneal injection of calcein (6 mg/kg body weight) was performed 7 and 3 days before death. The specimens were stained for tartrate-resistant acid phosphatase (TRACP).

12.2.3   Flow Cytometry

We performed flow cytometric analyses on tibial bone marrow cells 7 days after unloading and normal loading. The protein expressions of PECAM-1, angiopoietin-1, angiopoietin-2, Flk-1, and vascular endothelial cadherin (VE-cadherin, CD144) were monitored. Bone marrow cells were obtained from tibias in each mouse and then used for staining and flow cytometric analyses using the methods described previously

(Sakai et al. 1998). Dead cells were excluded by staining with propidium iodide (Sigma, St. Louis, MO, USA). Stained cells were analyzed on a flow cytometer (EPICS XL; Coulter, Tokyo, Japan). Quantification of cell surface antigen on a single cell was calculated using standard beads, QIFKIT (DAKO Japan, Kyoto, Japan). PE- or FITC-labeled donkey anti-goat IgG was used as a negative control.

12.2.4 Cell Culture

We performed primary tibial bone marrow cell cultures. Marrow cultures were initiated using the method of Maniatopoulos et al. (1988). Bone marrow was flushed out from the proximal cut end, a single-cell suspension was prepared by repeated aspiration, and all bone marrow cells were counted in a cell counter (F-820; Sysmex, Kobe, Japan). Bone marrow cells were used for the colony-forming units-fibroblastic (CFU-f) assay, as described previously (Sakai et al. 1996). We used anti-PECAM-1 antibody to evaluate the shutdown effect on tibial bone marrow capacity for bone cell development. To determine the effect of PECAM-1 on osteoblasts, we cocultured osteoblastic cell line MC3T3E-1 (Riken Cell Bank, Tsukuba, Japan) with or without PECAM-1-expressing endothelial cell line EOMA (American Type Culture Collection, Manassas, VA, USA) or nonexpressing endothelial cell line ISOS-1 (Cell Resource Center for Biomedical Research, Tohoku University, Sendai, Japan). We also reconfirmed PECAM-1 action on alkaline phosphatase (ALP) production in osteoblastic cell line, using anti-PECAM-1 antibody.

12.2.5 ALP Production of Cultured Cells

Flow cytometric analysis of MC3T3-E1 cells was performed using a FACScan (Becton Dickinson, Mountain View, CA, USA), following the procedure described previously (Tanaka et al. 1992). Cytoplasmic antigens of MC3T3-E1 cells pretreated with cell permeabilization kit (Caltag, Burlingame, CA) were stained by anti-ALP mAb in FACS media for 30 min at 4°C. After washing the cells three times with FACS media, they were further incubated with FITC-labeled goat anti-mouse IgG Ab for 30 min at 4°C. The mAbs-stained cells were detected using FACScan. ALP production on a single cell was quantified using standard beads, QIFKIT (DAKO Japan) as described previously (Tanaka et al. 1998, 1999).

12.2.6 Quantitative Real-Time Reverse-Transcriptase-Polymerase Chain Reactions

Bone marrow cells were obtained from tibias of 7-day unloaded and loaded mice and then used for isolating total RNA and quantitative reverse-transcriptase-polymerase chain reactions (RT-PCR) analyses using the methods described previously (Tanaka et al. 2004). Briefly, the bone marrow was flushed out from the proximal cut end with 2 ml of phosphate buffered saline. The bone marrow cells were frozen in liquid nitrogen. Total RNA was extracted using an acid guanidinium thiocyanate–phenol–chloroform method after homogenizing (Chomczynskin and Sacchi 1987). First-strand cDNA was reverse-transcribed from total RNA using Moloney murine leukemia reverse transcriptase (SuperScript; Life Technologies, Rockville, MD, USA) and oligo (dT) 12–18 primer (Life Technologies). Specific PCR primers were designed from published sequences of murine genes. Quantitative RT-PCR analysis was performed using an Optical System Interface software (version 3.0; Bio-Rad, Hercules, CA). PCR reactions were performed in four independent experiments. The expression levels of mRNA were normalized using β-actin as housekeeping gene and expressed as a relative value to the baseline control over the time course.

12.2.7 Statistical Analysis

Results are expressed as the mean ± standard error of the mean (SEM). Differences between groups were examined for statistical significance using the Mann–Whitney's U test. A p value less than 0.05 denoted the presence of a statistically significant difference.

12.3 Results

12.3.1 Bone Volume After Unloading and Reloading

Trabecular bone volume and bone formation were rapidly reduced to approximately 50% of the normal control level within 7 days and stabilized at that level in the subsequent 7 days of unloading (Fig. 12.1). Seven-day unloading reduced mRNA expressions of osterix and osteocalcin in bone marrow cells. Fourteen-day reloading after 7-day unloading completely restored trabecular bone volume, but this did not happen after 14-day unloading. Bone formation rapidly recovered immediately after reloading.

12.3.2 Bone Volume and Formation in Disrupted p53 Gene

Trabecular bone volume and bone formation in wild-type were significantly reduced after 7-day unloading, but in p53 knockout mice, bone volume and bone formation were not reduced at all (Fig. 12.2). In a culture study, 7-day unloading in wild-type decreased ALP-positive CFU-f, but not in p53 knockout mice (Fig. 12.3).

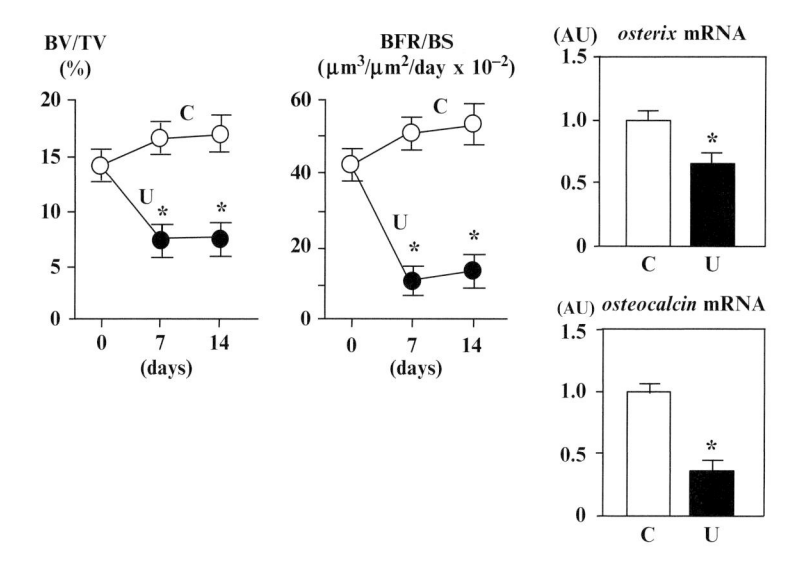

Fig. 12.1 Bone volume and bone formation after unloading. mRNA expressions were evaluated in tibial bone marrow cells after 7-day unloading. *BV/TV* one volume/tissue volume, *BFR/BS* bone formation rate/bone surface, *C* control, *U* unloading. Data are expressed as the mean±SEM. *$p<0.05$ versus C by Mann–Whitney's *U* test

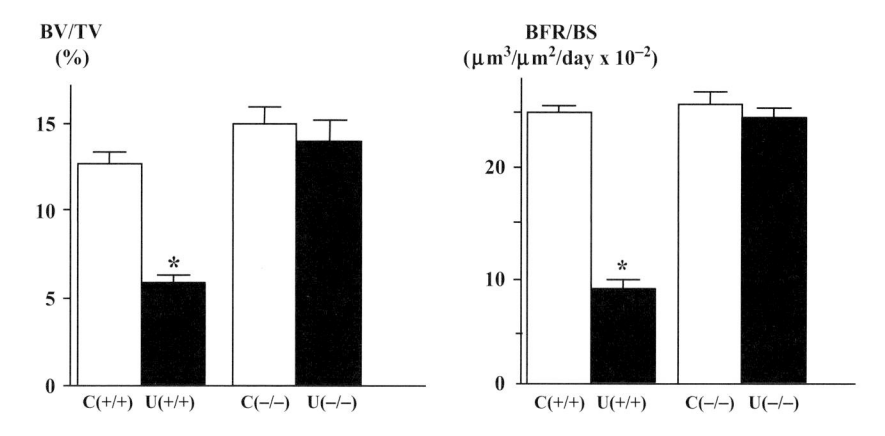

Fig. 12.2 Disruption of the p53 gene alleviated the reduction of bone volume and bone formation after 7-day unloading. *BV/TV* Bone volume/tissue volume, *BFR/BS* bone formation rate/bone surface, *C* control, *U* unloading. (+/+) wild-type mice, (–/–) p53 gene knockout mice. Data are expressed as the mean±SEM. *$p<0.05$ versus C by Mann–Whitney's *U* test. (cf. Sakai et al. 2002)

There was no difference among four groups in total CFU-f. In wild-type mice, unloading increased p53 gene expression in bone marrow cells. We detected apoptotic cells using terminal dUTP nick end labeling (TUNEL) staining. In wild-type mice, unloading increased apoptotic osteocytes and trabecular osteoblasts. In p53 knockout mice, there was no increase in apoptotic osteocytes and osteoblasts.

172 A. Sakai and T. Nakamura

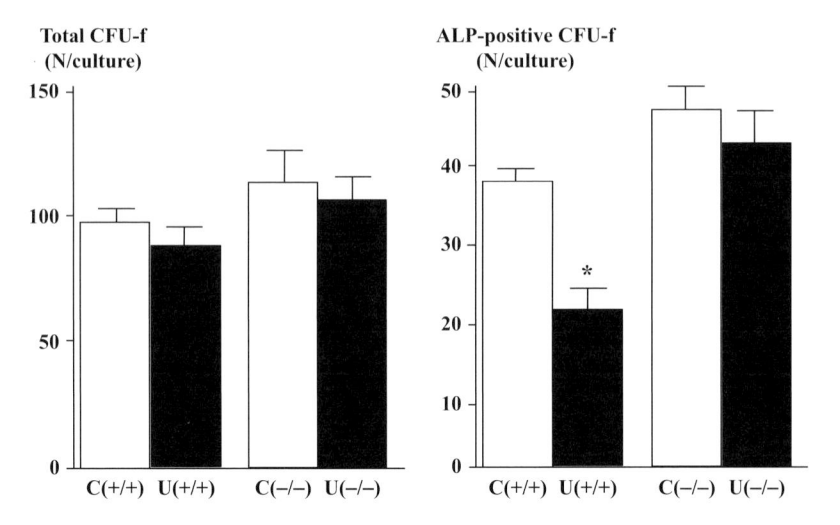

Fig. 12.3 ALP-positive CFU-f was reduced by 7-day unloading, but not in disrupted p53 gene status. *CFU-f* Colony-forming units-fibroblastic, *ALP* alkaline phosphatase, *C* control, *U* unloading. (+/+) wild-type mice, (–/–) p53 gene knockout mice. Data are expressed as the mean±SEM. *$p < 0.05$ versus C by Mann–Whitney's *U* test. (cf. Sakai et al. 2002)

12.3.3 PECAM-1 Expression After Unloading

Seven-day unloading significantly reduced the percentage of PECAM-1-positive cells in bone marrow cells compared with that of normal loading (Fig. 12.4). There were no significant differences in angiopoietin-1, angiopoietin-2, Flk-1, and VE-cadherin between unloading and loading. The expression level of PECAM-1 was lower after 3-day reloading and similar after 5-day reloading compared to the respective values at days 10 and 12 of loading. On the other hand, the PECAM-1 expression was maintained in humeral bone marrow cells obtained from loaded forelimbs as internal bone marrow control after unloading and reloading. The expression level of PECAM-1 mRNA in unloading was significantly lower at 7 days, compared to that in loading.

12.3.4 Bone Marrow Cell Development

The formation of ALP-positive CFU-f was significantly reduced under the medium condition of supplementation of anti-PECAM-1 antibody, compared with that of IgG. ALP production by cultured mouse osteoblastic cell line MC3T3-E1 was enhanced in the presence of mouse endothelial cell line EOMA, positive for PECAM-1 expression, but not in the presence of mouse endothelial cell line ISOS-1, negative for PECAM-1 expression. Anti-PECAM-1 antibody dose-dependently inhibited the increase in ALP production of MC3T3-E1 cocultured with EOMA (Fig. 12.5).

Percentage of bone marrow cells with PECAM-1 expression (%)

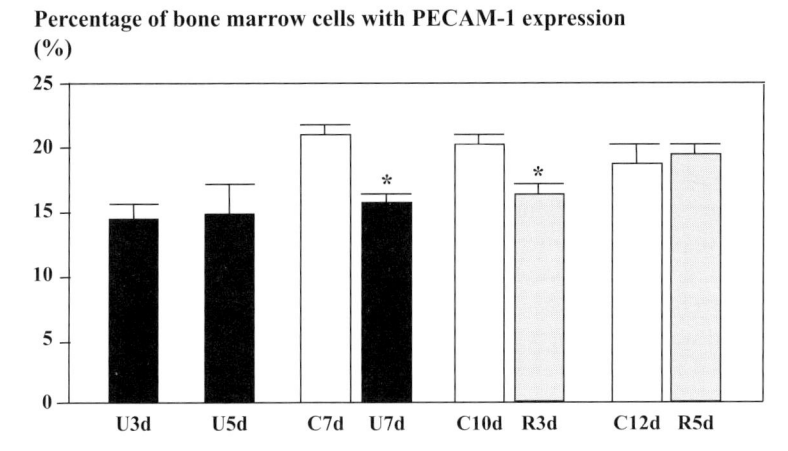

Fig. 12.4 PECAM-1 expression of tibial bone marrow cells after 7-day unloading and subsequent 5-day reloading. *PECAM-1* Platelet endothelial cell adhesion molecule-1, *C* control, *U* unloading, *R* reloading, *d* day. Data are expressed as the mean ± SEM. *$p < 0.05$ versus C by Mann–Whitney's *U* test. (cf. Sakuma-Zenke et al. 2005)

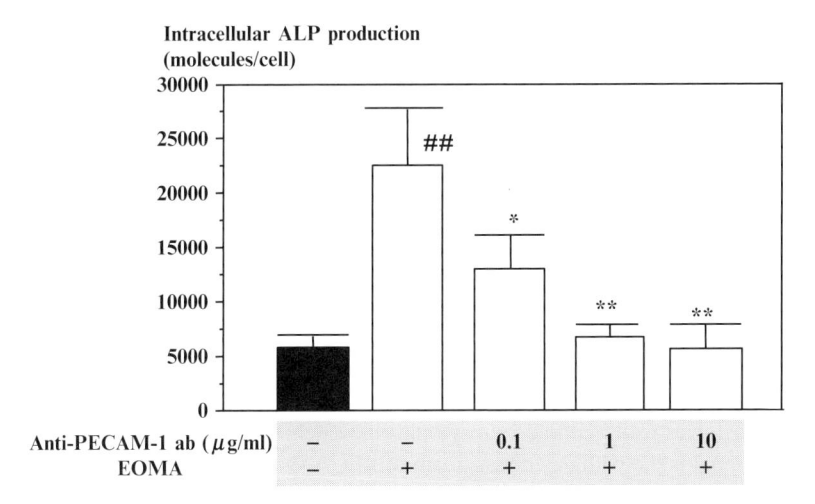

Fig. 12.5 Anti-PECAM-1 antibody dose-dependently reduced the increase in intracellular ALP production of MC3T3-E1 cocultured with EOMA. *ALP* Alkaline phosphatase, *PECAM-1* platelet endothelial cell adhesion molecule-1, *ab* antibody. Data are expressed as the mean ± SEM. ## $p < 0.01$ versus ab (–) and EOMA (–), *, ** $p < 0.05$, 0.01 versus ab (–) and EOMA (+) by Mann–Whitney's *U* test. (cf. Sakuma-Zenke et al. 2005)

12.4 Discussion

We demonstrated that disruption of p53 gene preserves trabecular bone mass and bone formation after unloading (Sakai et al. 2002) and that decreased osteogenic potential after unloading is related to reduction of PECAM-1 expression in bone marrow cells (Sakuma-Zenke et al. 2005).

Trabecular bone loss due to skeletal unloading was related to facilitation of intracellular p53–p21 signaling in bone marrow cells. Unloading significantly increased the percentage of apoptotic cells relative to that in loaded groups in p53$^{+/+}$ mice. However, there was no significant difference between the unloaded and loaded groups in p53$^{-/-}$ mice. Our results indicate that unloading induces p53-related apoptosis of osteoblasts and osteocytes. These results are in agreement with the results of our previous studies showing the presence of TUNEL-positive osteoblasts and osteocytes in p53$^{+/+}$ mice after limb immobilization, but not in p53$^{-/-}$ mice (Okazaki et al. 2004). It is also reported that cell death in the femur of glucocorticoid-treated rabbits, affecting osteoblasts and osteocytes, comprised up to half of the bone volume and was consistent with apoptosis (Weinstein et al. 1998).

PECAM-1 is located on endothelial cell surfaces in bone where it can be affected by unloading and mediate the mechanical sensing signals in bone. Sho et al. (2001) reported that apoptosis of endothelial cells was induced by a reduction in blood flow. Osawa et al. (1997) reported rapid tyrosine phosphorylation of PECAM-1 in endothelial cells exposed to fluid shear stress. Osmotic changes also induced similar PECAM-1 and extracellular signal-regulated kinase (ERK) phosphorylation with nearly identical kinetics (Fujiwara et al. 2001; Osawa et al. 2002). Duncan et al. (1999) reported that PECAM-1-disrupted mice are viable and undergo normal vascular development, suggesting that PECAM-1 is not critical for vasculogenesis. In our experiment, PECAM-1 could play a role in mediating mechanical sensing signals in bone, but not in vascularization. Thus, we consider that PECAM-1 on responded endothelial cells serves as a mechanosensor and directly or indirectly activates osteogenic potential through endothelial cell migration (Rainger et al. 1999), adhesion (Schnittler et al. 1997), and differentiation (Arihiro and Inai 2001).

PECAM-1 acts as an adhesion molecule as well as a regulatory molecule of integrin (Dangerfield et al. 2002; Tanaka et al. 1992). PECAM-1 on endothelial cells signals osteoblastic cells through heterophylic binding with $\alpha_v\beta_3$ integrin (Buckley et al. 1996; Piali et al. 1995; Rainger et al. 1999) and CD38 (Deaglio et al. 1998; Fernandez et al. 1998). Human CD38 is a cell surface molecule involved in the regulation of lymphocyte adhesion to endothelial cells (Deaglio et al. 1998), and is also found in bone tissue (Fernandez et al. 1998). PECAM-1 and CD38 cognate interactions modulate heterotypic adhesion as well as implement cytoplasmic calcium fluxes. We propose that $\alpha_v\beta_3$ integrin signal is the pathway downstream of PECAM-1 expression.

Sakata et al. (2003, 2004) reported that unloading suppressed osteoblast development by inhibiting the activation of insulin-like growth factor-I (IGF-I) signal pathway through downregulation of $\alpha_v\beta_3$ integrin. IGF-1 enhanced bone formation at the normal loading, but did not at the unloading (Fig. 12.6). This result is in agreement with that of ALP-positive CFU-f. IGF-1 enhanced cell proliferation at the normal loading, but did not at the unloading. This enhanced cell proliferation was inhibited by echistation, inhibitor of $\alpha_v\beta_3$ integrin signal. IGF-1 enhanced phosphorylation of IGF-1 receptor at the normal loading, but did not at the unloading. This enhanced phosphorylation of IGF-1 receptor was dose-dependently inhibited by echistation. IGF-1 activated Ras and enhanced phosphorylation of ERK1/2 at the normal loading, but did not at the unloading. Unloading increased apoptosis of osteoblasts and osteocytes, independently of IGF-1 signal (Fig. 12.7). On the other hand, IGF-1 enhanced phosphorylation of Akt both at the loading and unloading.

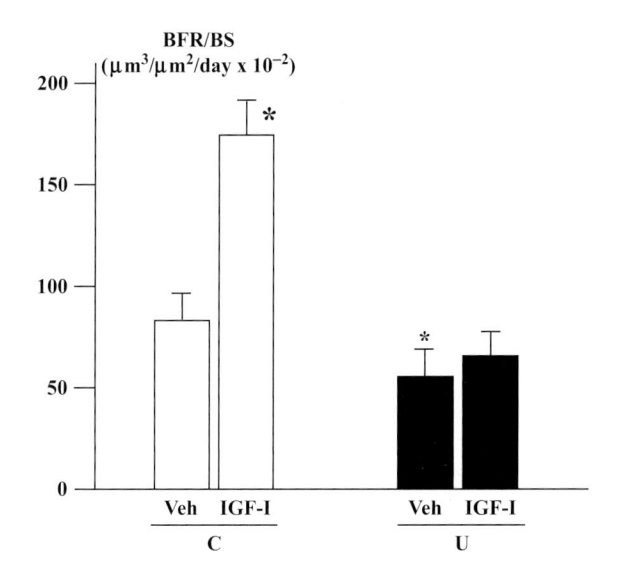

Fig. 12.6 IGF-I did not enhance bone formation under unloading condition. *BFR/BS* Bone formation rate/bone surface, *Veh* vehicle, *IGF-I* insulin-like growth factor-I, *C* control, *U* unloading. Data are expressed as the mean ± SEM. *$p < 0.05$ versus Veh/C by Mann–Whitney's *U* test. (cf. Sakata et al. 2003)

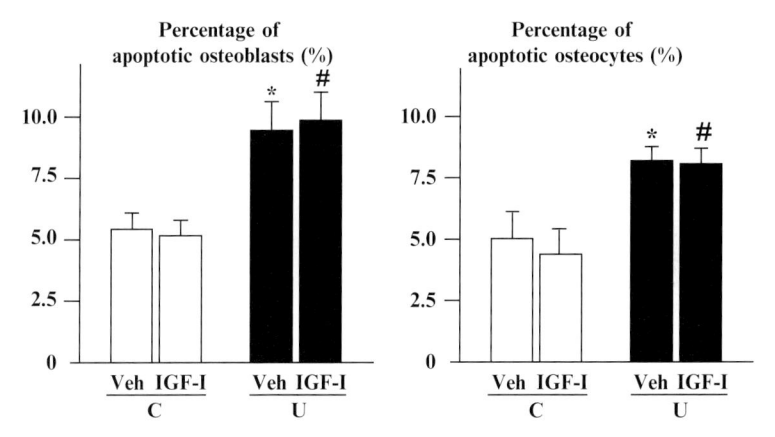

Fig. 12.7 Seven-day unloading induced apoptosis in osteoblasts and osteocytes independently of IGF-I signal. *Veh* Vehicle, *IGF-I* insulin-like growth factor-I, *C* control, *U* unloading. Data are expressed as the mean ± SEM. *$p < 0.05$ versus Veh/C, #$p < 0.05$ versus IGF-I/C by Mann–Whitney's *U* test. (cf. Sakata et al. 2004)

In conclusion, unloading inhibits the activation of IGF-I signal pathway, through downregulation of PECAM-1 and $\alpha_v\beta_3$ integrin, leading to decreased proliferation and differentiation of osteoblasts and their precursors. Independently of IGF-1 signal, apoptotic signal through the p53 gene is facilitated. As a result bone formation is rapidly reduced after unloading. These results are summarized as a scheme in Fig. 12.8. Recently, the Wnt/β-catenin signal pathway is recognized as an important regulator of bone mass and bone cell functions (Bonewald and Johnson 2008). Hind limb unloading yields a significant increase in *Sost* expression in the tibia (Robling et al. 2008). Sclerostin is the protein product of the *Sost* gene and a potent inhibitor of bone formation via inhibition of Wnt/β-catenin signaling (Fig. 12.9). Modulation of sclerostin

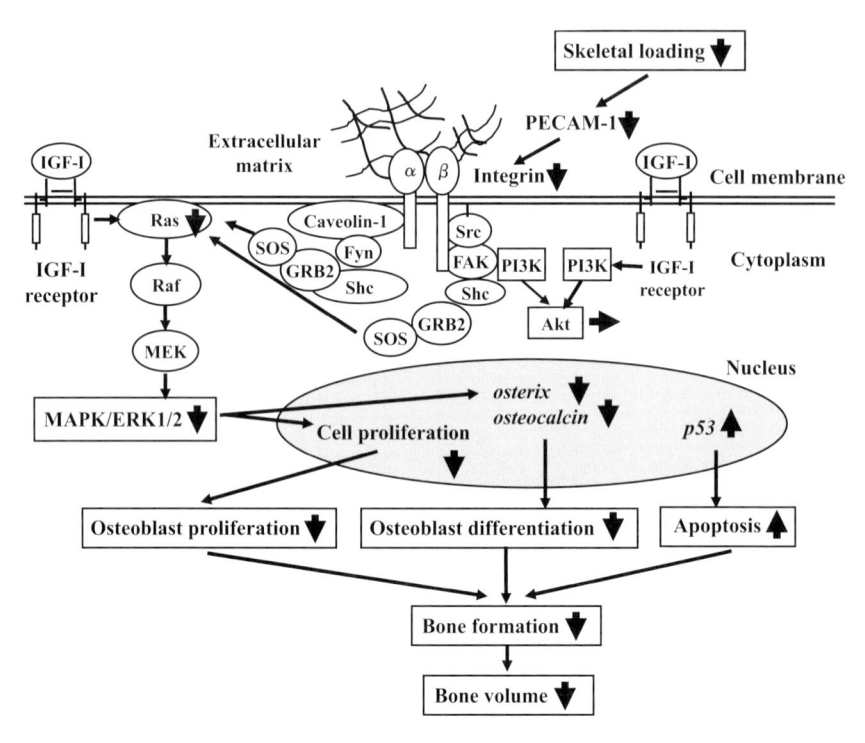

Fig. 12.8 Schema of signal pathway after skeletal unloading. *PECAM-1* Platelet endothelial cell adhesion molecule, *IGF-I* insulin-like growth factor-I

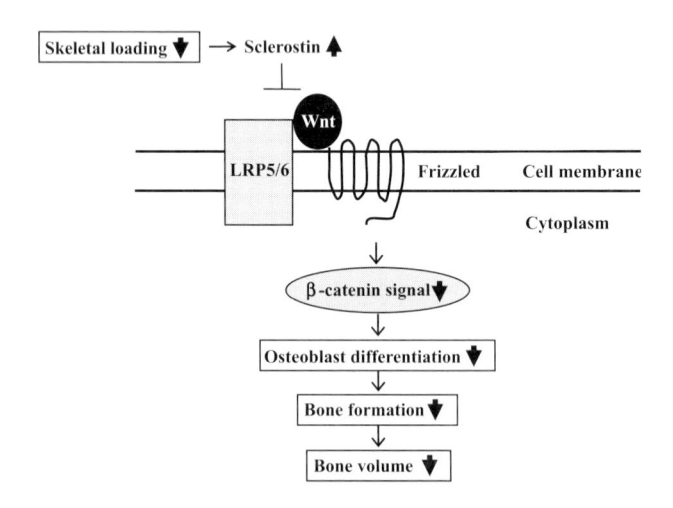

Fig. 12.9 Wnt/β-catenin signal regulates osteoblast development after unloading. Unloading reduced Wnt/β-catenin signal by enhancing sclerostin expression in osteocytes. *LRP* Low-density lipoprotein receptor-related protein. (cf. Lin et al. 2009)

levels appears to be a finely tuned mechanism by which osteocytes coordinate regional and local osteogenesis in response to decreased mechanical stimulation (Lin et al. 2009). Sclerostin plays an essential role in mediating bone response to mechanical unloading through inhibiting Wnt/β-catenin signaling.

References

Arihiro K, Inai K (2001) Expression of CD31, Met/hepatocyte growth factor receptor and bone morphogenetic protein in bone metastasis of osteosarcoma. Pathol Int 51:100–106

Bonewald LF, Johnson ML (2008) Osteocytes, mechanosensing and Wnt signaling. Bone 42:606–615

Buckley CD, Doyonnas R, Newton JP et al. (1996) Identification of $\alpha_v\beta_3$ as a heterotypic ligand for CD31/PECAM-1. J Cell Sci 109:437–445

Chomczynskin P, Sacchi N (1987) Single-step method of RNA isolation by acid guanidinium thiocyanate–phenol–chloroform extraction. Anal Biochem 162:156–159

Dangerfield J, Larbi KY, Huang MT et al. (2002) PECAM-1 (CD31) homophilic interaction up-regulates $\alpha_6\beta_1$ on transmigrated neutrophils in vivo and plays a functional role in the ability of α_6 integrins to mediate leukocyte migration through the perivascular basement membrane. J Exp Med 196:1201–1211

Deaglio S, Morra M, Mallone R et al. (1998) Human CD38 (ADP-ribosyl cyclase) is a counter-receptor of CD31, an Ig superfamily member. J Immunol 160:395–402

Duncan GS, Andrew DP, Takimoto H et al. (1999) Genetic evidence for functional redundancy of platelet/endothelial cell adhesion molecule-1 (PECAM-1): CD31-deficient mice reveal PECAM-1-dependent and PECAM-1-independent functions. J Immunol 162:3022–3030

Fernandez JE, Deaglio S, Donati D et al. (1998) Analysis of the distribution of human CD38 and of its ligand CD31 in normal tissues. J Biol Regul Homeost Agents 12:81–91

Fujiwara K, Masuda M, Osawa M et al. (2001) Is PECAM-1 a mechanoresponsive molecule? Cell Struct Funct 26:11–17

Kostenuik PJ, Halloran BP, Morey-Holton ER et al. (1997) Skeletal unloading inhibits the in vitro proliferation and differentiation of rat osteoprogenitor cells. Am J Physiol 273:E1133–E1139

Lin C, Jiang X, Dai Z et al. (2009) Sclerostin mediates bone response to mechanical unloading via antagonizing Wnt/beta-catenin signaling. J Bone Miner Res 24:1651–1661

Maniatopoulos C, Sodek J, Melcher AH (1988) Bone formation in vitro by stromal cells obtained from bone marrow of young rats. Cell Tissue Res 254:317–330

Morey ER, Baylink DJ (1978) Inhibition of bone formation during space flight. Science 201:1138–1141

Okazaki R, Sakai A, Ootsuyama A et al. (2004) Trabecular bone mass and bone formation are preserved after limb immobilization in p53 null mice. Ann Rheum Dis 63:453–456

Osawa M, Masuda M, Harada N et al. (1997) Tyrosine phosphorylation of platelet endothelial cell adhesion molecule-1 (PECAM-1, CD31) in mechanically stimulated vascular endothelial cells. Eur J Cell Biol 72:229–237

Osawa M, Masuda M, Kusano K et al. (2002) Evidence for a role of platelet endothelial cell adhesion molecule-1 in endothelial cell mechanosignal transduction: is it a mechanoresponsive molecule? J Cell Biol 158:773–785

Piali L, Hammel P, Uherek C et al. (1995) CD31/PECAM-1 is a ligand for $\alpha_v\beta_3$ integrin involved in adhesion of leukocytes to endothelium. J Cell Biol 130:451–460

Rainger GE, Buckley CD, Simmons DL et al. (1999) Neutrophils sense flow-generated stress and direct their migration through $\alpha_v\beta_3$-integrin. Am J Physiol 276:H858–H864

Robling AG, Niziolek PJ, Baldridge LA et al. (2008) Mechanical stimulation of bone in vivo reduces osteocyte expression of Sost/sclerostin. J Biol Chem 283:5866–5875

Sakai A, Nakamura T, Tsurukami H et al. (1996) Bone marrow capacity for bone cells and trabecular bone turnover in immobilized tibia after sciatic neurectomy in mice. Bone 18:479–486

Sakai A, Nishida S, Okimoto N et al. (1998) Bone marrow cell development and trabecular bone dynamics after ovariectomy in ddy mice. Bone 23:443–451

Sakai A, Sakata T, Tanaka S et al. (2002) Disruption of the p53 gene results in preserved trabecular bone mass and bone formation after mechanical unloading. J Bone Miner Res 17:119–127

Sakata T, Sakai A, Tsurukami H et al. (1999) Trabecular bone turnover and bone marrow cell development in tail-suspended mice. J Bone Miner Res 14:1596–1604

Sakata T, Halloran BP, Elalieh HZ et al. (2003) Skeletal unloading induces resistance to insulin-like growth factor I on bone formation. Bone 32:669–680

Sakata T, Wang Y, Halloran BP et al. (2004) Skeletal unloading induces resistance to insulin-like growth factor-I (IGF-I) by inhibiting activation of the IGF-I signaling pathways. J Bone Miner Res 19:436–446

Sakuma-Zenke M, Sakai A, Nakayamada S et al. (2005) Reduced expression of platelet endothelial cell adhesion molecule-1 in bone marrow cells in mice after skeletal unloading. J Bone Miner Res 20:1002–1010

Schnittler HJ, Puschel B, Drenckhahn D (1997) Role of cadherins and plakoglobin in interendothelial adhesion under resting conditions and shear stress. Am J Physiol 273:H2396–H2405

Sho E, Sho M, Singh TM et al. (2001) Blood flow decrease induces apoptosis of endothelial cells in previously dilated arteries resulting from chronic high blood flow. Arterioscler Thromb Vasc Biol 21:1139–1145

Tanaka Y, Albelda SM, Horgan KJ et al. (1992) CD31 expressed on distinctive T cell subsets is a preferential amplifier of β_1 integrin-mediated adhesion. J Exp Med 176:245–253

Tanaka Y, Mine S, Figdor CG et al. (1998) Constitutive chemokine production results in activation of leukocyte function-associated antigen-1 on adult T-cell leukemia cells. Blood 91:3909–3919

Tanaka Y, Minami Y, Mine S et al. (1999) H-Ras signals to cytoskeletal machinery in induction of integrin-mediated adhesion of T cells. J Immunol 163:6209–6216

Tanaka M, Sakai A, Uchida S et al. (2004) Prostaglandin E2 receptor (EP4) selective agonist (ONO-4819.CD) accelerates bone repair of femoral cortex after drill-hole injury associated with local upregulation of bone turnover in mature rats. Bone 34:940–948

Uchida S, Sakai A, Kudo H et al. (2003) Vascular endothelial growth factor is expressed along with its receptors during the healing process of bone and bone marrow after drill-hole injury in rats. Bone 32:491–501

Vico L, Novikov VE, Very JM et al. (1991) Bone histomorphometric comparison of rat tibial metaphysics after 7-day tail suspension vs. 7-day spaceflight. Aviat Space Environ Med 62:26–31

Weinstein RS, Jilka RL, Parfitt AM et al. (1998) Inhibition of osteoblastogenesis and promotion of apoptosis of osteoblasts and osteocytes by glucocorticoids. Potential mechanisms of their deleterious effects on bone. J Clin Invest 102:274–282

Wlodarski P, Wasik M, Ratajczak MZ et al. (1998) Role of p53 in hematopoietic recovery after cytokine treatment. Blood 91:2998–3006

Yoshihara H, Arai F, Hosokawa K et al. (2007) Thrombopoietin/MPL signaling regulates hematopoietic stem cell quiescence and interaction with the osteoblastic niche. Cell Stem Cell 1:685–697

Chapter 13
Mechanosensing in Bone and the Role of Glutamate Signalling

Tim Skerry

13.1 Introduction

The response of bone to mechanical loading is much more complex than it appears at first. The widely-held idea that the adaptive mechanism in the skeleton leads to inhibition of resorption and induction of formation in response to loading, and the opposite processes as a consequence of disuse is very simplistic indeed because the goal of adaptive (re)modelling is not a particular amount of bone, but instead, the more tenuous concept of a more appropriate structure to resist the demands of changed loading circumstances. This means that in many circumstances a change in activity that requires a different load-bearing structure may be met with one of several outcomes each of which answers adequately the needs of the circumstances in question. Some of those solutions could involve a change in bone mass, and some the rearrangement of existing mass with or without addition or removal of bone. The complexity of the adaptive process is also influenced heavily by other factors so that the ultimate adaptive response is almost invariably a compromise between purely mechanical demands and the effects of different competing or complimentary physiological influences. Even these physiological balances are coloured by the effects of pathological factors that influence the ability of skeletal cells to perform the ideal actions, whether those are alterations in the amount of bone material formed by osteoblasts, or removed by osteoclasts, or the rate of change of formation or resorption, or the physical properties and sites of formation or removal of the material in question. This means that a full understanding of the adaptive mechanisms of the skeleton includes information at the level of mechanotransduction, but also at a series of levels of higher complexity which in concert account for the observed physiological changes of bone in response to loading or disuse. Much current research in the field of bone biology is focused upon the early stages of the transduction and signalling in response to very simple mechanical

T. Skerry (✉)
Mellanby Bone Centre, School of Medicine and Biomedical Sciences,
University of Sheffield, Beech Hill Road, Sheffield S10 2RX, UK
e-mail: t.skerry@sheffield.ac.uk

M. Noda (ed.), *Mechanosensing Biology*,
DOI 10.1007/978-4-431-89757-6_13, © Springer 2011

stimuli. This is not wrong, but too often it is taken for research that will explain the adaptive process rather than what it is, namely research into a small component of an early part of the process.

The purpose of this article is to describe some of the higher level features of bone's adaptive response to loading. Because of my own personal interest in the way that the skeleton has the ability to retain and modify its behaviour after repeated bouts of loading (whether in the short medium or long term) I will also review briefly some of the aspects of a signalling system in bone cells that utilises the excitatory amino acid glutamate. While our understanding of this system is not complete, it seems likely that it may be involved in the ways that the skeleton responds to brief loads, repeated potentiating loading or loading regimens with different rates or frequencies of application. The attractiveness of signalling systems that have known functions in the central nervous system is that mechanisms by which they underlie memory are well understood. It is clear that whether or not glutamate signalling contributes to such effects in bone, some biological mechanism or mechanisms do allow bone to retain information of previous loading.

In this article, loading and disuse will not be considered as separate processes, but as different aspects of a single spectrum of habitual activity in which complete disuse is the zero point reflecting the amount of bone present as a result of a purely genetically driven developmental process, and "normal" bone mass is the result of the superimposition on that "baseline" skeleton of a higher level of function. A useful analogy might be temperature, where we all accept the arbitrary nature of the Celsius scale in which temperatures different from our own are neither hot nor cold in an absolute sense, where the only true defining point is when all molecular movement ceases at $-273°C$. In the context of the skeleton, this means that changes in the components of habitual activity (whether they are due to altered magnitude, rate, duration or any other loading variable) in either direction, are not signals specific to induce atrophy or hypertrophy but are merely different in the degree to which the skeletal mass, architecture and strength are required to exceed that which is genetically determined. Whether the adaptive response to a new set of loading parameters is bone formation or resorption (to put it simplistically) depends only on the starting point before the change in habitual activity occurred.

13.2   Modes of Loading Affecting the Skeleton

In order to understand the processes involved in adaptive changes in bone, it is useful to consider the principles behind Donald Rumsfeldt's much maligned speech about "knowns and unknowns" (Rumsfeldt 2002). There are indeed *known* contributors to adaptive mechanisms in bone, *known unknowns*: mechanisms where there is some evidence for a role in skeletal mechanosensing but where we do not yet understand how they work, and *unknown unknowns*: mechanisms which must exist to explain established observations in physiology but for which we have no idea as to their specific cellular and molecular basis.

Fundamental to understanding of the details of functional adaptation in bone are the concepts of stress and strain, although detailed knowledge of engineering or maths is not necessary. Briefly, stress is force per unit area (Bradley and Sandifer 2009), so it is possible to calculate the stress on a bone by knowledge of its cross-sectional area and the load applied. However the more meaningful parameter for the skeleton is strain, which is a measure of the amount a bone deforms as a result of a stress. Strain is a simple concept because it is the ratio of the change in a given dimension as a result of the application of force, divided by the original dimension. So for example a human femur of length 400 mm might shorten by 0.3 mm when its owner stands, and the strain would be given by $0.3/400 = 0.00075$. For convenience, strains in relatively stiff materials are quoted in microstrain or strain $\times 10^6$, so that in this example, the bone deformation is 750 microstrain. It is important to emphasise that strain is a ratio, and as such, has no units. The common fashion of using the Greek letters mu and epsilon to denote microstrain has led to a misconception that these are units of strain, when that is not the case. Strains experienced by bones are rarely simple as they comprise strains with different directions that change during locomotion or movement. In the human tibia for example, strains we can measure on different surfaces of the bone include those in compression, tension and torsion. The full range of strains throughout a bone will be considerably more complex because of non-uniform geometry and material properties of skeletal elements (Currey 1984; Lanyon 1987). During activity, the magnitudes, rates and directions of those strains change. Even in the what is probably one of the "purest" example of dynamic bone strain in vivo, in the forelimbs of gibbons during brachiation (a form of locomotion in which the animals swing from limb to limb in trees), there are considerable variations in many parameters (Swartz et al. 1989).

13.3 Components of Osteogenic Strains: Strain Magnitude

It is widely understood that high loads such as those induced by lifting heavy weights are more potent in inducing adaptive changes in the skeleton than low loads and this appears to be entirely logical (Rubin and Lanyon 1985). If the effect of loading on bone strain magnitude is investigated, then it is interesting to see that in the long bones of most vertebrates, strains during peak physiological activity rarely exceed 2–3,000 microstrain (Hylander and Johnson 1997). Increases in peak strain magnitude above those levels as a result of support of heavier loads lead to adaptive modelling comprising either inhibition of bone resorption (Hillam and Skerry 1995), architectural changes such as straightening of a curved bone (Mosley et al. 1997), or bone apposition (or combinations of those effects) so that the same load causes less strain after adaptation. It is therefore entirely reasonable to think that cells signalling processes in the skeleton can discriminate between high and low bone strains. In this circumstance, I refer to bone strain as a determinant of cells signalling without implying what are the specific effects that induce cellular mechanotransduction. It is well known that both cellular deformation (Duncan and Misler 1989) and fluid flow over cells (Shivaram et al. 2010) have effects upon them, and while there are exponents for both direct cellular strain and fluid flow as the regulator of the adaptive response, it is likely that both contribute to signal transduction. However whole bone strain is the parameter we can measure in bones in vivo and ex vivo and as it induces both the cellular strain and intracortical fluid flow, it is the most useful measure of physiological effects of loading.

13.4 Strain Rate and Frequency

The magnitude of the applied force is not the sole component of a loading regimen that regulates its effect on bone. The rate of application of loads and the induced strain rate are also powerful determinants (O'Connor et al. 1982; Turner et al. 1995a). It has been shown experimentally that bones experience a range of different rates of strain (Milgrom et al. 2000; Foldhazy et al. 2005), and that under laboratory conditions, high strain rates (over 100,000 microstrain per second) are potently osteogenic while lower rates (<4,000 microstrain per second) but the same magnitude, may be less effective or completely ineffective in stimulating adaptive changes (Mosley and Lanyon 1998). These data appear to be relevant to physiology as we have showed that protection of a bone from loading in a sheep calcaneus model is associated with disuse bone loss which cannot be restored by the reintroduction of moderate rate and magnitude strain events by treadmill exercise (Skerry and Lanyon 1995). In this case, the physiological strains not reintroduced by that exercise were occasional high rate high magnitude deformations seen when sheep are startled and stamp or twitch their hind legs. Observations of similarly disparate strain magnitudes and rates in human bones during a range of activities suggests that strain rate is an important determinant of the effectiveness of bone loading to induce changes (Hillam et al. 1995). This means that whatever the signal transduction and activation mechanism is in bone cells, it must be capable of discriminating between short and long durations of mechanical stimuli. If we extrapolate from the experiments that show differences between loading that induces the same strain at different rates, then the skeleton is certainly able to perceive events that occur in a time of about 10 ms, and it is very likely that even more transitory events are sensed. At present, it is not known whether such temporal discriminatory power is the result of a rapid transduction system which can activate a more sustained signalling response, or a signalling system that responds in or near to real time to short duration events, or a combination of those two.

It is much less clear how the frequency of the application of dynamic loading affects its potency on the skeleton. This is largely because it is hard to separate strain frequency and rate in loading regimens. As the frequency of application of loads rises, so does the rate under most experimental conditions, and this means that effects of physiological and supraphysiological frequencies of loading could be due to complex interactions of transduction and onward signalling (Turner et al. 1995b). At the same time as frequency and rate rise, it becomes difficult to maintain the range of dynamic strain magnitudes, so that different studies are hard to interpret in an integrated way. The effectiveness of loading at frequencies in the range of 30 Hz at strain magnitudes in ranges lower than 50 microstrain certainly suggests that there may be a tradeoffs between different components of a loading regimen (Rubin et al. 2001). Further to this, it has been shown that the application of ultrasound to intact bones induces some but not all of the changes seen in response to more physiological levels of loading (Perry et al. 2009). In studies in vivo it is not possible to measure the strains induced by ultrasound, but it is certain that they would be very low indeed.

13.5 Strain Direction

There is relatively little specific information on way that changes in the distribution of strain alter adaptive changes, but intuitively it would be very surprising if this did not occur. The largest body of evidence that direction has an influence comes from experiments in which artificially applied loads are imposed on bones of animals in vivo. In all cases where strains have been measured, the artificial imposition regimen does not mimic physiological strain distributions. In the isolated avian ulna preparation that was developed by Rubin and Lanyon in the 1980s, the application of strains that were of the same magnitude as those that occurred when the birds flapped their wing vigorously, but with an altered distribution, led to profound adaptive responses (Rubin and Lanyon 1987). Since then, models in rodents imposing 4 point or cantilevered bending or axial compression on the tibiae have replicated those effects where physiological magnitudes of strain but very different directions alter adaptive processes (RaabCullen et al. 1994; Gross et al. 2002). Interestingly in the rodent ulna end loading model, where the artificial regimen induced bending strains in the bone that were more similar to physiological locomotor bending strains, higher magnitudes of strain were needed to affect modelling processes (Torrance et al. 1994; Hillam and Skerry 1995). However these experiments show only that strain directions/ distributions different from those experienced habitually induce short–term adaptive changes. One source of experimental data on the effects of prolonged changed strain distribution is as far as I know unexplored, but very tractable. Patients who have low lumbar disc protrusions with unilateral effects on sciatic nerve function lose motor supply to several muscles in the lower limb, often including the lumbrical muscles responsible for arching of the foot and some lateral parts of the gastrocnemius. These changes lead to altered gait and ground reaction force patterns (Skerry and Goodship, unpublished data). pQCT analysis of a single individual showed that there were changes in the cross-section of the distal tibia over the 2 years following these gait changes, despite unchanged body mass or habitual activity.

Few experiments have addressed the issue of altered strain direction directly however. Preliminary results from a study we have performed suggest that if the direction of strain induced by applied loading is the only change in a loading regimen, then that is not detectable by bones (Skerry and Peet 1997). We applied different loading waveforms to the ulnae of rats, so that some animals were loaded rapidly and unloaded rapidly, some were loaded rapidly and unloaded slowly, and some loaded slowly and unloaded fast, but in all cases the peak strain magnitude was identical. Unsurprisingly fast loading and unloading was a potent stimulus to form new bone. The effect of applying compression fast and releasing it slowly was less, but indistinguishable from slow loading and rapid release. This suggests that some forms of exercise such as weightlifting (where the weight is dropped once the lift has been approved) or archery where a bowstring is drawn over several seconds but released in an instant, may influence bone from rapid "unloading" inducing high rates of strain change as much or more than the low rate slow loading events.

13.6   Strain Regimen Duration, Repetition and Interruption

The final loading parameter to address in the context of its influence on the properties of a transduction/signalling mechanism is the duration of any mechanical stimulus and the way in which repetition and division alter its effects. It was shown some decades ago that in birds whose ulnae were protected from loading, the introduction of only 8 s of dynamic loading in each 24 h period of disuse was sufficient to maintain bone mass that was no different from normal controls. Increases in the number of load cycles showed that 72 s of osteogenic dynamic strain in each day was sufficient to produce a maximal osteogenic response that was not increased by greater durations/numbers of cycles (Rubin and Lanyon 1987). This suggests that in a response to a single bout of loading, the process responsible for sensing its effects saturates rapidly. There is logic in this. If a habituated skeleton is exposed to novel strain stimuli regularly, but only a few times daily, then to avoid fracture an adaptive response is needed.

Further studies have revealed though that if a single daily saturating number of load cycles is divided into two, three or four bouts of loading totalling the same single saturating number of load cycles in 1 day, then the osteogenic effect is greater as the number of bouts increases (Robling et al. 2002). It has also been shown that the insertion of rest periods of between 9 and 15 s between individual load cycles not only does not reduce the response from that which occurs to loading of the same total duration (but more cycles), but actually potentiates the response of bone compared to loading without rest periods between cycles (Srinivasan et al. 2002, 2007).

13.7   Site Specificity of Habitual Bone Strains

It is often suggested that adaptive changes in bone occur to reinstate some particular threshold level of strain magnitude in response to the altered level of activity. In the case of increased activity, it has been suggested that when strains exceed 1,500 microstrain, adaptation occurs to increase bone mass or change architecture, increasing stiffness so that habitual strains fall to previous levels (Frost 1987). The converse is suggested to occur if habitual strains fail to reach 1,000 microstrain, when bone loss or altered architecture leads to a more compliant structure that then experiences

1,000–1,500 microstrain after the changes. This is a useful way to illustrate the adaptive mechanism in a simple sense, but it is not sufficiently accurate if our goal is to understand fully the mechanisms underlying adaptive changes. A further level of complexity arises because different parts of the skeleton have different "expectations" of habitual bone strain. Many long bones experience similar magnitudes and rates of strain in the 1–3,000 microstrain range with rates as fast as 200,000 microstrain per second (Rubin and Lanyon 1984), but others never experience any such levels (Hillam et al. 1995; Rawlinson et al. 1995). The data on bone strains throughout the skeleton is incomplete in any species, but it is clear from the studies that are available that some parts of the skeleton such as the limbs and vertebral bodies of pigs and sheep (Lanyon and Smith 1970; Lanyon 1971, 1972, 1973; Lanyon et al. 1975), the tail vertebrae of rats (unpublished data), and the skulls of several species only experience much lower magnitudes and rates of strain (Hillam et al. 1995; Ravosa et al. 2000; Thomason et al. 2001). In the context of the skull, it is interesting that some parts such as the mandible, maxilla, temporal ridge and probably the occiput do experience strains in the hundreds of microstrain (Ross and Hylander 1996). The cranium however has been shown to experience only very low strains in all the studies so far performed. In the few experiments to measure cranial strains, it has been find that it is very hard to induce strains over 100 microstrain by reasonably physiological activities (Hillam et al. 1995). Even in birds, behavioural and anatomical arrangements have developed to reduce the consequences of pecking so that impacts of the beak are not transmitted in their entirety to the cranium (May et al. 1979). The reasons for this appear logical in that protection of the brain from potentially damaging trauma would be a driver for evolutionary changes. This leads us to the idea that instead of targeting strain magnitude, the adaptive process is likely to aim to optimise some "customary strain stimulus" which might include components of magnitude, rate, duration, frequency, rest periods modified by a site specific input that is different not only in different parts of the skeleton but also in different parts of the same bone, and probably on different surfaces of a single cortex or trabecula. At present there appears to be no information to explain the mechanisms behind such site specific differences in different bones although it has been shown that both osteoclasts and osteoblasts from different regions of the skeleton are not the same (Rawlinson et al. 1995; Perez-Amodio et al. 2004). It seems likely that osteocytes in different regions may provide additional levels of site specificity of mechanosensing.

13.8 Inferences from the Known Effects of Altered Loading

To sum up the inferences we can draw from existing data, we know several important facts

1. While strain magnitude is a potent influence on the osteogenicity of a loading regimen, other variables such as strain rate and frequency are as important.
2. Different parts of the skeleton have different "set points" for adaptive changes. It is not clear whether bones that habitually experience very low strains are insensitive

to their effects (i.e. resistant to disuse bone loss), or exquisitely sensitive to them with a similar feedback loop as other bones.
3. The skeleton can perceive very transient loading events and distinguish them from slower events, responding differently to the two.
4. Rapid repetition of strain changes desensitise bone to subsequent events.
5. In longer timescales, previous loading within a few hours potentiates the effect of subsequent bouts of loading.

13.9   The Role of Glutamate Signalling in the Skeleton's "Strain Memory"

From the observed responses of bone to different forms of loading, and specifically the 3rd, 4th and 5th conclusions above, it is necessary to postulate a system in the skeleton that can retain information on the effects of previous loading. Such a system or systems need to have the ability to perform both potentiation and desensitisation functions, just as those which occur in the CNS where they form the basis for learning and memory. When we discovered evidence for the expression in bone of many molecules involved in synaptic transmission (Mason et al. 1997), where potentiation and depression are well understood, the possibility of a similar process was an obvious idea. The expression and function of many "neurotransmitter" molecules has now been detected in bone (Patton et al. 1998; Skerry and Genever 2001; Bliziotes et al. 2002; Hinoi et al. 2004; Spencer et al. 2004; Igwe et al. 2009), and it is tempting to believe that there is proof that they account for retention of loading information by the skeleton. However, at present the case is far from proven.

   In the case of glutamate signalling in bone, despite many studies by ourselves and other groups showing that bone cells possess glutamate receptors, the molecular machinery necessary to release the excitatory amino acid, and many accessory molecules associated with long-term potentiation events in the CNS, there is little direct evidence for a physiological role for glutamate. Functional studies in vitro show activation of glutamate receptors on bone cells by physiological and pharmacological ligands (Laketic-Ljubojevic et al. 1999), and inhibition of that activation by chemical antagonists of the receptors (Peet et al. 1999; Itzstein et al. 2000). The same antagonists cause profound changes in both bone formation by osteoblasts and bone resorption by osteoclasts. However data in vivo are much less clear. Many chemical antagonists of glutamate receptors cross the blood brain barrier so are very poor for discriminatory physiological studies of their direct effects on bone because of their modification of many behavioural activities with impacts on the whole body. A few glutamate receptor antagonists are poorly able to cross into the brain yet have rather slight effects on bone physiology. In studies where we have administered those antagonists in conjunction with application of loading to bones in vivo, the changes have been subtle.

   The use of transgenic animals has the potential to clarify this issue, in that tissue-specific knockout of elements of the glutamate signalling system in bone should allow their role to be researched more thoroughly. While initial studies showed changes in bone formation during development (Martinez-Bautista et al. 2008), adults bone phenotypes are mild. There are several conclusions that could be drawn from the data so far obtained. First it is possible that the expression and function of glutamate receptors has no physiological significance, a conclusion that has been suggested but which seems at best puzzling (Gray et al. 2001). Second there may be considerable redundancy in strain memory signalling that involves glutamate, so that in its absence other compensations occur. Finally it may be that the role of glutamate signalling is highly specific in relation to mechanosensing. If any effect is to regulate differently the responses to loading at different rates/frequencies, or to change the response to divided bouts of loading or those with rest periods between cycles, then there are currently no experimental data published that confirm or refute such a possibility. There is little doubt that if the role of glutamate is highly specific, then the habitual exercise regimens of lab mice in cages are unlikely to uncover such effects during normal activity.

13.10 Conclusions

The response of the skeleton to loading is highly complex and while we understand a great deal about it, there are many unknowns that could have profound impacts on therapies for bone loss. Improved understanding of the subtle ways in which different aspects of a loading regimen provide specific influences on bone may give clues as to how we could impose safely, highly osteogenic but non-damaging stimuli on frail skeletons. Better understanding of the site-specific properties of bone's responses could also have high impact, as could studies on the way that the skeleton retains information on previous loading. While putative cognitive enhancers have been devised to improve brain processing abilities, it may be simpler to develop agents that modulate similar signalling processes in bone to make habitual activities in each of us the perceived equivalent of a vigorous workout.

Acknowledgments The work performed in my lab that I mention here was funded largely by grants from BBSRC, with additional support from ARC, MRC and Conacyt.

References

Bliziotes M, Gunness M et al (2002) The role of dopamine and serotonin in regulating bone mass and strength: studies on dopamine and serotonin transporter null mice. J Musculoskelet Neuronal Interact 2(3):291–295

Bradley R, Sandifer C (2009) Cauchy's Cours d'analyse (sources and studies in the history of mathematics and physical sciences). Springer, New York

Currey JD (1984) Can strains give adequate information for adaptive bone remodeling? Calcif Tissue Int 36:S118–S122

Duncan R, Misler S (1989) Voltage-activated and stretch-activated Ba2+ conducting channels in an osteoblast-like cell line (UMR 106). FEBS Lett 251:17–21

Foldhazy Z, Arndt A et al (2005) Exercise-induced strain and strain rate in the distal radius. J Bone Joint Surg Br 87(2):261–266

Frost HM (1987) The mechanostat: a proposed pathogenic mechanism of osteoporoses and the bone mass effects of mechanical and nonmechanical agents. Bone Miner 2:73–85

Gray C, Marie H et al (2001) Glutamate does not play a major role in controlling bone growth. J Bone Miner Res 16:742–749

Gross TS, Srinivasan S et al (2002) Noninvasive loading of the murine tibia: an in vivo model for the study of mechanotransduction. J Bone Miner Res 17(3):493–501

Hillam RA, Skerry TM (1995) Inhibition of bone resorption and stimulation of formation by mechanical loading of the modeling rat ulna in vivo. J Bone Miner Res 10(5):683–689

Hillam RA, Jackson M et al (1995) Regional differences in human bone strain in-vivo. J Bone Miner Res 10:S443

Hinoi E, Takarada T et al (2004) Glutamate signaling in peripheral tissues. Eur J Biochem 271(1):1–13

Hylander WL, Johnson KR (1997) In vivo bone strain patterns in the zygomatic arch of macaques and the significance of these patterns for functional interpretations of craniofacial form. Am J Phys Anthropol 102:203–232

Igwe JC, Jiang X et al (2009) Neuropeptide Y is expressed by osteocytes and can inhibit osteoblastic activity. J Cell Biochem 108(3):621–630

Itzstein C, Espinosa L et al (2000) Specific antagonists of NMDA receptors prevent osteoclast sealing zone formation required for bone resorption. Biochem Biophys Res Commun 268:201–209

Laketic-Ljubojevic I, Suva LJ et al (1999) Functional characterization of N-methyl-D-aspartic acid-gated channels in bone cells. Bone 25(6):631–637

Lanyon LE (1971) Strain in sheep lumbar vertebrae recorded during life. Acta Orthop Scand 42(1):102–112

Lanyon LE (1972) In vivo bone strain recorded from thoracic vertebrae of sheep. J Biomech 5(3):277–281

Lanyon LE (1973) Analysis of surface bone strain in the calcaneus of sheep during normal locomotion. Strain analysis of the calcaneus. J Biomech 6(1):41–49

Lanyon LE (1987) Functional strain in bone tissue as an objective, and controlling stimulus for adaptive bone remodelling. J Biomech 20:1083–1093

Lanyon LE, Smith RN (1970) Bone strain in the tibia during normal quadrupedal locomotion. Acta Orthop Scand 41(3):238–248

Lanyon LE, Hampson WGJ et al (1975) Bone deformation recorded in vivo from strain gauges attached to the human tibial shaft. Acta Orthop Scand 46:256–268

Martinez-Bautista S, Wang N et al (2008) Mice lacking NMDA receptor expression in osteoblasts have pronounced vertebral bone abnormalities. J Bone Miner Res 23:S87

Mason DJ, Suva LJ et al (1997) Mechanically regulated expression of a neural glutamate transporter in bone: a role for excitatory amino acids as osteotropic agents? Bone 20:199–205

May PR, Fuster JM et al (1979) Woodpecker drilling behavior. An endorsement of the rotational theory of impact brain injury. Arch Neurol 36(6):370–373

Milgrom C, Finestone A et al (2000) In-vivo strain measurements to evaluate the strengthening potential of exercises on the tibial bone. J Bone Joint Surg Br 82(4):591–594

Mosley JR, Lanyon LE (1998) Strain rate as a controlling influence on adaptive modeling in response to dynamic loading of the ulna in growing male rats. Bone 23(4):313–318

Mosley JR, March BM et al (1997) Strain magnitude related changes in whole bone architecture in growing rats. Bone 20(3):191–198

O'Connor JA, Lanyon LE, MacFie H (1982) The influence of strain rate on adaptive bone remodelling. J Biomech 15(10):767–781

Patton AJ, Genever PG et al (1998) Expression of an NMDA type receptor by human and rat osteoblasts and osteoclasts suggests a novel glutamate signaling pathway in bone. Bone 22:645–649

Peet NM, Grabowski PS et al (1999) The glutamate receptor antagonist MK801 modulates bone resorption in vitro by a mechanism predominantly involving osteoclast differentiation. FASEB J 13:2179–2185

Perez-Amodio S, Beertsen W et al (2004) (Pre-)osteoclasts induce retraction of osteoblasts before their fusion to osteoclasts. J Bone Miner Res 19(10):1722–1731

Perry MJ, Parry LK et al (2009) Ultrasound mimics the effect of mechanical loading on bone formation in vivo on rat ulnae. Med Eng Phys 31(1):42–47

RaabCullen DM, Akhter MP et al (1994) Periosteal bone formation stimulated by externally induced bending strains. J Bone Miner Res 9:1143–1152

Ravosa MJ, Johnson KR et al (2000) Strain in the galago facial skull. J Morphol 245(1):51–66

Rawlinson SC, Mosley JR et al (1995) Calvarial and limb bone cells in organ and monolayer culture do not show the same early responses to dynamic mechanical strain. J Bone Miner Res 10(8):1225–1232

Robling AG, Hinant FM et al (2002) Improved bone structure and strength after long-term mechanical loading is greatest if loading is separated into short bouts. J Bone Miner Res 17(8):1545–1554

Ross CF, Hylander WL (1996) In vivo and in vitro bone strain in the owl monkey circumorbital region and the function of the postorbital septum. Am J Phys Anthropol 101(2):183–215

Rubin CT, Lanyon LE (1984) Dynamic strain similarity in vertebrates; an alternative to allometric limb bone scaling. J Theor Biol 107:321–327

Rubin CT, Lanyon LE (1985) Regulation of bone mass by mechanical strain magnitude. Calcif Tissue Int 37:411–417

Rubin CT, Lanyon LE (1987) Osteoregulatory nature of mechanical stimuli: function as a determinant for adaptive remodeling in bone. J Orthop Res 5:300–310

Rubin C, Turner AS et al (2001) Anabolism. Low mechanical signals strengthen long bones. Nature 412(6847):603–604

Rumsfeldt D (2002) The unknown. Department of Defence News Briefing.

Shivaram GM, Kim CH et al (2010) Novel early response genes in osteoblasts exposed to dynamic fluid flow. Philos Trans A Math Phys Eng Sci 368(1912):605–616

Skerry TM, Genever PG (2001) Glutamate signalling in non-neuronal tissues. Trends Pharmacol Sci 22(4):174–181

Skerry TM, Lanyon LE (1995) Interruption of disuse by short-duration walking exercise does not prevent bone loss in the sheep calcaneus. Bone 16:269–274

Skerry TM, Peet NM (1997) "Unloading" exercise increases bone formation in rats. J Bone Miner Res 12(9):6

Spencer GJ, Hitchcock IS et al (2004) Emerging neuroskeletal signalling pathways: a review. FEBS Lett 559(1–3):6–12

Srinivasan S, Weimer DA et al (2002) Low-magnitude mechanical loading becomes osteogenic when rest is inserted between each load cycle. J Bone Miner Res 17(9):1613–1620

Srinivasan S, Ausk BJ et al (2007) Rest-inserted loading rapidly amplifies the response of bone to small increases in strain and load cycles. J Appl Physiol 102(5):1945–1952

Swartz SM, Bertram JEA et al (1989) Telemetered in vivo strain analysis of locomotor mechanics of brachiating gibbons. Nature 342(6247):270–272

Thomason JJ, Grovum LE et al (2001) In vivo surface strain and stereology of the frontal and maxillary bones of sheep: implications for the structural design of the mammalian skull. Anat Rec 264(4):325–338

Torrance AG, Mosley JM et al (1994) Noninvasive loading of the rat ulna in vivo induces a strain related modeling response uncomplicated by trauma of periosteal pressure. Calcif Tissue Int 54(3):241–247

Turner C, Owan HI et al (1995a) Mechanotransduction in bone – role of strain-rate. Am J Physiol 269:E438–E442

Turner CH, Yoshikawa T et al (1995b) High frequency components of bone strain in dogs measured during various activities. J Biomech 28(1):39–44

Chapter 14
Osteoclast Biology and Mechanosensing

Géraldine Pawlak, Virginie Vives, Emmanuelle Planus,
Corinne Albiges-Rizo, and Anne Blangy

14.1 Introduction

Bone matrix is composed of a mineral phase, which guarantees its stiffness, and collagen fibers, which give elasticity and allow bone to absorb energy. These two components cooperate to ensure bone strength. Bone is a dynamic tissue undergoing remodeling throughout life. This turnover allows the tissue to prevent and repair its own damage and to respond to external forces. It ensures permanent adaptation of the skeleton to the mechanical and physiological conditions. Bone dynamics relies on the cooperative action of two differentiated cell types: the osteoclasts to degrade bone and the osteoblasts to lay down new bone.

Osteoclasts are multinucleated cells specialized for the degradation of bone and mineralized cartilage. They arise from the fusion of precursors from the myeloid hematopoietic lineage, which also gives rise to dendritic cells and various types of macrophages (Boyle et al. 2003). The main cytokines involved in the differentiation of osteoclasts are the ligands of two transmembrane receptors: receptor activator for nuclear factor κB ligand (RANKL) and macrophage colony stimulating factor (M-CSF). They respectively bind to the membrane receptors RANK and c-fms. Osteoclast formation also involves the membrane receptors: osteoclast-associated receptor (OSCAR) and triggering receptor expressed on myeloid cells 2 (TREM-2) (Cella et al. 2003; Kim et al. 2002; Paloneva et al. 2003). For signal transduction, these two receptors rely on adaptor molecules that contain an immunoreceptor tyrosine-based activation motif (ITAM) and function as coreceptors. OSCAR interacts with Fc receptor common γ subunit (FcRγ) (Ishikawa et al. 2004) whereas TREM-2 binds to DAP12 (DNAX-activating protein) (Paloneva et al. 2003). As yet,

G. Pawlak, V. Vives, and A. Blangy (✉)
Montpellier University, Centre de Recherche de Biochimie Macromoléculaire,
CNRS, 1919 route de Mende, 34297 Montpellier Cedex 5, France
e-mail: anne.blangy@crbm.cnrs.fr

G. Pawlak, E. Planus, and C. Albiges-Rizo
Institut Albert Bonniot, INSERM U823, CNRS ERL3148 Université Joseph Fourier,
Equipe DySAD, Site Santé, BP 170, 38042 Grenoble Cedex 9, France

the ligands for OSCAR and TREM-2 remain unknown. In vitro, the two cytokines RANKL and M-CSF are sufficient to obtain fully functional osteoclasts from purified monocytic precursors in a tissue culture dish (Boyle et al. 2003).

Adhesion of osteoclasts and osteoclast precursors to the extracellular matrix is mediated exclusively by podosomes. As is the case for all monocytic lineage cells, osteoclasts never assemble focal adhesions (Linder 2007). Podosomes are formed by a central column of F-actin fibers perpendicular to the substrate (core, Fig. 14.1a). It is surrounded by an actin cloud, a loose network of actin cables that run parallel to the substrate (cloud, Fig. 14.1a). Assembly of the core is controlled by CD44 whereas the cloud is organized by $\alpha v \beta 3$ integrin (Chabadel et al. 2007). Podosomes can assemble into various superstructures depending on osteoclast differentiation stage and substrate nature (Destaing et al. 2003; Luxenburg et al. 2007). During osteoclast differentiation on nonmineralized substrates, podosomes initially assemble as clusters, then rearrange into dynamic rings that expand and eventually fuse to form a unique pericellular podosome belt (Fig. 14.1b). When osteoclasts are seeded on mineralized matrices, podosomes get compacted to assemble into a broad actin ring or sealing zone (Lakkakorpi and Vaananen 1991; Saltel et al. 2004). Such specific adhesion structure is necessary for bone resorption. The sealing zone surrounds the apical membrane of the osteoclast; the ruffled border, which is the real site of bone resorption (Fig. 14.1c). Osteoclast activity results in the local dissolution of the mineral and the subsequent digestion of the collagen fibers. This leads to the formation of a resorption pit or resorption lacuna. At the ruffled border, protons are secreted to dissolve the mineral components of the bone matrix by acidification. The mineral phase of bone is predominantly made of calcium phosphate: hydroxyapatite. Proteases are also released to degrade the proteins of the bone matrix, predominantly type 1 collagen (Fig. 14.1c). At the bottom of the resorption pit, incompletely digested collagen fibers remain (Fig. 14.1d). The degraded material is internalized by the osteoclast and transported from the apical membrane toward the secretion zone at the basal membrane of the osteoclast (Mulari et al. 2003; Vaananen and Laitala-Leinonen 2008).

On mineralized substrates, osteoclasts alternate between resorption and migration phases. When they migrate, osteoclasts do not resorb bone and the sealing zone disassembles (Kanehisa and Heersche 1988). Osteoclast morphology is significantly changing throughout differentiation and between resorbing and nonresorbing phases. This involves extensive reorganization of the cytoskeleton and the plasma membrane. In particular, organization of podosomes is very dynamic. They assemble as clusters, rings, belts or sealing zones. The formation of podosome superstructures in osteoclasts is controlled by RhoGTPases (Ory et al. 2008), in particular RhoA (Chellaiah et al. 2000; Destaing et al. 2005) and Wrch1/RhoU (Brazier et al. 2009), and by the tyrosine kinases Src and Pyk2 (Destaing et al. 2008; Gil-Henn et al. 2007).

Osteocytes arise from the terminal differentiation of osteoblasts and are embedded within the mineralized matrix of bones. These star-shaped cells are widely accepted as the major bone mechanosensors that signal towards osteoclasts and osteoclast precursors to stimulate bone resorption (Bonewald 2006; Sims and Gooi 2008; Skerry 2008). The self-ability of osteoclasts to sense and respond to

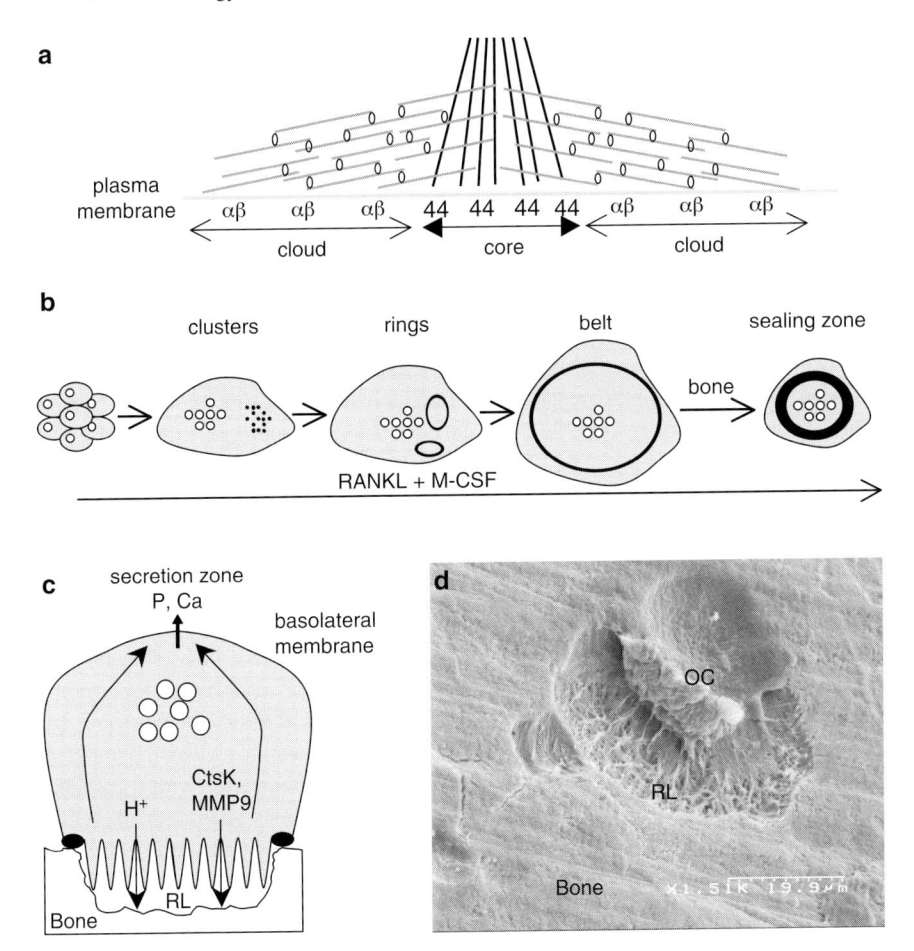

Fig. 14.1 Organization of adhesion structures in osteoclasts. (**a**) Schematic side view of a podosome. The actin fibers run perpendicular to the substrate in the podosome core (*black lines*) and parallel in the cloud (*dark gray lines*). MyosinII (*open ovals*) and $\alpha v\beta 3$ ($\alpha\beta$) integrin are in the podosome cloud and CD44 (*44*) in the core (adapted from Chabadel et al. 2007). (**b**) Dynamics of podosome organization during osteoclast differentiation and function (adapted from Destaing et al. 2003). During osteoclast differentiation stimulated by RANKL and M-CSF, mononucleated precursors fuse to form polykaryons. During osteoclast maturation, podosomes in clusters (*black dots*) organize as rings that expand and fuse to form a peripheral belt (*thin black rings*). On bone, podosomes assemble the sealing zone (*thick black ring*). *White dots* Nuclei. (**c**) Schematic organization of a resorbing osteoclast. The sealing zone made of podosomes (*black ovals*) surrounds osteoclast apical membrane differentiated into a ruffled border. At this site, osteoclasts secrete protons (*H+*) and proteases: Cathepsin K (*CtsK*) and matrix metalloprotease (*MMP9*) to resorb bone. This generates a resorption lacuna (*RL*) at the bone surface. Bone degradation products, including Calcium (Ca) and Phosphate (P), are internalized and transferred by transcytosis toward the secretion zone of the basal membrane. (**d**) Scanning electron micrograph of an osteoclast in a RL. The resorbing activity of a mouse monocyte derived osteoclast (*OC*) on bone leaves a RL with some collagen fibers sill visible at the bottom. *Scale bar* 19.9 μm

modifications of the physical properties of their environment has scarcely been studied so far. As a consequence, the direct influence of physical forces on osteo-clast biology has not been much explored. The aim of this article is to give an overview of early to very recent evidence that osteoclasts can sense and respond to modifications of their physical environment in a cell-autonomous fashion. We will also speculate on the potential molecular mechanisms controlling osteoclast mechanosensitive response.

The mechanosensitive properties of osteoclasts may have important impacts on their biology, especially to regulate differentiation and resorption. They may also have consequences for bone pathologies; in particular diseases that involve exacer-bate osteoclast activity including age-related osteoporosis or tumor metastasis to the bone. Finally, osteoclast response to mechanical stimulation should be kept in mind for the design of resorbable biomaterials for bone repair and substitution.

14.2 Podosomes: Recently Proven Mechanosensors

Adhesion sites are considered as the main cellular structures involved in substrate sensing. In focal adhesions, integrins act as mechanosensitive receptors. They are able to translate modifications of extracellular matrix physical properties: topography, rigidity or anisotropy, into intracellular signaling (Geiger et al. 2009; Schwartz and DeSimone 2008). In particular, integrin $\alpha v\beta 3$ can sense matrix rigidity (Jiang et al. 2006; Kostic and Sheetz 2006). Podosomes share many components with focal adhe-sions and integrin $\alpha v\beta 3$ is the major integrin in osteoclasts (Block et al. 2008). It localizes to the podosome cloud (Chabadel et al. 2007; Destaing et al. 2008). As opposed to focal adhesions that are stable for 30–90 min, podosomes are highly dynamic structures: their life span in osteoclasts is only a few minutes (Destaing et al. 2003, 2008). Until recently, the unstable nature of podosomes appeared contradictory with any ability to sense and transmit forces between molecular complexes inside and outside the cell. Thus, podosomes have not so far been considered as good candidate mechanosensors in contrast to focal adhesions (Geiger and Bershadsky 2002).

However, several recent studies contradict this postulate. A study on self-organized podosome rosettes in fibroblasts showed that stiffer substrates induced lower podo-some density with a longer life span (Collin et al. 2006). Rosettes expand and contract owing to podosome dissolution and reformation. The velocity of rosettes expansion and shrinking decreased when substrate rigidity increased (Collin et al. 2006). This is the first demonstration that individual podosomes or self-organized podosome rosettes can be affected by the physical properties of the extracellular matrix. It was further shown that podosome rings could exert myosin II-dependent traction forces on extracellular matrix proteins (Collin et al. 2008). Forces increased with matrix stiffness, showing that podosomes were true mechanosensors. This study also reported that podosome movements were stimulated by local forces applied at the cell surface. When stress increased, displacement of podosomes increased (Collin et al. 2008).

These two reports established that podosomes were bona fide mechanosensors able to transmit mechanical stresses inside-out and outside-in.

Like osteoclasts, macrophages contact the extracellular matrix exclusively through podosomes (Linder 2007). Alveolar macrophages were shown to adapt their morphology to substrate stiffness (Fereol et al. 2006). Macrophages plated on glass were flat and spread. When plated on soft polyacrylamide gels, macrophages were more rounded and less spread. The maximal surface area of macrophages in contact with the substrate increased with polyacrylamide gel stiffness. These cells whose adhesion to the substrate resides exclusively on podosomes were shown to exhibit low internal tension and high sensitivity to substrate stiffness and to be able to adapt their shape to matrix stiffness (Fereol et al. 2009).

Altogether, these findings suggest that osteoclasts and their precursors should be able to sense the physical properties of their environment using podosomes as mechanosensors.

14.3 Importance of Mechanical Signals During Osteoclast Differentiation

Mechanical stimuli have an essential role in the maintenance of bone mass. Reduced weight bearing, due for instance to motor paralysis, prolonged bed rest or during space flights, provokes bone loss (LeBlanc et al. 2007; Rubin et al. 2006). Although osteoclasts are responsible for bone resorption, most studies explored the effect of mechanical stress on bone-forming osteoblasts (see Part 3 of this book: "Skeletal Response").

14.3.1 Osteoclast Differentiation: The Bone Touch

Osteoclasts differentiate from hematopoietic precursors of the monocyte–macrophage lineage, present in the circulating blood, the bone marrow or other hematopoietic tissues such as spleen or fetal liver (Roodman 1999). Using a model system in rats to induce osteoclast formation in vivo, an early study showed that mononucleated precursors are recruited from the vascular layer and migrate towards the bone surface (Baron et al. 1986). Once in contact with bone, precursors fuse and give rise to fully differentiated osteoclasts (Baron et al. 1986). This founder study established that, although osteoclast specific markers such as tartrate resistant acid phosphatase (TRAP) are expressed before cells contact the bone, fusion of mononuclear precursors occurs only after their attachment onto the bone surface. Further histological observations on tissues from sites of bone resorption showed that, under physiological and pathological conditions, cells expressing the full morphological and functional properties of mature osteoclasts are restricted to the bone surface (Shen et al. 2006).

14.3.2   The Need for an Appropriate Experimental System

When reaching the bone surface, differentiating osteoclast precursors make contact with osteoblasts-derived lining cells and osteocytes and thus receive novel chemical and physical signals. During differentiation, precursor cells leave the soft environment of the vascular layer to interact with the bone, a tissue of much higher rigidity. Elastic moduli of vascular and epithelial layers are in the order of kilopascal, whereas bone elastic modulus is several orders of magnitude higher, reaching 10–25 GPa (Rho et al. 1997; Turner et al. 1999). Osteoclast differentiation is completed only after precursors contact the bone. Therefore, a rigid substrate may be necessary to provide essential mechanical signals. Bone microenvironment is too complex to allow dissection of the respective contributions of chemical and physical cues to osteoclasts differentiation. This precludes distinguishing in vivo between mechanical signals received by the osteoclasts themselves from those transmitted to osteoclasts by other bone mechanosensitive cells such as osteoblasts or osteocytes. Before the identification of RANKL as the essential cytokine for osteoclastogenesis, osteoclasts were produced in coculture systems with osteoblasts or total bone marrow cells. These cells later proved to provide the RANKL necessary for osteoclast differentiation. Some studies have explored osteoclast differentiation in response to mechanical strain in vitro in coculture. These studies concluded that this strain repressed osteoclastogenesis (Rubin et al. 1997, 1999). However, it was later found that mechanical stimulation inhibited RANKL secretion by osteoblasts and bone marrow cells. Therefore, reduced osteoclast formation in response to strain could not be attributed to a mechanosensitive response of the precursors themselves (Rubin et al. 2000).

From identification of RANKL as the essential cytokine in osteoclastogenesis (Lacey et al. 1998; Takahashi et al. 1999), in vitro methods were developed to differentiate osteoclasts from purified precursors isolated from hematopoietic organs such as bone marrow, blood or spleen. These precursors differentiate into osteoclasts with bone resorbing properties in the presence of M-CSF and RANKL within 1–2 weeks (Boyle et al. 2003; Lacey et al. 1998). Model cell lines are also used to generate osteoclasts in vitro (Hentunen et al. 1999), such as the popular RAW264.7 mouse macrophage cell line. RAW264.7 cells differentiate into bone resorbing osteoclasts in the presence of RANKL within 4–5 days (Hsu et al. 1999). These in vitro culture systems now allow exploration of mechanical stimuli direct impact on osteoclast differentiation.

14.3.3   Osteoclast Differentiation Requires a Stiff Substrate

Differentiation of osteoclasts in vitro is usually performed on polystyrene plates or glass coverslips. Polystyrene and glass are very stiff substrates, with an elastic modulus around 3 GPa and 50 GPa, respectively. This is in the range of bone stiffness (10–25 GPa), but is much more rigid than the vascular layer (several kilopascal).

When differentiated in vitro, osteoclast precursors are in contact with a stiff matrix from the very beginning of the differentiation process, whereas in vivo, their commitment into the osteoclast lineage starts in the soft environment of the vascular layer before they contact bone (Baron et al. 1986). Miyamoto and colleagues studied osteoclastic differentiation of precursors kept in semisolid nonadherent conditions. Precursors committed to the osteoclast lineage and expressed specific markers such as TRAP, but were unable to fuse (Miyamoto et al. 2000). After cells were transferred back to adherent conditions, they formed multinucleated osteoclasts, in agreement with the earlier in vivo observations of Baron and colleagues. The in vitro study further established that integrin $\alpha v \beta 3$ is essential for anchorage-dependent multinuclear osteoclast formation (Miyamoto et al. 2000). Although osteoclast precursors expressed integrin $\alpha v \beta 3$, and even if its ligand vitronectin was abundant in the culture medium, precursors failed to fuse when kept in semisolid conditions. Thus, multinucleated osteoclast formation defect in these conditions was very likely due to the absence of $\alpha v \beta 3$ integrin physical stimulation.

This suggests that two types of signals are necessary for osteoclastic differentiation. On one hand, chemical stimulation provided by cytokines irreversibly commits the hematopoietic precursors to the osteoclast lineage and induces the expression of specific markers such as integrin $\beta 3$, TRAP or cathepsin K. On the other hand, a mechanical stimulation resulting from the contact with a rigid substrate, i.e., the bone in vivo, allows already committed yet mononucleated precursors to fuse through an integrin $\alpha v \beta 3$-dependent mechanism. To support this hypothesis, we performed osteoclastic differentiation of RAW264.7 cells on soft 10 kPa polyacrylamide gels similar to those shown to inhibit alveolar macrophage spreading (Fereol et al. 2006). We found that RAW264.7 cells induced for osteoclastic differentiation by RANKL were not able to form multinucleated cells when plated on polyacrylamide gels (Fig. 14.2a). This result is in line with the observations of Miyamoto and colleagues. Interestingly, when cells were differentiated on glass for 3 days and then transferred as mononucleated preosteoclasts onto soft polyacrylamide gels, they were able to fuse with the same efficiency as cells kept on glass throughout differentiation (Fig. 14.2b, c). These observations further sustain the hypothesis that a rigid substrate is needed to induce the part of the osteoclastic differentiation program involved in the fusion process. In addition to the observations by Miyamoto and colleagues, we found that the rigid substrate was not necessary for the fusion process itself, as fusion indexes were comparable between perfusion osteoclasts replated on 10-kPa polyacrylamide gels and on glass (Fig. 14.2b, c). This suggests that contacts with a stiff substrate induce an irreversible priming signal necessary for subsequent precursor fusion.

Requirement of a rigid substrate for osteoclast precursor fusion accounts for the in vivo observation that multinucleated fully mature osteoclasts are only observed in contact with bone or mineralized cartilage (Baron et al. 1986; Shen et al. 2006). Mechanical signals are known to affect gene expression (Wang et al. 2007, 2009) and cell differentiation (Engler et al. 2006). The above observations strongly suggest that osteoclast differentiation involves expression of a specific set of genes responding to mechanical stimulation transduced by integrin $\alpha v \beta 3$.

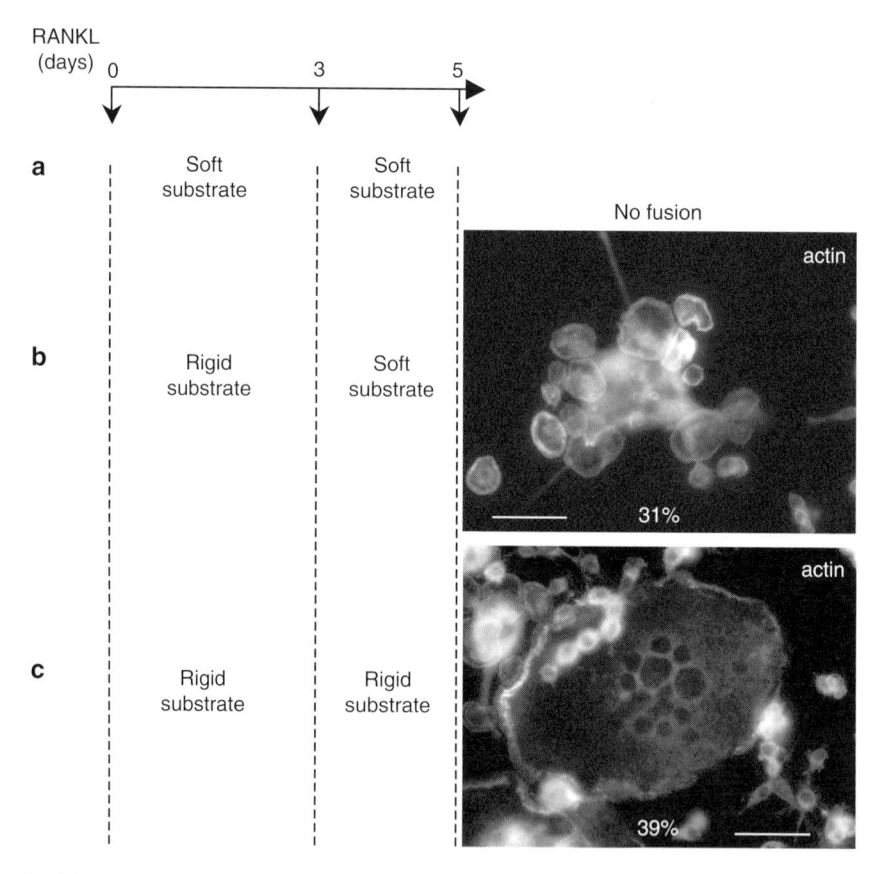

Fig. 14.2 Osteoclast precursors do not fuse when differentiated on a soft substrate. RAW264.7 cells were seeded on vitronectin-coated 10 kPa polyacrylamide gels (**a**) or glass (**b, c**) in the presence of RANKL. After 3 days, preosteoclasts were still mononucleated. They were left on polyacrylamide gels (**a**) or transferred from glass onto 10 kPa polyacrylamide gels (**b**) or back onto glass (**c**). Differentiation was pursued for two more days in the presence of RANKL. F-actin was stained with rhodamine-labeled phalloidin (*bottom images*) and DNA was stained with Hoechst 33257 (not shown). The efficiency of osteoclast differentiation was assessed by measuring the fusion index (number of nuclei in cells with more than three nuclei/total number of nuclei) in each condition (% in bottom images). *Scale bar* 20 μm

14.3.4 Strain Inhibits Osteoclast Differentiation

Mechanical stimulation is a crucial factor for the maintenance of bone mass, as shown for instance by space flight and bed rest studies (LeBlanc et al. 2007; Rubin et al. 2006). Bones that bear weight adapt their structure to the mechanical environment: loading results in increased bone formation by osteoblasts and decreases bone degradation by osteoclasts whereas unloading produces the reverse effects. Osteocytes and osteoblasts are bone cells essential for sensing changes in loading and communicating the need for adapting bone mass and structure (Teti and Zallone 2009).

Osteoclasts are rather considered as effectors of bone resorption instructed and regulated by osteoblasts and osteocytes in this context. Nonetheless, a few recent reports suggest that osteoclast precursors can respond to mechanical stimuli in a cell-autonomous fashion.

As mentioned above, Rubin and colleagues showed that mechanical stimulation repressed osteoclastogenesis. They used a coculture system in which the RANKL necessary for osteoclast differentiation was provided by osteoblasts and its production was inhibited by mechanical stimulation (Rubin et al. 1999, 2000). Therefore, a direct effect of mechanosignals on osteoclastogenesis remained to be established. A recent study explored the effect of mechanical strain on RANKL-induced differentiation using purified osteoclast precursors (Kadow-Romacker et al. 2009). Precursors were allowed to differentiate on a dentine substrate submitted daily to short micromechanical stimulation, mimicking physiological bone microstrains. For a certain duration and frequency, microstrains significantly decreased osteoclast differentiation (Kadow-Romacker et al. 2009). Therefore, osteoclast precursors appear to be sensitive to mechanical stimulation, which intended to inhibit osteoclastogenesis.

14.3.5 Microgravity Stimulates Osteoclast Differentiation

A few reports examined the effects of reduced mechanical stimulation on osteoclast precursor differentiation. To simulate weightlessness, hindlimb unloading can be provoked in rodent by tail suspension. This treatment results in trabecular bone loss (Morey-Holton and Globus 2002). Osteoclast precursors purified from unloaded mouse bone marrow were found to produce the same number of osteoclasts as compared to normal weight-bearing mice. Nevertheless, these osteoclasts exhibited increased spreading suggesting that osteoclast precursors were sensitive to unloading (Ishijima et al. 2007).

Ground-based rotational devices such as the clinostat (Cogoli 1993) or the rotary cell culture system (RCCS) developed by NASA (Mitteregger et al. 1999) were also used to simulate microgravity. Two reports briefly explored its incidence on osteo-clastic differentiation of RAW264.7 cells. When cells were subjected to microgravity for 24 h in a RCCS prior to differentiation under normal gravity conditions, the expression of TRAP was increased and the number of osteoclasts produced was twice as high as with cells kept under normal gravity (Saxena et al. 2007). In apparent contradiction to these findings, RAW264.7 cells grown in a Clinostat throughout differentiation showed no modification of TRAP and reduction of integrin $\beta 3$ expression. No quantification of the number of osteoclasts was given, but the authors claimed that the size of osteoclasts was smaller when differentiated under microgravity (Makihira et al. 2008). Finally, a very recent study analyzed osteoclastic differentiation of mouse bone marrow-derived precursors during a real space flight. Differentiation of precursors was more efficient when submitted to microgravity than when kept at normal gravity. Osteoclasts were larger and expression of integrin $\beta 3$, TRAP, Cathepsin K and Calcitonin receptor genes was increased (Tamma et al. 2009).

This tends to confirm the observations of Saxena and colleagues indicating that reduced gravity stimulates osteoclastogenesis.

Several in vitro experiments suggest that osteoclast precursors may directly participate in bone mass adaptation to loading by sensing and responding to mechanical signals. Nevertheless, only a few studies are available so far. Understanding the exact effects of mechanical stimulation and gravity on osteoclast differentiation will require further examination.

14.4 Importance of Mechanical Signals for Bone Resorption by Osteoclasts

Little is known about the mechanisms that determine where on the bone an osteoclast starts resorbing, how alternation between resorbing and migrating phases is regulated, or what controls the duration of individual osteoclast resorbing activity. As osteoclast precursors can sense and respond to mechanical signals during differentiation, mechanical signals may also regulate individual osteoclast resorbing function.

14.4.1 Mature Osteoclasts Sense and Respond to Matrix Stiffness

Macrophage spreading was shown to increase with substrate stiffness (Fereol et al. 2006). Similar to alveolar macrophages, osteoclasts adhere to the substrate by means of podosomes. We used vitronectin-coated glass or vitronectin-coated soft polyacrylamide gels identical to those used by Fereol and colleagues and tested whether osteoclasts were sensitive to substrate stiffness. We compared mature osteoclast spreading and actin organization after plating for 6 h on different substrates. Similar to alveolar macrophages, we observed that spreading of osteoclasts increased when plated on substrates with increasing stiffness (Fig. 14.3). Whereas osteoclasts were spread with a regular circular shape on glass, they showed reduced spreading with numerous pericellular membrane processes on 10-kPa gels and they failed to spread on even softer 0.5-kPa gels (Fig. 14.3a). Similarly, osteoclasts do not spread when seeded on osteoblastic, fibroblastic or epithelial cell layers (Saltel et al. 2006, 2008). Analyzing actin organization (Fig. 14.3b), we found that osteoclasts exhibited a peripheral podosome belt on glass whereas they showed numerous actin rings on 10-kPa gels (see also Fig. 14.2b). On softer 0.5-kPa gels, osteoclasts failed to assemble any recognizable adhesion structures. Instead, actin organized in patches. Similarly, membrane protrusions in osteoclasts undergoing transmigration through an osteoblastic cell layer were shown to contain actin patches and to be devoid of podosomes. Only when protrusions contacted the underlying glass substrate did podosomes assemble (Saltel et al. 2006, 2008).

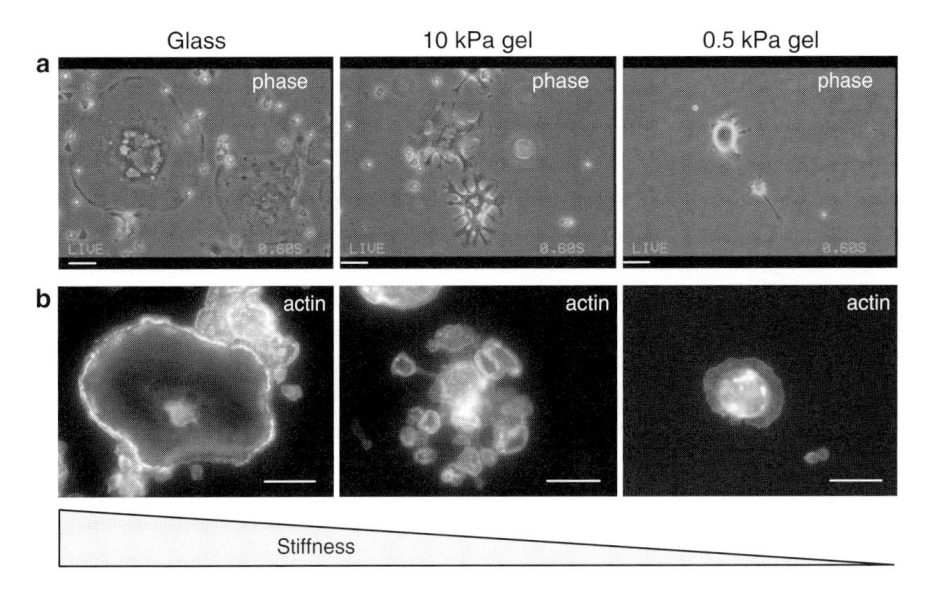

Fig. 14.3 Osteoclast morphology and actin organization depends on substrate stiffness. Osteoclasts differentiated from RAW264.7 cells were seeded either onto vintronectin-coated glass or onto soft (10-kPa or 0.5-kPa) polyacrylamide gels. After 6 h, live osteoclasts were imaged by phase contrast (**a**) and then fixed. F-actin was stained with rhodamine-labeled phalloidin to visualize adhesion structures by fluorescent microscopy (**b**). *Scale bars* 20 μm

These observations suggest that actin organization in mature osteoclast is dependent on matrix stiffness. Podosome belt, which mimics the sealing zone on nonmineralized substrates (Luxenburg et al. 2007), is unable to assemble on a soft matrix. Therefore, the podosome compaction step, necessary for the assembly of a functional sealing zone, may require not only the mineral components of bone but also its stiffness. These may participate in the mechanism of sealing zone disassembly at the end of the resorption process and explain the alternate migration and resorption phases of the osteoclast: when the mineral matrix has been dissolved and the collagen fibers resorbed by the osteoclast, the resulting "melted" bone may not be stiff enough to sustain a packed organization of podosomes, leading to the dissolution of the sealing zone.

14.4.2 Bone Resorption: The Influence of Mechanical Stimulation

Bone adapts to loading and remodels in response to the mechanical forces that provoke its deformation (Skerry 2008). Physiological mechanical stimulations between 50 and 1,500 microstrains (με) do not result in any modification of bone structure whereas higher strains increase bone formation and lower strains favor bone resorption (Kadow-Romacker et al. 2009; Qin et al. 1998).

Two studies examined the effect of microstrains on the resorbing activity of purified osteoclasts seeded on dentine disks subjected to cyclic deformation. In the first report, osteoclasts on dentine submitted to continuous stretching cycles for 24 h exhibited higher resorption activity than osteoclasts seeded on unstretched dentine (Kurata et al. 2001). Nevertheless, the bending conditions used to produce stretching (1,730 με at a frequency of 1 Hz) were likely to provoke modifications in the structure of dentine. In fact, a recent report using a similar stimulation device mentions that bending dentine disks at 1,100 με at a frequency of 1 Hz led to loss of elastic rebond after only 1.6 min (Kadow-Romacker et al. 2009). Therefore, in the study by Kurata et al., dentine structure alteration may have occurred, such as microdamage, which could have stimulated osteoclast activity. Kadow-Romacker and colleagues used lower frequencies and few minutes daily treatments that maintained the integrity of dentine. They found a mild reduction in osteoclast resorbing activity when seeded on dentine subjected to cyclic deformation (Kadow-Romacker et al. 2009). In line with this latter study, strain was also shown to increase the apoptosis of osteoclasts (MacQuarrie et al. 2004). This suggests that strain intends to limit bone resorption by acting directly on osteoclasts.

Conversely, microgravity increases osteoclast resorbing activity, as it also favors osteoclast precursor differentiation. Tamma et al. compared bone resorption by purified mature osteoclasts kept at normal gravity or subjected to gravitational changes during a space flight. They showed that osteoclast activity was inversely proportional to gravity (Tamma et al. 2009). In line with this conclusion, hindlimb unloading was found to increase osteoclast surface (Ishijima et al. 2007), but association with increased resorbing capacity was not tested.

Bones experience frequent and variable mechanical strains. Their effects on osteoblasts have been extensively studied whereas little has been done so far to explore their influence on osteoclasts. The few studies above suggest that osteoclasts can respond to mechanical strains and gravity. Osteoclasts seem to increase their activity when mechanical stimulation diminishes. This is in line with increased bone resorption observed in vivo due to motor paralysis, long bed rest or a space flight. More studies are needed to definitely establish the direct influence of mechanical forces on osteoclasts bone resorbing activity.

14.4.3 Can Osteoclasts Recognize and Adapt to Topographical Features?

Bone is an anisotropic tissue composed of type 1 collagen fibrils embedded in calcium phosphate crystals. At the nanometer scale, bone surface is woven with mineral plates attached to collagen fibrils (Bozec et al. 2005; Hassenkam et al. 2004).

Microdamage that is generated by bone load modifies its surface topography and targets the sites for bone remodeling (Chapurlat and Delmas 2009; Hassenkam et al. 2006). Microcracks locally sever osteocytes and cause their apoptosis. This signals the location and size of damage and stimulates resorption by osteoclasts. It has been shown that osteocyte apoptosis is a key signal to activate local osteoclast recruitment and then resorption (Cardoso et al. 2009; Martin and Seeman 2008). Nevertheless, the subsequent mechanisms controlling osteoclast activity are not known. In particular, how local topography modifications due to damage influence osteoclast activity remains to be established.

The presence of minerals is essential to induce osteoclast polarization and resorption (Chambers et al. 1984). Osteoclasts can resorb protein-free calcium phosphate substrates (Gomi et al. 1993), and the nature of the calcium phosphate influences resorption (Doi et al. 1999). One explanation may be that different calcium phosphate crystals display variable surface topographies. But they also have different acid solubilities, which are known to influence resorption (Ramaswamy et al. 2005). Assessing purely topographical effects on osteoclast activity would require comparisons between surfaces of identical chemistry. For instance, the comparison of osteoclastic resorption between sintered hydroxyapatite disks (Gomi et al. 1993) and bone slices (Matsunaga et al. 1999) abraded with different grinding papers showed that resorption increases with roughness. But these two studies were done using coculture osteoclastic differentiation systems. The effect of roughness on resorption by osteoclasts appeared to result from the indirect stimulation of the osteoblasts present in the culture (Matsunaga et al. 1999). Similar studies with purified osteoclasts would be helpful to understand the importance of topography on osteoclastic bone resorption.

Several studies on sealing zone organization argue in favor of an influence of surface topography on osteoclast activity. Nacre is a resorbable biomaterial that has been used for bone grafting in animals (Berland et al. 2005). Nacre is made of aragonite tablets embedded in chitin. On nacre cut parallel to the tablets, osteoclasts assembled a large sealing zone following the geometry of the tablets. Conversely, on nacre cut perpendicular to the tablets, the sealing zone was made of several short lines along the edges of nacre tablets (Duplat et al. 2007). This suggests that the sealing zone can orient according to the three-dimensional structure of the substrate, which should influence the area resorbed by the osteoclast. Geblinger and colleagues compared the dynamics of the sealing zones of osteoclasts plated onto bone or on protein-free calcite (Geblinger et al. 2009). On bone, osteoclasts assembled several small sealing zones. After disassembly, these tended to reform at the same position. Other areas of osteoclast membrane in contact with the bone always remained devoid of sealing zone formation. Conversely, on protein-free calcite, osteoclasts formed large sealing zones encompassing to the whole cell area. They conclude that osteoclasts can recognize specific areas on bone, which are to be resorbed. They also claim that surface topography itself could dramatically influence osteoclast behavior (Geblinger et al. 2009). Yet, their statement relies on unpublished observations, and further studies will be necessary to identify the local characteristics of the bone surface that may stimulate osteoclast activity.

14.5   Molecular Pathways Involved in Osteoclast Response to Mechanical Stimuli

There is now increasing evidence that osteoclasts are mechanosensitive cells. The recent demonstration that podosomes are true mechanosensors reinforces the idea that osteoclasts translate physical properties of the substrate into intracellular signaling (Collin et al. 2008). Due to the important consequences of mechanotransduction in many biological processes, molecular mechanisms involved in the transduction of mechanosignals have recently received much attention in various cell types. These studies allow speculating on what signaling pathways may be involved in osteoclast responses to mechanical stimuli.

Integrins are major force-bearing adhesion receptors that have an essential role in cell sensing and responding to the physical properties of the substrate (Puklin-Faucher and Sheetz 2009; Schwartz and DeSimone 2008). Rigidity response regulates cytoskeleton organization (Chou et al. 2009; Clark et al. 2007). Downstream of integrins, several actors in mechanotransduction signaling pathways have been identified, including docking protein p130Cas, a molecular sensor of mechanical forces (Sawada et al. 2006), protein tyrosine kinases and phosphatases (Giannone and Sheetz 2006; Kostic and Sheetz 2006), and Rho GTPases that are the major regulators of actin dynamics (Chen 2008; Katsumi et al. 2004). Some of these have essential functions in osteoclasts and could be involved in mechanotransduction during their differentiation or bone resorption.

14.5.1   Transduction of Mechanical Signal Through Integrin αvβ3, p130Cas and Src Family Tyrosine Kinases in Osteoclasts?

Integrin αvβ3 is a known mechanoreceptor that can sense and respond to matrix rigidity. This response involves phosphorylation of p130Cas (Sawada et al. 2006) and activation of Src family tyrosine kinases to allow cell spreading (Jiang et al. 2006; Kostic and Sheetz 2006).

In osteoclasts, a rigid substrate is required for precursor cell fusion (Fig. 14.2) via an integrin αvβ3-dependent pathway (Miyamoto et al. 2000). On soft substrates, mature osteoclasts spread less and do not assemble a podosome belt (Fig. 14.3). Integrin αvβ3 is essential for organization of the actin cloud around the podosome core and Src for podosome belt organization (Chabadel et al. 2007; Destaing et al. 2008). Moreover, activation of αvβ3 through adhesion onto vitronectin induces the phosphorylation of p130Cas in osteoclast precursors (Nakamura et al. 2003). Integrin αvβ3, p130Cas and Src have been shown to localize to the podosome cloud and to the sealing zone in osteoclasts (Chabadel et al. 2007; Lakkakorpi et al. 1999; Luxenburg et al. 2006). Therefore, all the molecular machinery necessary to sense and transduce mechanical signals is present in osteoclast podosome clouds (Fig. 14.4a).

As podosomes are mechanosensitive adhesion structures (Collin et al. 2008), mechanical stimulation of osteoclast podosomes is likely to activate integrin αvβ3,

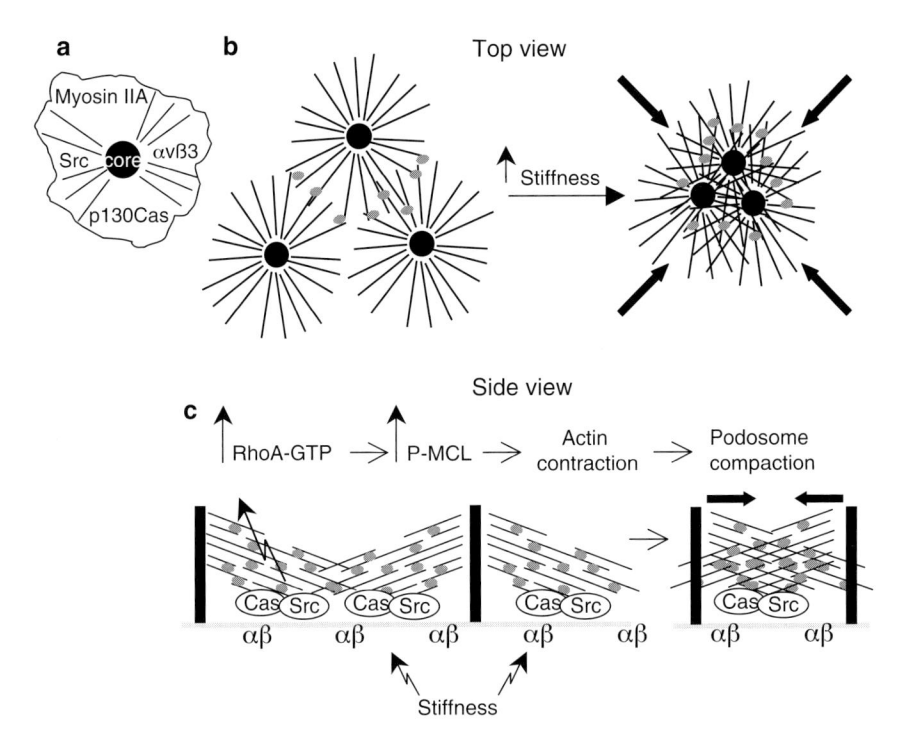

Fig. 14.4 Model mechanism of mechanotransduction at osteoclast podosomes. (**a**) Schematic top view of osteoclast podosomes. The podosome core contains F-actin fibers perpendicular to the substrate (*black disk*). It is surrounded by a cloud of F-actin fibers parallel to the substrate (*black lines*) where integrin αvβ3, p130Cas, Src and Myosin IIA also localize. (**b**) Tentative model for the molecular mechanisms driving podosome reorganization in response to mechanical stimulation. Substrate stiffness activates integrin αvβ3 (*αβ*) that transmits the mechanical signal to the molecular sensor p130Cas (*Cas*) and to the tyrosine kinase Src. Downstream, the activation of RhoA (*RhoA-GTP*) increases myosin light chain phosphorylation (*P-MLC*). Thereby activated, Myosin IIA (*gray dots*) induces actin network contraction in the clouds. This brings podosome cores closer and results in podosome compaction

induce p130Cas phosphorylation, and activate Src family kinases. This pathway would be essential in various key features of osteoclast biology: during differentiation to stimulate integrin αvβ3-dependent osteoclast precursor fusion (Miyamoto et al. 2000) and during bone resorption for the reorganization of podosome superstructures (Fig. 14.2), including the sealing zone.

14.5.2 Effectors of Mechanical Signals in Osteoclasts: Myosin IIA and Rho Family GTPases

Actomyosin contractility is a well-characterized intracellular effector to transduce mechanical signals downstream of integrins, and myosin II is the major motor protein that regulates actomyosin contractility. The Rho family RhoA GTPase is known to stimulate the activating phosphorylation of myosin regulatory light chain (MLC) and promote actomyosin contraction (Clark et al. 2007). RhoA is activated by mechanical signals (Sarasa-Renedo et al. 2006; Zhang et al. 2007) and regulates osteoblast response to extracellular matrix rigidity (Khatiwala et al. 2009). In osteo-clasts, RhoA localizes to podosomes. It controls their organization and is essential for sealing zone formation (Chellaiah et al. 2000; Ory et al. 2008). Myosin IIA localizes to the osteoclast sealing zone and is also present in the podosome cloud (Krits et al. 2002; McMichael et al. 2009). Podosome belt and sealing zone forma-tion requires important compaction of podosomes: podosome cores move closer to each other while actin fibers become denser in the cloud (Luxenburg et al. 2007). Putting these observations together, one could postulate that downstream of integrin $\alpha v \beta 3$, p130Cas and Src, RhoA transduces mechanical signals to Myosin IIA. By generating contractility forces on radial actin fibers in podosome clouds, this would provoke actin cytoskeleton contraction leading to podosome compaction and finally allow the assembly of the podosome belt or the sealing zone (Fig. 14.4b, c).

The Rho family Wrch1/RhoU GTPase is essential for osteoclast differentiation (Brazier et al. 2006). Wrch1 localizes to the podosome cloud (Ory et al. 2007) and regulates signaling downstream of integrin $\alpha v \beta 3$ and influences podosome organi-zation (Brazier et al. 2009). Wrch1 activity was also shown to regulate Myosin light chain phosphorylation in HeLa cells (Chuang et al. 2007). Therefore Wrch1 could participate with RhoA in the regulation of Myosin IIA activity downstream of integrin $\alpha v \beta 3$ in osteoclasts.

14.6   Conclusion

Several lines of evidence support the importance of mechanical signals in various aspects of osteoclast biology, although mechanotransduction has been scarcely explored so far as a regulator of osteoclast functions. Important advances were made recently that allowed better understanding of osteoclast biology and identification of molecular mechanotransduction signaling pathways. Novel experimental devices were also developed to apply controlled forces to individual cells. All these advances now provide an appropriate experimental environment to address how mechanical forces directly influence osteoclast biology.

Osteoclasts are responsible for physiological and pathological bone degradation and understanding their physiology is a major challenge. This is an important need given the increasing incidence of osteolytic diseases such as age-related osteoporo-sis and bone metastasis; all the more so as modulating osteoclast activity has proved to be an efficient therapeutic approach. Osteoclast response to mechanical stimuli is also expected to have major consequences on the outcome of bone substitutes used for engraftment. With the global population aging, the incidence of bone defects due to fractures, tumors or infection is increasing. Therefore, the need for bone replacement rises. Due to autograft limitation, engineering of new biomaterials to be used as bone substitutes is very active. Resorption of implants stimulates bone formation and allows optimal biointegration of the graft (Spence et al. 2008).

The chemical composition of bone implants has an important impact in their resorption rate (Spence et al. 2009). The mechanosensing properties of precursor and mature osteoclasts are also likely to modulate their interaction with implant surface and internal structures, which will impact on their resorbability and then integration within neighboring bone. A better understanding of osteoclast response to mechanical properties of biomaterials should help to engineer more effective bone substitutes.

Acknowledgments This work was supported by research grants from the Fondation pour la Recherche Médicale (DVO20081013473) and the Association pour la Recherche sur le Cancer (3897). Virginie Vives is a recipient of a post doctoral fellowship from the Association pour la Recherche sur le Cancer. The authors are grateful to Cécile Gauthier-Rouvière, Gilles Gadea, Gilles Uzé and Philippe Fort for critical reading of the manuscript.

References

Baron R, Neff L, Van Tran P, Nefussi JR, Vignery A (1986) Kinetic and cytochemical identification of osteoclast precursors and their differentiation into multinucleated osteoclasts. Am J Pathol 122:363–378

Berland S, Delattre O, Borzeix S, Catonne Y, Lopez E (2005) Nacre/bone interface changes in durable nacre endosseous implants in sheep. Biomaterials 26:2767–2773

Block MR, Badowski C, Millon-Fremillon A, Bouvard D, Bouin AP, Faurobert E, Gerber-Scokaert D, Planus E, Albiges-Rizo C (2008) Podosome-type adhesions and focal adhesions, so alike yet so different. Eur J Cell Biol 87:491–506

Bonewald LF (2006) Mechanosensation and transduction in osteocytes. Bonekey Osteovision 3:7–15

Boyle WJ, Simonet WS, Lacey DL (2003) Osteoclast differentiation and activation. Nature 423:337–342

Bozec L, de Groot J, Odlyha M, Nicholls B, Nesbitt S, Flanagan A, Horton M (2005) Atomic force microscopy of collagen structure in bone and dentine revealed by osteoclastic resorption. Ultramicroscopy 105:79–89

Brazier H, Stephens S, Ory S, Fort P, Morrison N, Blangy A (2006) Expression profile of RhoGTPases and RhoGEFs during RANKL-stimulated osteoclastogenesis: identification of essential genes in osteoclasts. J Bone Miner Res 21:1387–1398

Brazier H, Pawlak G, Vives V, Blangy A (2009) The Rho GTPase Wrch1 regulates osteoclast precursor adhesion and migration. Int J Biochem Cell Biol 41:1391–1401

Cardoso L, Herman BC, Verborgt O, Laudier D, Majeska RJ, Schaffler MB (2009) Osteocyte apoptosis controls activation of intracortical resorption in response to bone fatigue. J Bone Miner Res 24:597–605

Cella M, Buonsanti C, Strader C, Kondo T, Salmaggi A, Colonna M (2003) Impaired differentiation of osteoclasts in TREM-2-deficient individuals. J Exp Med 198:645–651

Chabadel A, Banon-Rodriguez I, Cluet D, Rudkin BB, Wehrle-Haller B, Genot E, Jurdic P, Anton IM, Saltel F (2007) CD44 and beta3 integrin organize two functionally distinct actin-based domains in osteoclasts. Mol Biol Cell 18:4899–4910

Chambers TJ, Thomson BM, Fuller K (1984) Effect of substrate composition on bone resorption by rabbit osteoclasts. J Cell Sci 70:61–71

Chapurlat RD, Delmas PD (2009) Bone microdamage: a clinical perspective. Osteoporos Int 20:1299–1308

Chellaiah MA, Soga N, Swanson S, McAllister S, Alvarez U, Wang D, Dowdy SF, Hruska KA (2000) Rho-A is critical for osteoclast podosome organization, motility, and bone resorption. J Biol Chem 275:11993–12002

Chen CS (2008) Mechanotransduction – a field pulling together? J Cell Sci 121:3285–3292

Chou SY, Cheng CM, Leduc PR (2009) Composite polymer systems with control of local substrate elasticity and their effect on cytoskeletal and morphological characteristics of adherent cells. Biomaterials 30:3136–3142

Chuang YY, Valster A, Coniglio SJ, Backer JM, Symons M (2007) The atypical Rho family GTPase Wrch-1 regulates focal adhesion formation and cell migration. J Cell Sci 120:1927–1934

Clark K, Langeslag M, Figdor CG, van Leeuwen FN (2007) Myosin II and mechanotransduction: a balancing act. Trends Cell Biol 17:178–186

Cogoli A (1993) The effect of hypogravity and hypergravity on cells of the immune system. J Leukoc Biol 54:259–268

Collin O, Tracqui P, Stephanou A, Usson Y, Clement-Lacroix J, Planus E (2006) Spatiotemporal dynamics of actin-rich adhesion microdomains: influence of substrate flexibility. J Cell Sci 119:1914–1925

Collin O, Na S, Chowdhury F, Hong M, Shin ME, Wang F, Wang N (2008) Self-organized podosomes are dynamic mechanosensors. Curr Biol 18:1288–1294

Destaing O, Saltel F, Geminard JC, Jurdic P, Bard F (2003) Podosomes display actin turnover and dynamic self-organization in osteoclasts expressing actin-green fluorescent protein. Mol Biol Cell 14:407–416

Destaing O, Saltel F, Gilquin B, Chabadel A, Khochbin S, Ory S, Jurdic P (2005) A novel Rho-mDia2-HDAC6 pathway controls podosome patterning through microtubule acetylation in osteoclasts. J Cell Sci 118:2901–2911

Destaing O, Sanjay A, Itzstein C, Horne WC, Toomre D, De Camilli P, Baron R (2008) The tyrosine kinase activity of c-Src regulates actin dynamics and organization of podosomes in osteoclasts. Mol Biol Cell 19:394–404

Doi Y, Iwanaga H, Shibutani T, Moriwaki Y, Iwayama Y (1999) Osteoclastic responses to various calcium phosphates in cell cultures. J Biomed Mater Res 47:424–33

Duplat D, Chabadel A, Gallet M, Berland S, Bedouet L, Rousseau M, Kamel S, Milet C, Jurdic P, Brazier M, Lopez E (2007) The in vitro osteoclastic degradation of nacre. Biomaterials 28:2155–2162

Engler AJ, Sen S, Sweeney HL, Discher DE (2006) Matrix elasticity directs stem cell lineage specification. Cell 126:677–689

Fereol S, Fodil R, Labat B, Galiacy S, Laurent VM, Louis B, Isabey D, Planus E (2006) Sensitivity of alveolar macrophages to substrate mechanical and adhesive properties. Cell Motil Cytoskeleton 63:321–340

Fereol S, Fodil R, Laurent VM, Balland M, Louis B, Pelle G, Henon S, Planus E, Isabey D (2009) Prestress and adhesion site dynamics control cell sensitivity to extracellular stiffness. Biophys J 96:2009–2022

Geblinger D, Geiger B, Addadi L (2009) Surface-induced regulation of podosome organization and dynamics in cultured osteoclasts. Chembiochem 10:158–165

Geiger B, Bershadsky A (2002) Exploring the neighborhood: adhesion-coupled cell mechanosensors. Cell 110:139–142

Geiger B, Spatz JP, Bershadsky AD (2009) Environmental sensing through focal adhesions. Nat Rev Mol Cell Biol 10:21–33

Giannone G, Sheetz MP (2006) Substrate rigidity and force define form through tyrosine phosphatase and kinase pathways. Trends Cell Biol 16:213–223

Gil-Henn H, Destaing O, Sims NA, Aoki K, Alles N, Neff L, Sanjay A, Bruzzaniti A, De Camilli P, Baron R, Schlessinger J (2007) Defective microtubule-dependent podosome organization in osteoclasts leads to increased bone density in Pyk2(−/−) mice. J Cell Biol 178:1053–1064

Gomi K, Lowenberg B, Shapiro G, Davies JE (1993) Resorption of sintered synthetic hydroxyapatite by osteoclasts in vitro. Biomaterials 14:91–96

Hassenkam T, Fantner GE, Cutroni JA, Weaver JC, Morse DE, Hansma PK (2004) High-resolution AFM imaging of intact and fractured trabecular bone. Bone 35:4–10

Hassenkam T, Jorgensen HL, Lauritzen JB (2006) Mapping the imprint of bone remodeling by atomic force microscopy. Anat Rec A Discov Mol Cell Evol Biol 288:1087–1094

Hentunen TA, Jackson SH, Chung H, Reddy SV, Lorenzo J, Choi SJ, Roodman GD (1999) Characterization of immortalized osteoclast precursors developed from mice transgenic for both bcl-X(L) and simian virus 40 large T antigen. Endocrinology 140:2954–2961

Hsu H, Lacey DL, Dunstan CR, Solovyev I, Colombero A, Timms E, Tan HL, Elliott G, Kelley MJ, Sarosi I, Wang L, Xia XZ, Elliott R, Chiu L, Black T, Scully S, Capparelli C, Morony S, Shimamoto G, Bass MB, Boyle WJ (1999) Tumor necrosis factor receptor family member RANK mediates osteoclast differentiation and activation induced by osteoprotegerin ligand. Proc Natl Acad Sci USA 96:3540–3545

Ishijima M, Tsuji K, Rittling SR, Yamashita T, Kurosawa H, Denhardt DT, Nifuji A, Ezura Y, Noda M (2007) Osteopontin is required for mechanical stress-dependent signals to bone marrow cells. J Endocrinol 193:235–243

Ishikawa S, Arase N, Suenaga T, Saita Y, Noda M, Kuriyama T, Arase H, Saito T (2004) Involvement of FcRgamma in signal transduction of osteoclast-associated receptor (OSCAR). Int Immunol 16:1019–1025

Jiang G, Huang AH, Cai Y, Tanase M, Sheetz MP (2006) Rigidity sensing at the leading edge through alphavbeta3 integrins and RPTPalpha. Biophys J 90:1804–1809

Kadow-Romacker A, Hoffmann JE, Duda G, Wildemann B, Schmidmaier G (2009) Effect of mechanical stimulation on osteoblast- and osteoclast-like cells in vitro. Cells Tissues Organs 190:61–68

Kanehisa J, Heersche JN (1988) Osteoclastic bone resorption: in vitro analysis of the rate of resorption and migration of individual osteoclasts. Bone 9:73–79

Katsumi A, Orr AW, Tzima E, Schwartz MA (2004) Integrins in mechanotransduction. J Biol Chem 279:12001–12004

Khatiwala CB, Kim PD, Peyton SR, Putnam AJ (2009) ECM compliance regulates osteogenesis by influencing MAPK signaling downstream of RhoA and ROCK. J Bone Miner Res 24(5):886–898

Kim N, Takami M, Rho J, Josien R, Choi Y (2002) A novel member of the leukocyte receptor complex regulates osteoclast differentiation. J Exp Med 195:201–209

Kostic A, Sheetz MP (2006) Fibronectin rigidity response through Fyn and p130Cas recruitment to the leading edge. Mol Biol Cell 17:2684–2695

Krits I, Wysolmerski RB, Holliday LS, Lee BS (2002) Differential localization of myosin II isoforms in resting and activated osteoclasts. Calcif Tissue Int 71:530–538

Kurata K, Uemura T, Nemoto A, Tateishi T, Murakami T, Higaki H, Miura H, Iwamoto Y (2001) Mechanical strain effect on bone-resorbing activity and messenger RNA expressions of marker enzymes in isolated osteoclast culture. J Bone Miner Res 16:722–730

Lacey DL, Timms E, Tan HL, Kelley MJ, Dunstan CR, Burgess T, Elliott R, Colombero A, Elliott G, Scully S, Hsu H, Sullivan J, Hawkins N, Davy E, Capparelli C, Eli A, Qian YX, Kaufman S, Sarosi I, Shalhoub V, Senaldi G, Guo J, Delaney J, Boyle WJ (1998) Osteoprotegerin ligand is a cytokine that regulates osteoclast differentiation and activation. Cell 93:165–176

Lakkakorpi PT, Vaananen HK (1991) Kinetics of the osteoclast cytoskeleton during the resorption cycle in vitro. J Bone Miner Res 6:817–826

Lakkakorpi PT, Nakamura I, Nagy RM, Parsons JT, Rodan GA, Duong LT (1999) Stable association of PYK2 and p130(Cas) in osteoclasts and their co-localization in the sealing zone. J Biol Chem 274:4900–4907

LeBlanc AD, Spector ER, Evans HJ, Sibonga JD (2007) Skeletal responses to space flight and the bed rest analog: a review. J Musculoskelet Neuronal Interact 7:33–47

Linder S (2007) The matrix corroded: podosomes and invadopodia in extracellular matrix degradation. Trends Cell Biol 17:107–117

Luxenburg C, Parsons JT, Addadi L, Geiger B (2006) Involvement of the Src-cortactin pathway in podosome formation and turnover during polarization of cultured osteoclasts. J Cell Sci 119:4878–4888

Luxenburg C, Geblinger D, Klein E, Anderson K, Hanein D, Geiger B, Addadi L (2007) The architecture of the adhesive apparatus of cultured osteoclasts: from podosome formation to sealing zone assembly. PLoS One 2:e179

MacQuarrie RA, Fang Chen Y, Coles C, Anderson GI (2004) Wear-particle-induced osteoclast osteolysis: the role of particulates and mechanical strain. J Biomed Mater Res B Appl Biomater 69:104–112

Makihira S, Kawahara Y, Yuge L, Mine Y, Nikawa H (2008) Impact of the microgravity environment in a 3-dimensional clinostat on osteoblast- and osteoclast-like cells. Cell Biol Int 32:1176–1181

Martin TJ, Seeman E (2008) Bone remodelling: its local regulation and the emergence of bone fragility. Best Pract Res Clin Endocrinol Metab 22:701–722

Matsunaga T, Inoue H, Kojo T, Hatano K, Tsujisawa T, Uchiyama C, Uchida Y (1999) Disaggregated osteoclasts increase in resorption activity in response to roughness of bone surface. J Biomed Mater Res 48:417–423

McMichael BK, Wysolmerski RB, Lee BS (2009) Regulated proteolysis of nonmuscle myosin IIA stimulates osteoclast fusion. J Biol Chem 284:12266–12275

Mitteregger R, Vogt G, Rossmanith E, Falkenhagen D (1999) Rotary cell culture system (RCCS): a new method for cultivating hepatocytes on microcarriers. Int J Artif Organs 22:816–822

Miyamoto T, Arai F, Ohneda O, Takagi K, Anderson DM, Suda T (2000) An adherent condition is required for formation of multinuclear osteoclasts in the presence of macrophage colony-stimulating factor and receptor activator of nuclear factor kappa B ligand. Blood 96:4335–4343

Morey-Holton ER, Globus RK (2002) Hindlimb unloading rodent model: technical aspects. J Appl Physiol 92:1367–1377

Mulari MT, Zhao H, Lakkakorpi PT, Vaananen HK (2003) Osteoclast ruffled border has distinct subdomains for secretion and degraded matrix uptake. Traffic 4:113–125

Nakamura I, Rodan GA, le Duong T (2003) Distinct roles of p130Cas and c-Cbl in adhesion-induced or macrophage colony-stimulating factor-mediated signaling pathways in prefusion osteoclasts. Endocrinology 144:4739–4741

Ory S, Brazier H, Blangy A (2007) Identification of a bipartite focal adhesion localization signal in RhoU/Wrch-1, a Rho family GTPase that regulates cell adhesion and migration. Biol Cell 99:701–716

Ory S, Brazier H, Pawlak G, Blangy A (2008) Rho GTPases in osteoclasts: orchestrators of podosome arrangement. Eur J Cell Biol 87:469–477

Paloneva J, Mandelin J, Kiialainen A, Bohling T, Prudlo J, Hakola P, Haltia M, Konttinen YT, Peltonen L (2003) DAP12/TREM2 deficiency results in impaired osteoclast differentiation and osteoporotic features. J Exp Med 198:669–675

Puklin-Faucher E, Sheetz MP (2009) The mechanical integrin cycle. J Cell Sci 122:179–186

Qin YX, Rubin CT, McLeod KJ (1998) Nonlinear dependence of loading intensity and cycle number in the maintenance of bone mass and morphology. J Orthop Res 16:482–489

Ramaswamy Y, Haynes DR, Berger G, Gildenhaar R, Lucas H, Holding C, Zreiqat H (2005) Bioceramics composition modulate resorption of human osteoclasts. J Mater Sci Mater Med 16:1199–1205

Rho JY, Tsui TY, Pharr GM (1997) Elastic properties of human cortical and trabecular lamellar bone measured by nanoindentation. Biomaterials 18:1325–1330

Roodman GD (1999) Cell biology of the osteoclast. Exp Hematol 27:1229–1241

Rubin J, Biskobing D, Fan X, Rubin C, McLeod K, Taylor WR (1997) Pressure regulates osteoclast formation and MCSF expression in marrow culture. J Cell Physiol 170:81–87

Rubin J, Fan X, Biskobing DM, Taylor WR, Rubin CT (1999) Osteoclastogenesis is repressed by mechanical strain in an in vitro model. J Orthop Res 17:639–645

Rubin J, Murphy T, Nanes MS, Fan X (2000) Mechanical strain inhibits expression of osteoclast differentiation factor by murine stromal cells. Am J Physiol Cell Physiol 278:C1126–C1132

Rubin J, Rubin C, Jacobs CR (2006) Molecular pathways mediating mechanical signaling in bone. Gene 367:1–16

Saltel F, Destaing O, Bard F, Eichert D, Jurdic P (2004) Apatite-mediated actin dynamics in resorbing osteoclasts. Mol Biol Cell 15:5231–5241

Saltel F, Chabadel A, Zhao Y, Lafage-Proust MH, Clezardin P, Jurdic P, Bonnelye E (2006) Transmigration: a new property of mature multinucleated osteoclasts. J Bone Miner Res 21:1913–1923

Saltel F, Chabadel A, Bonnelye E, Jurdic P (2008) Actin cytoskeletal organisation in osteoclasts: a model to decipher transmigration and matrix degradation. Eur J Cell Biol 87:459–468

Sarasa-Renedo A, Tunc-Civelek V, Chiquet M (2006) Role of RhoA/ROCK-dependent actin contractility in the induction of tenascin-C by cyclic tensile strain. Exp Cell Res 312:1361–1370

Sawada Y, Tamada M, Dubin-Thaler BJ, Cherniavskaya O, Sakai R, Tanaka S, Sheetz MP (2006) Force sensing by mechanical extension of the Src family kinase substrate p130Cas. Cell 127:1015–1026

Saxena R, Pan G, McDonald JM (2007) Osteoblast and osteoclast differentiation in modeled microgravity. Ann N Y Acad Sci 1116:494–498

Schwartz MA, DeSimone DW (2008) Cell adhesion receptors in mechanotransduction. Curr Opin Cell Biol 20:551–556

Shen Z, Crotti TN, McHugh KP, Matsuzaki K, Gravallese EM, Bierbaum BE, Goldring SR (2006) The role played by cell-substrate interactions in the pathogenesis of osteoclast-mediated peri-implant osteolysis. Arthritis Res Ther 8:R70

Sims NA, Gooi JH (2008) Bone remodeling: multiple cellular interactions required for coupling of bone formation and resorption. Semin Cell Dev Biol 19:444–451

Skerry TM (2008) The response of bone to mechanical loading and disuse: fundamental principles and influences on osteoblast/osteocyte homeostasis. Arch Biochem Biophys 473:117–123

Spence G, Phillips S, Campion C, Brooks R, Rushton N (2008) Bone formation in a carbonate-substituted hydroxyapatite implant is inhibited by zoledronate: the importance of bioresorption to osteoconduction. J Bone Joint Surg Br 90:1635–1640

Spence G, Patel N, Brooks R, Rushton N (2009) Carbonate substituted hydroxyapatite: resorption by osteoclasts modifies the osteoblastic response. J Biomed Mater Res A 90:217–224

Takahashi N, Udagawa N, Suda T (1999) A new member of tumor necrosis factor ligand family, ODF/OPGL/TRANCE/RANKL, regulates osteoclast differentiation and function. Biochem Biophys Res Commun 256:449–455

Tamma R, Colaianni G, Camerino C, Di Benedetto A, Greco G, Strippoli M, Vergari R, Grano A, Mancini L, Mori G, Colucci S, Grano M, Zallone A (2009) Microgravity during space-flight directly affects in vitro osteoclastogenesis and bone resorption. FASEB J 23:2549–2554

Teti A, Zallone A (2009) Do osteocytes contribute to bone mineral homeostasis? Osteocytic osteolysis revisited. Bone 44:11–16

Turner CH, Rho J, Takano Y, Tsui TY, Pharr GM (1999) The elastic properties of trabecular and cortical bone tissues are similar: results from two microscopic measurement techniques. J Biomech 32:437–441

Vaananen HK, Laitala-Leinonen T (2008) Osteoclast lineage and function. Arch Biochem Biophys 473:132–138

Wang JH, Thampatty BP, Lin JS, Im HJ (2007) Mechanoregulation of gene expression in fibroblasts. Gene 391:1–15

Wang N, Tytell JD, Ingber DE (2009) Mechanotransduction at a distance: mechanically coupling the extracellular matrix with the nucleus. Nat Rev Mol Cell Biol 10:75–82

Zhang SJ, Truskey GA, Kraus WE (2007) Effect of cyclic stretch on beta1D-integrin expression and activation of FAK and RhoA. Am J Physiol Cell Physiol 292:C2057–C2069

Index